Evolution of Social Insect Colonies

CW00920189

Oxford Series in Ecology and Evolution

Edited by Robert M. May and Paul H. Harvey

The Comparative Method in Evolutionary Biology
Paul H. Harvey and Mark D. Pagel

The Causes of Molecular Evolution
John H. Gillespie

Dunnock Behaviour and Social Evolution
N. B. Davies

Natural Selection: Domains, Levels, and Challenges
George C. Williams

Behaviour and Social Evolution of Wasps: The Communal Aggregation Hypothesis
Yosiaki Itô

Life History Invariants: Some Explorations of Symmetry in Evolutionary Ecology
Eric L. Charnov

Quantitative Ecology and the Brown Trout
J. M. Elliott

Sexual Selection and the Barn Swallow
Anders Pape Møller

Ecology and Evolution in Anoxic Worlds
Tom Fenchel and Bland J. Finlay

Anolis Lizards of the Caribbean: Ecology, Evolution, and Plate Tectonics
Jonathan Roughgarden

From Individual Behaviour to Population Ecology
William J. Sutherland

Evolution of Social Insect Colonies: Sex Allocation and Kin Selection
Ross H. Crozier and Pekka Pamilo

Evolution of Social Insect Colonies

Sex Allocation and Kin Selection

Ross H. Crozier

School of Biological Science
University of New South Wales
and
School of Genetics and Human Variation
La Trobe University, Melbourne

and

Pekka Pamilo

Department of Genetics
University of Helsinki
and
Department of Genetics
Uppsala University

Oxford New York Tokyo
OXFORD UNIVERSITY PRESS
1996

Oxford University Press, Walton Street, Oxford OX2 6DP
Oxford New York
Athens Auckland Bangkok Bombay
Calcutta Cape town Dares Salaam Delhi
Florence Hong Kong Istanbul Karachi
Kuala lumpur Madras Madrid Melbourne
Mexico City Nairobi Paris Singapore
Taipei Tokyo Toronto

and associated companies in
Berlin Ibadan

Oxford is a trade mark of Oxford University Press

Published in the United States
by Oxford University Press Inc., New York

© Ross H. Crozier and Pekka Pamilo, 1996

All rights reserved. No part of this publication may be reproduced, stored in a retrieval system, or transmitted, in any form or by any means, without the prior permission in writing of Oxford University Press. Within the UK, exceptions are allowed in respect of any fair dealing for the purpose of research or private study, or criticism or review, as permitted under the Copyright, Designs and Patents Act, 1988, or in the case of reprographic reproduction in accordance with the terms of licences issued by the Copyright Licensing Agency. Enquiries concerning reproduction outside those terms and in other countries should be sent to the Rights Department, Oxford University Press, at the address above.

This book is sold subject to the condition that it shall not, by way of trade or otherwise, be lent, re-sold, hired out, or otherwise circulated without the publisher's prior consent in any form of binding or cover other than that in which it is published and without a similar condition including this condition being imposed on the subsequent purchaser.

A catalogue record for this book is available from the British Library

Library of Congress Cataloging-in-Publication Data
Crozier, R.H. (Rossiter Henry), 1943-
Evolution of Social Insect Colonies/Ross H. Crozier and Pekka Pamilo.
(Oxford series in ecology and evolution)
Includes bibliographical references and index.
1. Insect societies. 2. Sex allocation. 3. Kin selection (Evolution) I. Pamilo, Pekka. II. Title. III. Series.
QL496. C76 1996 595.7051—dc20 95-32182

ISBN 0 19 854943 1 (Hbk.)
ISBN 0 19 854942 3 (Pbk.)

Typeset by AMA Graphics Ltd., Preston, Lancs
Printed in Great Britain by
Bookcraft Ltd, Midsomer Norton, Avon

Acknowledgements

We have benefited enormously from the comments, data, and advice of numerous colleagues who have discussed the subject matter of this work with us during its gestation, made lengthy by the disruptions caused by each of us, in turn, moving between institutions. For such comments we thank Alfred Buschinger, James Cook, Mark Elgar, Wille Fortelius, Penny Kukuk, Christian Peeters, Risa Rosenberg, Michael Schwarz, Perttu Seppä, Lotta Sundström and most especially Joan Herbers, David Queller and Rainer Rosengren, who made extremely detailed comments on an early version. Our work on evolutionary genetics has been supported by the Australian Research Council and the Finnish and Swedish Natural Science Research Councils; without this support we would not have gathered the experience making this effort possible. We also thank our editors, Robert May and Paul Harvey, for retaining interest in this project.

Contents

1

Introduction

1.1 The problems

Social insects fascinate because of the finely coordinated organization of their societies and because of their great ecological success in many biotopes (Wilson, 1990). Studies on insect social organization have traditionally focused on understanding colony-level phenomena. The emphases in such studies have been on understanding colony growth, the division of labour between colony members, the coordination of functions between individuals, and the communication mechanisms that make this finely tuned coordination possible (e.g. Oster and Wilson, 1978; Moritz and Southwick, 1992; Hölldobler, 1995). Hölldobler and Wilson (1990:29) emphasize the importance of series-parallel task processing in the enormous success of social insects (i.e. individual ants can specialize in components of social life, and do not each have to perform the entire range of tasks necessary for their colony's success). Recently, however, it has been realized that these successful societies do not always function in complete harmony but rather that the cooperating individuals can have different interests and that conflicts may occur. Aggression between competing conspecific colonies has been known for a long time, but such conflicts can alternate with cooperation for the same colonies, as exemplified in the famous (for Europeans) wood ant wars (Mabelis, 1979; Rosengren and Pamilo, 1983). Conflicts exist not only between colonies but within them. Individuals which coexist and cooperate differ genetically and by age and reproductive capacity and therefore selection may result in them favouring different outcomes of colony activity.

The two views and research traditions, one emphasizing cooperation and coordination, and the other conflicts, tend to be concerned with different colony activities (Pamilo, 1991a). The two traditions are to some extent concerned with different levels of analysis and need not therefore be regarded as conflicting (Sherman, 1988). In any case, study of ultimate causation, such as selection at the individual and colony levels, should illuminate studies of proximate causation, such as the communication systems within and between colonies.

Social insects have a special place in evolutionary biology, as already noted by Darwin (1859; pages 140–2 in the 1979 edition) long before us. The problem from the evolutionary point of view is that only some of the individuals reproduce. This division between reproductive and non-reproductive sections of the community provides us with two major problems. First, how did such a life-pattern arise in the first place? Why did selection lead to the abandonment of reproduction by some individuals if it is through reproduction that the genes for such behaviour are passed on? This problem, if we agree that the evolution of sociality has been an adaptive change, leads immediately to further questions, such as: for whom is this pattern adaptive? what is the unit of selection leading to this pattern? The question of the units of selection arises unavoidably in certain parts of this book, but it is not our intent to enter deeply into terminological discussions (Lewontin, 1970; Hull, 1980; Dawkins, 1982; Crozier, 1987a), but rather to uncover the processes involved, however one chooses to name them. These processes frequently involve conflicts between units at different levels in the hierarchy of the levels of selection which it is convenient to distinguish in social insects.

Even though the worker individuals in social insect colonies do not normally reproduce, they still play a major role in the evolution of colonial life. They, of course, make the colony function by building, supporting and defending it. They not only gather resources to the colony, they also deliver them amongst the colony members. By these means they can control the allocation of resources in the colony and further their own interests. This leads to the second major evolutionary question, the conflicts between different individuals in the colonies. One part of this confrontation is the general parent–offspring conflict, the divergent fitness interests of parents and offspring create disagreements between these parties over the allocation of parental investment (Trivers, 1974). Although the theory of parent–offspring conflict is well formulated, there is little empirical evidence of its importance in nature. Mock and Forbes (1992) consider social insects to provide one of the best opportunities to test the theory. More generally, the kin selection theory of social evolution predicts that workers should try to increase their own evolutionary success by favouring closely related individuals. It appears that in many cases the interests of queens and workers, as well as those of different kin groups coexisting in a colony, do not agree with each other. One of the major questions in the biology of social insects is to examine who are actually running colonies, the queens or the workers. We should examine in which ways the workers can push their own interests, and how the predicted social conflicts can affect the evolution of colonial characteristics.

We are especially interested in the evolution of social life and reproductive decisions in social insects. Very little is known about genetic variation for any trait relevant to these characteristics, even though evolutionary theories are largely based on explicit genetic models. But there are other avenues to explore experimentally. First, the phenotypic predictions of the models can be examined and the various hypotheses tested without anything being known about the genetic architecture of the traits concerned. Second, it is possible to use genetic markers for estimating relevant genetic parameters pertaining to colony and population structure, and then to use these estimates when formulating hypotheses to be tested. If selection is assumed to be weak, these estimates based on markers can be used instead of those unknown values pertaining to loci actually under selection imposed by the behavioural correlates of the social life-pattern. In fact, we especially seek to emphasize the role of genetic studies in deriving data relevant to testing theories of social evolution and sex allocation.

The function of a colony is to produce more reproductives and colonies. That production measures its success in the evolutionary sense. However, a colony is not genetically homogeneous, and we need to consider how this success is partitioned among the colony members.

We should also note that, at least in perennial species, the colony cannot use all of its resources in the production of new reproductives: it has to maintain itself and perhaps to grow further. This clearly leads to a problem of resource allocation between growth and reproduction at each stage of its existence. Another allocation problem concerns sex allocation, because the male and female offspring from a colony may carry different sets of genes and may be of different genetic value to the various colony members.

Sex allocation is not merely the relative apportionment of resources into the production of males versus females, but the relative apportionment into male versus female reproductive *function*. For example, in many species new colonies are formed by budding or fission of existing colonies, and the workers in such founding groups are then part of the investment in female function. The complexity resulting from the social insect life-pattern can be organized around the following major questions (Pamilo, 1991a).

1. Who reproduces in the colony?

2. How are resources allocated between reproductive and non-reproductive functions (i.e. between the production of reproductives and colony growth and maintenance)?

3. How are resources allocated between male and female functions?

All three questions include aspects which are specific to social insects. In other words, eusociality and reproductive resource allocation are interrelated in a complex way. The three questions are not distinct, but rather the answer to one will affect the answer to the others. Illustrating the interrelationships between these questions, and trying to answer them, will be the task of this book.

1.2 What are social insects?

Few insects are not social to some degree, if we include courtship and mating as social activities. Many also show parental care. When talking of the 'social insects', we generally refer to species with specific types of cooperation. This cooperation can take various forms, and this allows classification of the levels of sociality. One view of the various levels of sociality is shown in Table 1.1.

The **eusocial** insects exemplify the various principles we wish to present, and so we will confine ourselves to considering them. Eusocial insects are those with a reproductive division of labour, cooperative brood care, and the presence of non-reproductive helpers (often distinguished morphologically as 'workers') of a later generation to the reproductives (Michener, 1969, 1974:372; Wilson, 1971:464). The greatest evolutionary puzzle and interest focus on the reproductive specialization. There are cases outside the traditional social insects that are often called eusocial because they have evolved the sterile caste, even though they may not otherwise fit the strict definition of the term.

To focus the book yet further, we will largely confine ourselves to examples drawn from the 'traditional' eusocial insects, the ants, termites and eusocial bees and wasps, even though it is now known that some aphids (Aoki, 1982, 1987; Itô, 1989), beetles (Kent and Simpson, 1992), thrips (Crespi, 1992a), and spiders (Vollrath, 1986) may also meet the definition of eusociality. The eusocial insects thus fall into five major taxonomic groups: the hymenopteran suborder Aculeata (six families with eusocial species), the order Isoptera (nine families), and single families in the orders Homoptera, Thysanoptera, and Coleoptera. This distribution is shown in Table 1.2, and we discuss the 'non-traditional' cases briefly further on in this chapter. There is also noticeable variety of the subsocial and parasocial species found in the families of aculeate Hymenoptera (Eickwort, 1981). This taxonomic distribution indicates that eusociality has

Table 1.1. Levels of social organization in insects, after Wilson (1971:5), Michener (1969, 1974:38), and Starr (1984).

	Continued care of young	Cooperative brood care	Reproductive division of labour	Colonies with at least two adult generations	Egg-layers morphologically differentiated
Solitary	–	–	–	No colonies	–
Subsocial	+	–	–	–	–
Parasocial					
Communal	+	–	–	–	–
Quasisocial	+	+	–	±	–
Semisocial	+	+	+	–	–
Eusocial					
Primitive	+	+	+	+	–
Advanced	+	+	+	+	+

Table 1.2. Taxonomic distribution of eusocial insect species (modified from Pamilo, 1991c; after Snelling, 1981; Myles and Nutting, 1988).

	Family	Subfamily	Eusocial species
Hymenoptera,	Aculeata		
	Anthophoridae		In seven genera in the tribe Allodapini
	Apidae	Apinae (honey-bees)	Six highly eusocial species
		Bombinae (bumble-bees)	300 primitively eusocial species and their social parasites.
		Euglossinae (orchid bees)	None
		Meliponinae (stingless bees)	200 eusocial species
	Halictidae		In six genera of the tribes Halictini and Augochlorini
	Sphecidae		*Microstigmus*
	Vespidae	Polisitinae	Over 500 species, all eusocial
		Stenogastrinae	Some primitively eusocial species
		Vespinae	ca. 80 species, all eusocial
	Formicidae	11 subfamilies	ca. 8800 described species, all highly eusocial or descended from highly eusocial species
	Many other families		None
Isoptera			(All species eusocial)
Lower termites	Hodotermitidae		16 species
	Indotermitidae		6 species
	Kalotermitidae		332 species
	Mastotermitidae		1 species
	Rhinotermitidae		204 species
	Serritermitidae		1 species
	Stylotermitidae		28 species
	Termopsidae		16 species
Higher termites	Termitidae		1685 species
Homoptera	Pemphigidae		Sterile soldiers in six genera
Coleoptera	Curculionidae		*Austroplatypus incompertus*
Thysanoptera	Phlaeothripidae		Subfertile soldiers in *Oncothrips*

evolved at least twelve times, perhaps more, in the Hymenoptera. Furthermore, some species may be described as having both eusocial and non-eusocial colonies, and it seems likely that the eusocial life-pattern may revert to a non-eusocial one under some circumstances (Michener, 1985, 1990; Packer,

1991). Michener (1990), when discussing the evolution in the Halictinae, suspects that 'eusocial behavior has arisen repeatedly, dozens or hundreds of times, and that reversion to solitary behavior is also easy'. Whether advanced eusocial bees (honey-bees and stingless bees) have evolved independently or not from primitively social forms is also controversial because morphological and molecular characters seem to support different phylogenies (Cameron, 1993).

The Hymenoptera, the order with most eusocial species, are holometabolous insects. The larvae of all eusocial species are relatively helpless and rely on being fed by the adults; either the eggs are laid in cells containing food stored for the larvae to eat ('mass provisioning'), or the larvae are fed continuously ('progressive provisioning'). The hymenopteran species have male-haploid sex determination. Unfertilized, haploid eggs develop into males, whereas the females are diploid and develop from fertilized eggs. The division of labour in hymenopteran societies involves only adult females (with rare and problematic exceptions). The females can specialize not only behaviourally but also morphologically, the most important distinction being between reproductive and non-reproductive individuals. The species differ widely in the complexity of the caste system. The size and complexity of the colonies also vary greatly, from a few individuals in primitively eusocial bees and wasps and in some ants to millions in for example army ants. The colonies also have different life cycles. They can be founded by a single female who thus has a period of solitary life (independent colony foundation), or she can be supported by workers when the colony fissions or buds or when she enters an existing colony (dependent colony foundation). Many temperate bees and wasps have normally annual colonies, only females overwinter and they establish new colonies independently. Honey-bees, stingless bees, swarm-founding polybine wasps, ants and termites have perennial colonies. In many species a colony can recruit new reproductive queens, either by replacing the old queens or having many coexisting queens in the same nest. We will later return to these complexities, examine how they affect the queen–worker conflicts within colonies and how these features themselves may have been affected by such conflicts.

All termites are eusocial. Their closest living relatives are generally regarded as the three species of the roach genus *Cryptocercus*, which are subsocial (Eickwort, 1981), but this view is not supported by a cladistic analysis of morphological and behavioural characters (Thorne, 1991). Furthermore, molecular sequence data indicate that the living termites may be diphyletic and have arisen from more than one cockroach stock (Vawter, 1991; cf. DeSalle *et al.*, 1992). The termites are commonly divided into lower and higher groups (Table 1.2). There are basic differences between higher and lower termites in the determination and diversity of castes and the kind of symbionts in their guts (ciliate protozoa in lower termites and bacteria in higher termites). Termites are hemimetabolous insects and all individuals except reproductives are immature (Noirot and Pasteels, 1987). In many lower termites, the working individuals are late larvae, nymphs, and pseudergates (individuals differing from nymphs in lacking wing buds) and in the families Rhinotermitidae, Mastotermitidae and Hodotermitidae they are true workers. In higher termites, the working individuals belong to a terminal caste of true workers, namely individuals incapable of further moults. In addition to these workers or worker equivalents, in both higher and lower termites there are some individuals which develop into soldiers, a specialized terminal caste. Both sexes of termites are diploid and take part in a broad range of colony activities, although in some species all soldiers are of one sex only (Noirot and Pasteels, 1987), which one it is depending on the species. In lower termites, the working individuals (nymphs and pseudergates) can moult into reproductives and do not represent an obligatorily sterile or terminal caste. The sterile castes are the soldiers and true workers which cannot reverse their specialization and develop into reproductives any longer. The sterile caste has evolved at least three times in termites (Noirot and Pasteels, 1987; Myles and Nutting, 1988). It is worth noting that even though the pseudergates and the working nymphs of lower termites have the potential of developing into reproductives, many of them die before they have a chance to do so and consequently we see these individuals as specialists in worker activities.

Aphids are also hemimetabolous insects. Various species in the Pemphigidae have soldiers during their thelytokous generations. The existence of these sterile soldiers led Itô (1989) to classify these aphids as eusocial, although it is arguable that the further requirement of an overlap of generations is met. Although thelytoky results in genetically identical siblings, the apparent lack of kin recognition can lead to mixing of aphid groups lowering the relatedness of interacting individuals (Aoki *et al.*, 1991; Sakata and Itô, 1991). Another example of a sterile caste in clonal groups is found in polyembryonic wasps where some siblings develop into precocious larvae that protect the rest of the brood at the expense of their own reproduction (Cruz, 1981). The story is somewhat more complicated than that because a female often lays two eggs in a host, one diploid and the other haploid. The precocious larvae are predominantly females and kill their own haploid brothers which develop from the haploid egg within the same host (Grbic *et al.*, 1992).

The ambrosia beetle *Austroplatypus incompertus* has been termed eusocial because of a prevalence of colonies with one reproductive female and several unmated other females (Kent and Simpson, 1992). Males leave the nest galleries before their sisters. Those females remaining behind lose the last segments of their tarsae and are apparently unable to leave.

Thrips are a further hemimetabolous group. They resemble the Hymenoptera in being male-haploid. In two species of the Australian gall-inhabiting genus *Oncothrips*, some individuals of both sexes have rapid development into a wingless form with enlarged armed forelegs (Crespi, 1992a). These micropterous individuals act as soldiers in that they attack invading insects of other species and usually kill them. The soldiers have reduced fecundity compared to the foundresses, and hence Crespi (1992a) argued that these species should be regarded as eusocial, resembling for example those species of eusocial Hymenoptera in which workers can reproduce.

Spiders are, of course, not insects, but a group of terrestrial arthropods which have independently achieved high sociality on several occasions (Buskirk, 1981). Social spider colonies are highly inbred populations with no obvious adaptations reducing the fecundity of any class of individuals. However, Vollrath (1986) argued that *Anelosimus eximius* should be regarded as eusocial because a large proportion of the females remain unmated, yet still help catch prey consumed by all. Vollrath argues that the unmated females can be regarded as 'helpers' to the mated ones, and rests his case for eusociality on this point. Strictly speaking, it is doubtful that *A. eximius* qualifies as eusocial because of the overlap of generations criterion, although this point remains to be empirically tested.

Moving further away from insects and arthropods, we meet eusocial colonies, surprisingly, in mammals. The species considered as eusocial are all subterranean mole rats, and it seems that eusociality has evolved several times in that group (Sherman *et al.*, 1991; Burda and Kawalika, 1993).

The 'traditional' eusocial insects possess two contrasting genetic systems. The males of Hymenoptera are (usually) haploid and impaternate, whereas the females are diploid and have two parents. Male-haploidy leads to major asymmetries in relatedness and genetic value between the two sexes. In termites, both sexes are diploid and all individuals have two parents, so that the two sexes present essentially familiar patterns of relatedness (although the sex-chromosome translocation systems in many species causes same-sex siblings to be more closely related than opposite-sex ones, Section 1.8). It is the asymmetries between the sexes in patterns of relatedness and genetic value that gives the sex allocation patterns of eusocial Hymenoptera their special complexity and their pride of place in this book.

A *role* is a definable set of activities. Reproduction is one of the two chief roles in the division of labour characterizing eusocial insects. The other chief role is the worker role (Fig. 1.1). Workers feed and otherwise assist and protect their colony's reproductives in the production of further reproductives. If the workers are able to assist preferentially some of the colony's offspring rather than others, their choice of whom to assist can differ from that of the colony's reproductives. This difference in choice can lead to conflict over resource allocation in the colony.

Associated with the division of roles in insect societies is caste, a concept very important but somewhat ambiguous in the social insect literature (Peeters and Crozier, 1988). Under one definition, castes are assemblages of individuals already distinct morphologically at the start of adult life. Castes then differ in role but each caste may have more than one role. Under another usage (e.g. Oster and Wilson, 1978:19), caste is completely defined by role: a caste is an assemblage of individuals that need differ from other castes only in their role.

Discussions about the meaning of caste and role are generally irrelevant to this book, because the morphologically defined castes generally agree with the

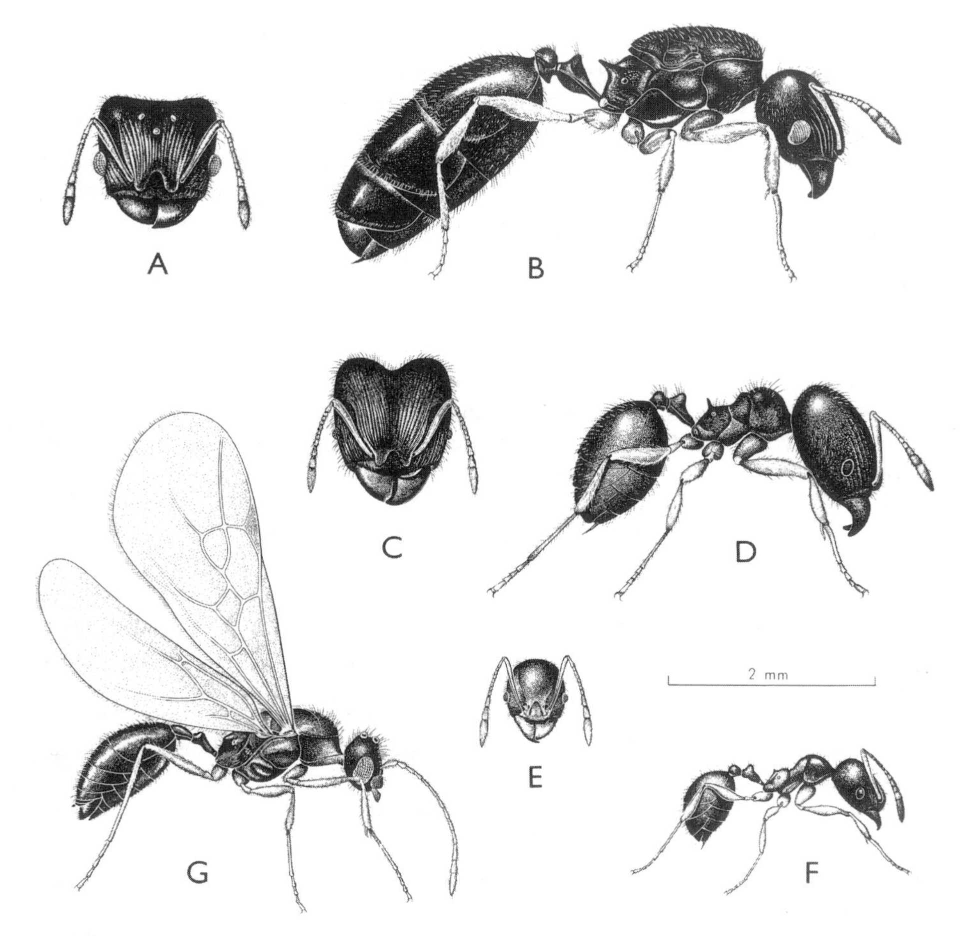

Fig. 1.1 A well-developed caste system: colony members in an Australian species of the ant genus *Pheidole*. The winged individual is a male, whereas the individual with a large bulky thorax is a queen (who has removed her wings after mating). The remaining individuals are workers. (From Naumann, I. D. (ed.) 1991, *The insects of Australia*, Vol. 2, 2nd edn, Melbourne University Press, by permission.)

functionally defined roles. There are, however, some exceptions. In primitively eusocial species, there may be no clear morphological distinction between the reproductive females and the worker females, and the distinction between the two is based on behaviour (e.g. Yanega, 1992). Very much the same situation, but evolved secondarily, is observed in some ponerine ants in which true (morphological) queens have been replaced by, or coexist with, mated workers as occupiers of the mated egg-layer role. These mated workers have been termed *gamergates* (Peeters and Crewe, 1984). Gamergates are morphologically workers, but they fill the egg-layer role occupied in other species by queens.

The distinction between workers and reproductives is not absolute. In many species, workers are not sterile although they do have reduced reproductive capacities. In some ants, for example, workers lay some or all of the eggs destined to yield males (Wilson, 1971:333; Crozier, 1975:67–8; Brian, 1983:201; Bourke, 1988; see also Section 4.2), as also happens with some bees (Machado *et al.*, 1984). Very often the reproductive capacities of workers are suppressed in the presence of reproductives; removal of the reproductives often leads to reproductive activation of the workers. Workers in orphaned hymenopteran colonies often lay unfertilized eggs (Wilson, 1971:305; Crozier, 1974; Velthuis, 1985) and may in some bees also mate and produce diploid eggs (Section 5.1). In termites, removal of the reproductives can lead to the conversion of some working individuals to reproductives (Wilson, 1971:188–96).

The essence of the eusocial insects is that they live in colonies. In the analogy of Oster and Wilson (1978:21–2), these colonies are fortified factories whose products are reproductives concerned with establishing further colonies. Colonies of the same population are in reproductive competition with each other; such competition often extends to aggressive territoriality (Wilson, 1971:447–52). The colonies themselves are genetically heterogeneous, consisting of genetically different individuals, and this heterogeneity can lead to reproductive competition within colonies. The genetic diversity present within social insect colonies stems first of all from the fact that colony members are usually derived from eggs resulting from normal meiosis in the reproductives, and that colonies with more than one mated egg-layer (polygyny) and multiple mating by females (polyandry) may occur. The production of males from worker-laid eggs adds further to this complexity (Section 4.2).

In social Hymenoptera, colonies may have one to many queens or gamergates, depending on the species (Wilson, 1971:331–3; West-Eberhard, 1978). These colonies are then termed polygynous (strictly speaking, polygyny refers to the presence of multiple queens and not to multiple mated workers, but for simplicity in this book we assume congruence of form and function). Queens may mate once to many times, i.e. they are either monandrous or polyandrous (Crozier and Page, 1985; Page, 1986; see also Sections 4.1, 4.3). The degree of polygyny and polyandry can also vary in termites (Thorne, 1982). Polygynous colonies can outlive any single individual, and this succession of reproductive generations poses new problems for resource allocation: should new sexuals be encouraged to leave and found new colonies, or to remain and strengthen the colony?

Whereas some eusocial species have secondarily lost the morphological queen caste and reverted to worker reproduction (gamergates), others have lost the worker castes and become obligatory social parasites. An analogous kind of parasitism, cleptoparasitism, is found in some solitary bee species which usurp and use the nests dug and provisioned by their host species. The socially parasitic ants, bees, and wasps use not only the nest of the host species but also its workers for raising their queens and males. This coexistence of two species in the same nest raises the interesting question of which individuals, those of the parasite or those of the host, determine how resources are allocated in the sexual production of the parasite (Trivers and Hare, 1976). Of course, in addition to the problem of sex allocation, the evolution of social parasitism is extremely interesting in terms of speciation, because the parasites are often, although not always, the closest living relatives of their hosts (e.g. Pamilo *et al.*, 1981; Carpenter *et al.*, 1993).

1.3 Male-haploidy is characteristic of Hymenoptera

Male-haploidy is a genetic system under which ordinary males are haploid and females are diploid. Recourse to biology textbooks and adding up the numbers of animal species known to be male-haploid leads to an estimate of about 20% of known animal species being male-haploid. In some insects and mites (Bull, 1983:153), both males and females arise from fertilized eggs, with the paternal genome being expelled from eggs destined to be males. The Hymenoptera, in common with various other insects, mites, ticks and rotifers (Bull, 1983:148), combine male-haploidy with *arrhenotoky*, the production of males from unfertilized eggs and females from fertilized eggs. Bull (1983:149) estimates that the combination of arrhenotoky and male-haploidy has evolved about 12 times.

While male-haploidy need not involve arrhenotoky, it is also true that arrhenotoky need not involve male haploidy. Nur (1972) found that both the coccid *Lecanium putmani* and one 'race' of *L. cerasifex* are arrhenotokous; both sexes are diploid with the diploid number restored in males by fusion of the products of the first cleavage division (males are thus completely homozygous, effectively as if haploid).

Male-haploidy and arrhenotoky form the basic genetic system of the Hymenoptera, but there are also many species in which *thelytoky*, the production of females from unfertilized eggs, has become established secondarily. Sometimes complex life cycles result, such as in gall wasps (Cynipidae), which have an alternation between arrhenotokous and thelytokous generations with a further predisposition of the thelytokous generation females to specialize in producing offspring of only one sex in the arrhenotokous generation. The arrhenotokous generation females themselves also tend to specialize in producing either male- or female-producers of the thelytokous generation (the population genetics of this system is considered by Crozier, 1975:62–6).

Laidlaw and Page (1986) make the interesting point that a hymenopteran female can be considered to be an hermaphrodite. Thus, a female produces two types of gametes: eggs which are fertilized and give rise to new females, and gametes which perform a male function by developing into drones, making multiple copies of themselves in the drones' testes and fertilizing eggs. Under this convention, males are packages of gametes produced by females.

Male-haploidy leads to several important consequences concerning social evolution. First, it allows unmated females to produce sons. This is particularly important for workers that have lost the ability to mate but still have the option of direct reproduction by laying haploid eggs. Second, it makes it possible for an egg-laying female to control the sex ratio of her brood by deciding whether to fertilize an egg or not. Third, the male-haploid sex determination makes the degrees of genetic relatedness among close relatives depart in an interesting way from those characterizing diploid organisms. This fact has been central in the explanations of social evolution in the hymenopteran insects (Hamilton, 1964b) and it forms the cornerstone in the hypothesis of the queen–worker conflicts.

A male-haploid pedigree is shown in Fig. 1.2 and the connections in this pedigree yield the values of relatedness shown in Table 1.3. Clearly, male-haploidy imposes a strong intrinsic bias towards asymmetrical relatednesses between individuals of different sex. A haploid male passes the same set of genes to all his daughters. The genetic differences among full sisters are caused only by segregation of the maternal alleles, there being a 50% chance

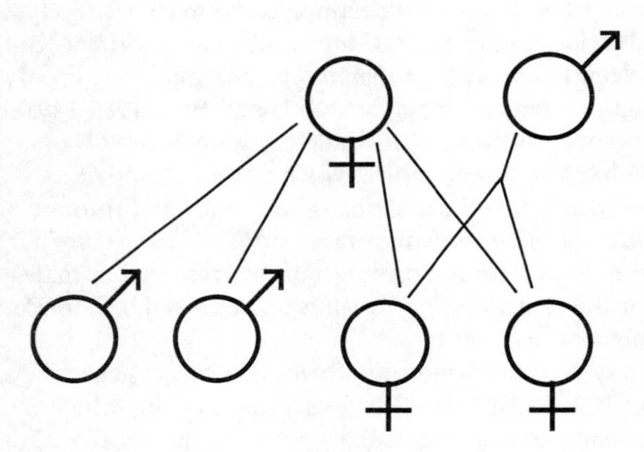

Fig. 1.2 A simple male-haploid pedigree. Normal males arise from unfertilized eggs and are haploid, whereas females arise from fertilized eggs. The joining of the line linking the male to his daughters reflects the fact that he passes on identical genes to every daughter.

Table 1.3. Pedigree coefficient of relatedness (g_{YX}) values for male-haploids. Relatednesses are given for individual Y to individual X.

X	Y								
	Mother	Father	Daughter	Son	Sister	Brother	Aunt	Niece	Nephew
Female	1/2	1	1/2	1	3/4	1/2	3/8	3/8	3/4
Male	1/2	0	1/2	0	1/4	1/2	3/8	1/8	1/4

of two sisters inheriting the same allele from the mother. As a result, two full sisters are expected to have 75% of their genes identical by descent, and we say that their genetic relatedness is $g = 0.75$. (Depending on the segregation of the maternal chromosomes, there is in practice some variation around the expected relatedness of 0.75 between sisters. If we take a single locus, two sisters either inherit the same maternal allele or different maternal alleles, and the relatedness at that locus is either 1.0 or 0.5, respectively.) The relatedness between a female and her daughters is 0.50. This comparison led Hamilton (1964b) to suggest that the hymenopteran females could benefit by raising highly related sisters ($g = 0.75$) rather than their own daughters ($g = 0.50$). Another important consequence is the difference between sisters and brothers. Whereas the sisters are highly related, a brother receives only maternal alleles, and the relatedness of a brother to a sister (e.g. to a worker in the colony) is $g = 0.50$. This difference means that a worker raising larvae in a colony would benefit more by raising closely related sisters than by raising brothers. This is the cause of the sex allocation conflict between queens and workers; workers are expected to prefer female-biased sex ratios while queens should favour equal investment in females and males (Trivers and Hare, 1976). We will later return to the definition and estimation of genetic relatedness in Chapter 2.

The fourth important consequence of male-haploidy concerns the reproductive values of males and females (Taylor, 1988). The males pass their genes only to daughters, whereas the females produce both daughters and sons. The females therefore contribute more to the gene pool of future generations, and we can say that they have a higher reproductive value than the males have. We will show in the next chapter how the coefficients of relatedness and the sex-specific reproductive values are defined, and we also show how they are taken into account in formal models of social evolution.

We should finally remark that Laidlaw and Page (1986) suggested that the terminology of hymenopteran relatives should be based on those degrees of genetic relatedness found in diploid organisms. The sisters of a diploid organism are related by 0.5, and therefore Laidlaw and Page suggested that hymenopteran females with the same mother and father should be called super sisters (relatedness of 0.75), and females with the same mother and whose fathers are brothers should be called full sisters (the relatedness being 0.5). Although this terminology is appealing, we do not use it here, but use the term full sisters to refer to female siblings having the same father and the term half sisters for siblings having different fathers.

1.4 Sex in Hymenoptera is commonly determined by heterozygosity

As seen above, male-haploidy has important evolutionary implications. We will therefore briefly discuss next the mechanisms of sex determination in the

Hymenoptera. Sex determination under arrhenotoky is a major departure from the chromosomal systems in which one or the other of the sex chromosomes (X or Y, W or Z) plays an overwhelming role (Bull, 1983:22). In male-haploids, there is no possibility of difference in dosage between one section of the genome and another, ruling out both sex chromosomes and dosage compensation, except as applied to nucleo-cytoplasmic balance (Crozier, 1975:73–4; Crozier, 1985:203).

In two hymenopterans, the parasitoid *Bracon hebetor* and the honey-bee *Apis mellifera*, there is abundant evidence that sex determination is due to a single multi-allelic locus (see Crozier, 1977a and Bull 1983:145–59 for general reviews), and there are others which appear to have same system (Table 1.4). Heterozygotes at this locus are females, while homozygotes and hemizygotes develop into males. In a panmictic population, a proportion $1/k$ of the diploid individuals are expected to be homozygous diploid males, where k is the effective number of alleles at the sex locus. The diploid males are part of the population's genetic load because they are either sterile or lead to sterile or low-fecundity offspring (their sperm are diploid, yielding triploid offspring). In honey-bees, workers eat the diploid drone larvae at an early age. Diploid males are known in a wide range of Hymenoptera (Table 1.4). In some cases, as noted above, pedigree analysis indicates the operation of heterozygosity-based sex determination. It is tempting to adduce the mere presence of diploid males as indicating such sex determination, but it is risky because diploid males are also known from highly inbred *Nasonia*, although these may not be due to homozygosity for sex-determining loci (reviewed by Crozier, 1977a; Luck *et al.*, 1993; Cook, 1993b).

Sex-locus alleles at high frequency are eliminated proportionately more often through occurring in diploid males than are low-frequency alleles, and the number of alleles segregating in the population depends on the balance between mutation, frequency-dependent selection favouring rare alleles, and the loss of alleles due to genetic drift (e.g. Yokoyama and Nei, 1979). In accordance with this expectation, estimates of the numbers of alleles at such loci are quite high: nine for *Bracon hebetor* (Whiting, 1961), 20.0 for *Melipona compressipes fasciculata* (Kerr, 1987), 10–13 for *Solenopsis invicta* (Ross *et al.*, 1988), and up to 18.9 for *Apis mellifer*a (Adams *et al.*, 1977).

Of course, the arguments in the preceding paragraph presuppose that the number of *possible* sex alleles is very much larger than that attainable in populations of normal size. The absence of molecular information on the mode of action of the sex locus makes it hard to determine the value of the upper limit to the number of sex alleles. Similar estimates for the number of alleles in *Apis* and *Melipona*, which have very different population sizes, led Kerr (1987) to suggest that the observed estimates may already be close to such a limit. It would be interesting to determine the allelism rates of sex alleles in isolated populations of honey-bees. If the limiting number is very large, then the allelism rate should be low. If the limiting number is small, then the allelism rate should be high.

Table 1.4. Species in which diploid males have been found in Hymenoptera. The occurrence of triploid (*Crematogaster* sp. ANIC 2) and tetraploid (*Camponotus* sp. ANIC 5) individuals in two ant species (Imai *et al.*, 1977) probably reflects the occurrence of diploid males in them also. It is possible that *Nasonia* diploid males never arise from inbreeding but from other causes, and the diploid males of *Diplolepis* have one set of chromosomes heterochromatinized, suggesting that they too do not arise from inbreeding. It therefore seems that diploid males are readily produced by inbreeding in ichneumonoid parasitoids, aculeates (with the single exception of *Goniozus nephantidis* (Cook, 1993a)), and in sawflies, but not in chalcidoid parasitoids (Cook, 1993b).

Family *Species*	Notes and references
SAWFLIES	
Diprionidae	
Athalia rosae	1 Naito and Suzuki (1991)
Neodiprion	
nigroscutum	2 Smith and Wallace (1971)
pinetum	Stouthamer et al. (1992)
PARASITOIDS: ICHNEUMONOIDEA	
Braconidae	
Bracon	
brevicornis	Speicher and Speicher (1940), in Cook (1993b)
hebetor	1 Whiting (1939)
serinopae	Clark *et al.* (1963)
Cotesia rubecula	Stouthamer *et al.* (1992)
Microplitis croceipes	Steiner and Teig (1989), in Stouthamer *et al.* (1992)
Ichneumonidae	
Batheplectes curculionis	Unruh *et al.* (1984), in Luck *et al.* (1993)
Diadromus pulchellus	1 Hedderwick *et al.* (1985)
PARASITOIDS: CHALCIDOIDEA	
Cynipidae	
Diplolepis rosae	Stille and Dävring (1980)
Pteromalidae	
Nasonia vitripennis	3 Whiting (1960), Macy and Whiting (1969)
BEES	
Apidae	
Apis cerana	2 Hoshiba *et al.* (1981)
Apis mellifera	1 Drescher and Rothenbuhler (1964), Woyke (1969)
Bombus atratus	2 Garofalo (1973), see Crozier (1977a)
Melipona	
compressipes	W. E. Kerr (pers. comm.).
quadrifasciata	Kerr (1974)
Trigona quadrangula	Kerr (1974)

Table 1.4. *Continued*

Halictidae		
Augochorella striata		Packer and Owen (1990), Mueller et al. (1994)
Lasioglossum zephyrum		Kukuk and May (1990)
SOCIAL WASPS		
Vespidae		
Liostenogaster flavolineata		J. E. Strassmann (pers. comm.)
Mischocyttarus immarginatus		J. E. Strassmann (pers. comm.)
ANTS		
Formicidae		
Doronomyrmex kutteri		Buschinger and Fischer (1991)
Epimyrma stumperi		Fischer (1987), in Loiselle et al. (1990)
Formica		
pressilabris		Pamilo and Rosengren (1984)
aquilonia		Pamilo *et al.* (1994)
lugubris		Pamilo *et al.* (1994)
polyctena		Pamilo *et al.* (1994)
rufa		Pamilo *et al.* (1994)
truncorum		Pamilo *et al.* (1994)
Harpagoxenus sublaevis		Fischer (1987), in Loiselle *et al.* (1990)
Lasius alienus		Pearson (1983b)
Leptothorax		
ambiguus		Herbers and Grieco (1994)
muscorum	2	Loiselle *et al.* (1990)
sp. A		Loiselle *et al.* (1990)
Pseudolasius sp. nr *emeryi*		Hung *et al.* (1972)
Rhytidoponera		
chalybaea		Ward (1978:239–42)
confusa		Ward (1978:239–42)
Solenopsis invicta	2	Ross and Fletcher (1985a)

1. Pedigree analyses demonstrate single-locus sex determination.
2. Pedigree analyses indicate that sex determination is probably by a single-locus scheme.
3. Pedigree analyses rule out single-locus sex determination and indicate that heterozygosity may not be involved at all.

Inbreeding data and pedigree analyses indicate the operation of one-locus sex determination in a range of Hymenoptera (Table 1.4). The operation of multiple-locus schemes has been suggested for some species, but the data for these cases do not distinguish them from the single-locus scheme (Crozier, 1977a). For example, Smith and Wallace (1971) inferred a single-locus scheme for the sawfly *Neodiprion nigroscutum*, but Kerr (1974), noting a deficiency of diploid males in some crosses, suggested that a two-locus multi-allele scheme might apply. Given the occurrence of similar deficiencies in *Bracon* crosses, known to be due to diploid male mortality (Whiting, 1943), it does not seem that the *Neodiprion* data can be used definitively to infer the number of sex loci.

In many other species of Hymenoptera, inbreeding does not lead to production of diploid males. Crozier (1971; see also Snell, 1935) suggested that in such species sex might be determined by two to several sex loci, with heterozygotes at one or more loci being females, the rest males. But even prolonged inbreeding does not produce diploid males in many other species, such as the much studied parasitoid *Nasonia vitripennis* (Bull, 1983:152). Although the operation of a multi-locus scheme would delay the production of diploid males under inbreeding for several generations, they are still expected to occur eventually. Cook (1993a) has used the expected production of diploid males under severe inbreeding to exclude the multiple-locus model under reasonable assumptions for the parasitoid *Goniozus nephantidis* (Bethylidae).

Sex determination in *Nasonia* and similar insects therefore remains problematic. According to Kerr (see 1974, 1975, 1987), sex determination in such cases is by a scheme in which female-determining loci have increased effects in diploids (are additive) whereas male-determining loci do not. Crozier (1975: 73–4) noted that there is no dosage difference between male- and female-determining loci between haploids and diploids in male-haploid species (a/b = 2a/2b), so that dosage compensation between nuclear genes cannot occur under this genetic system. There is, however, a difference between haploid and diploid zygotes in the ratio of nuclear DNA to cytoplasmic constituents, so that Kerr's scheme devolves to one of additivity and balance (Crozier, 1977a). Kerr suggests that during evolution the heterozygosity-based systems (which he regards as cases where additivity of the female-determining genes occurs only if they are heterozygous) replaced the nucleo-cytoplasmic scheme, and that such evolution is reversible if there is for some reason a greater cost to pure additivity as against heterozygosity-dependent additivity.

Poirié *et al.* (1992) suggest that sex in *Nasonia* is determined by imprinting, with a switch gene turned off in eggs but on in sperm. In this case, destruction by irradiation of the egg pronucleus and subsequent fertilization should yield haploid females whereas in *Bracon* this procedure yields haploid males (Whiting, 1946).

It is worth noting that sex determination in *Drosophila*, now fairly well understood (e.g. Keyes *et al.*, 1992), is essentially a matter of nucleo-cytoplasmic balance. The number of X-chromosomes in the embryonic genome is estimated by the relative activities of the embryo's numerator genes (especially *sis-a* and *sis-b*) compared to those of denominator genes such as *dpn* (Cline, 1993).

As Bull (1983:177–85) notes, rare alleles under the heterozygosity-based scheme are selected for because they reduce the proportion of diploid males. An allele allowing automatic development as a male if the individual is haploid, and as a female if the individual is diploid, would be favoured over an allele requiring complementation for female development. It is therefore more likely that whatever mechanism operates in inbreeding parasitoids is secondary, rather than ancestral. This conclusion is strengthened by the indication (Smith and Wallace, 1971) that a sawfly, *Neodiprion nigroscutum*, has a heterozygosity-based mechanism: the Symphyta, which include sawflies, are generally believed to be ancestral to all other Hymenoptera (Malyshev, 1968:10–25).

The occurrence of diploid males is potentially a serious complication in the estimation of sex allocation ratios. Diploid males arise from fertilized eggs 'intended' to be workers or queens, and hence may not represent investment in males; furthermore they are essentially dead genetically and lack function as males. In fact, diploid males might be partly included in the investment in females: they form part of the production of diploid zygotes from which future queens will be derived (the rest being workers in addition to the diploid males). Whether they should be counted as investment in males or females depends on whether or not they were raised as males or were 'thought of' as females. The answer depends on how the workers recognize the sex of the larvae they raise. The potential for error is shown well by the fire ant *Solenopsis invicta*. This species has two forms distinguished by the type of colonies they form: monogynous (single-queened colonies) and polygynous (multiple-queened colonies) (Ross and Fletcher, 1985b). In both forms, about 12% of the newly mated queens produce about half of their diploid progeny as males. These diploid males make up most of the males produced by such queens, and would grossly distort calculations of sex allocation ratios if not identified as being diploid.

Normal mature nests of monogynous *Solenopsis invicta* do not produce *any* diploid males, whereas a majority of the males of the polygynous form are diploid. This dramatic difference stems from the differing social system of the two forms: the monogynous colonies founded by those queens producing diploid males die out because of insufficient worker numbers, whereas queens who produce diploid males in polygynous nests are saved by the workers produced by other queens in the colony (Ross and Fletcher, 1986). It is interesting to note that apparently monogynous colonies of some *Formica* ants produce large numbers of diploid males (Pamilo *et al.*, 1994).

Because they do not occur naturally in mature colonies of monogynous *S. invicta*, diploid males should have negligible impact on sex allocation patterns in such populations. In polygynous colonies diploid males represent part of the cost of producing sexuals and paradoxically, if 'raised as females', they weight allocation towards the production of haploid males, which are then relatively cheaper (compared to females) than in monogynous colonies. It would be worthwhile to monitor the polygynous populations to see whether there are changes in the years ahead, because this form is spreading in North America and the population may not be at equilibrium.

The situation in *Bombus atratus* may be similar in some respects to that of monogynous *Solenopsis invicta*: a colony whose queen produces diploid males grows significantly more slowly than one whose queen does not (Plowright and Pallett, 1979).

The number of sex alleles can be estimated if diploid males are detected and it is assumed that the population is panmictic (Adams *et al.*, 1977). Under the further assumption of equilibrium, the number of sex alleles, $\hat{k}$, may be estimated by:

$$\hat{k} = \frac{1}{D}$$

where D is the proportion of all diploid individuals in the population which are males. Adams *et al.* (1977) used computer simulation in an apparent jackknife procedure to derive the confidence limits for $\hat{k}$. Computer simulation may be avoided if the number of matings per female is known. Adams *et al.* (1977) found $\hat{k}$ to be 18.9 for a Brazilian population, and that the best estimate for the number of males successfully mating with each queen in the same population is 17.3. Kerr (1987) found that $\hat{k} = 20.0$ for a population of the stingless bee *Melipona compressipes fasciculata*. For some other estimation methods, see Kerr (1986).

Ross and Fletcher (1985a) plausibly infer the operation of a single sex locus in the fire ant *S. invicta*. They also found that the diploid males are heavier than the haploids, as is also the case in *Neodiprion nigroscutum* (Smith and Wallace, 1971) and *Apis mellifera* (Woyke, 1978). Using a maximum likelihood procedure based on the frequencies of electrophoretically detectable diploid males and matched matings, the number of sex-alleles in North American *S. invicta* has been estimated to be 10.0–15.0 (Ross and Fletcher, 1985a; Ross *et al.*, 1988). The estimates from native fire ant populations in Argentina suggest that the number of sex alleles is much larger, 66–86 per population (Ross *et al.*, 1993). If sex is determined by two loci, instead of one, the number of alleles per locus can be much smaller.

If the population size remains constant for a long time, the number of sex alleles and their frequency distribution reach equilibrium values determined by selection, mutation, and stochastic processes (Yokoyama and Nei, 1979). Cornuet (1980) slightly modified their approach and showed that the expected number of sex alleles at equilibrium is

$$E(k) = \sqrt{-2N_e / \ln(u\sqrt{8\pi N_e})}$$

where u is the mutation rate to new sex-determining alleles and N_e is the effective population size. If the mutation rate lies between 10^{-7} and 10^{-5}, the expected number of sex alleles at equilibrium would thus be about 4–5 in populations with an effective size of 100, 13–18 with $N_e = 1000$ and 45–60 with $N_e = 10\,000$, assuming that the sex locus can in fact have so many alleles.

1.5 Male-haploidy allows maternal control of sex ratio

As noted by Hamilton (1967), arrhenotoky frees organisms from the sex-chromosome mechanics seen in creatures such as *Drosophila* and mammals, allowing females great flexibility to adjust the sex ratio. Because they provide sperm homogeneous with respect to sex determination, the ability of males to influence the sex ratio of the progeny produced by their mates has been largely eliminated in male-haploids. (Some exceptions may occur to this rule of paternal impotence, such as in the organ-pipe wasp *Trypoxylon*, in which males continue consorting with nest-provisioning females, and hence may have some influence over brood sex ratios (Brockmann and Grafen, 1989).) Instead of the behaviour of the sex-chromosomes at meiosis, the opening or closing of the spermathecal valve (Fig. 1.3), under the 'voluntary' control of the female, becomes the controlling element of sex determination. Rojas-Rousse and

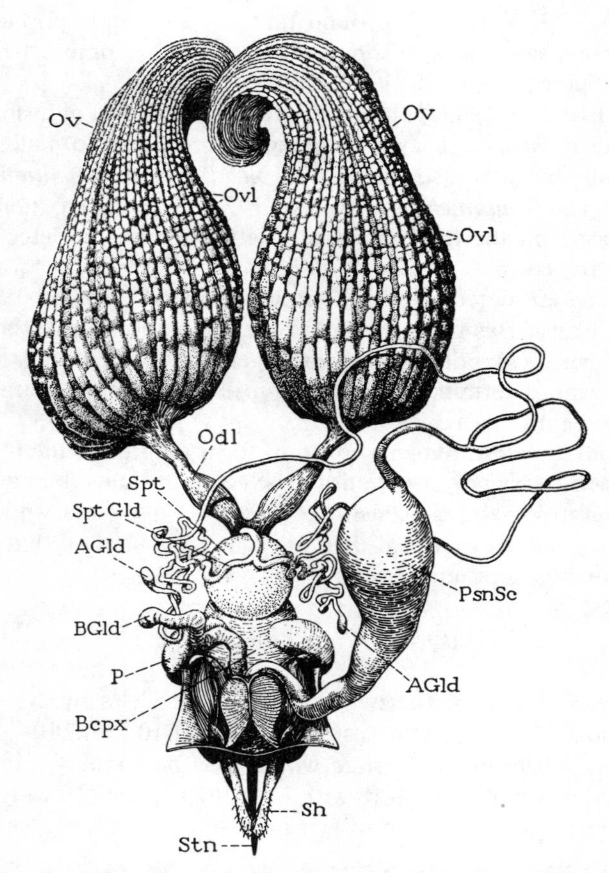

Fig. 1.3 Reproductive system of the honey-bee queen, showing the spermatheca (Spt). Other abbreviations are as follows: AGld, poison gland of sting; Bcpx, bursa copulatrix; BGld, accessory gland of sting; Odl, oviduct; Ov, ovary; Ovl, ovariole; P, lateral pouch of bursa copulatrix; PsnSc, poison gland reservoir; Spt Gld, spermathecal gland; Sh, sheath lobes of sting; Stn, shaft of sting. (Figure 104 of Snodgrass, R.E. 1956, *Anatomy of the honey bee*, Cornell University Press, Ithaca, NY. Reproduced with permission.)

Palevody (1981) discuss and compare spermathecal function in *Apis mellifera* and the parasitoid *Diadromus pulchellus*.

The ability of females to control the fertilization of the egg, and consequently the sex of the offspring, is clearly seen in that fertilization is not a random process. First, unfertilized eggs are generally produced during a shorter time interval than fertilized eggs. Second, whether an egg is fertilized or not is strongly dependent on where it is laid. In many parasitic wasps fertilization depends on the quality and size of the host, and on whether or not the host is already parasitized by other females (Charnov, 1982:48–66). In the honey bee, the sex of an offspring depends on whether or not the egg is laid in a drone cell.

A direct observation linking female behaviour to the subsequent sex of offspring comes from Gerber and Klostermeyer's (1970) observation that, in the bee *Megachile rotundata*, the female pauses during the laying of eggs later found to yield female offspring but not when she lays eggs later yielding male offspring. It is tempting to suggest that the pause in the production of female-producing eggs reflects the positioning of the egg and the opening of the spermathecal duct. The ability of *Nasonia vitripennis* females to adjust the sex ratios of their brood according to the number of females ovipositing on the same host (Section 1.6) is strong evidence for the ability to control sperm release in that species.

The placement of unfertilized eggs in drone cells and fertilized eggs in worker and queen cells indicates the queen's power of control over sex allocation in the honey-bee, but some rectification of 'mistakes' may go undetected because of workers eating eggs. This selective laying of fertilized and unfertilized eggs leads to the question of who really controls the sex allocation in the honey-bee colony: the queen who lays the eggs, or the workers who construct the brood combs?

It is this ability of females to exercise major control over the initial sex ratio, plus the asymmetries of relatedness under male-haploidy, that gives the study of sex allocation in Hymenoptera its special quality. In social Hymenoptera there is the added dimension of the strong influence of individuals other than the parents over sex allocation.

1.6 Male-haploids may show much genetic variation for the sex ratio

Studies on creatures with XY sex determination systems have shown that genetic variation for sex ratio in these cases is usually negligible, often undetectable (Maynard Smith, 1978:148–50; Williams, 1979; Toro and Charlesworth, 1982; Clutton-Brock and Iason, 1986). Williams (1979) concludes that the 1:1 sex ratio typical of most mammals and *Drosophila* is simply a consequence of sex chromosome mechanics. Charnov (1982:104) and Karlin and Lessard (1986:17) point out that the question is really what form of selection and past history led to fixation for such invariant ratios, and Karlin and Lessard (1986:17) point to the existence of a small level of variation in cattle (Bar Anon and Robertson, 1975) as showing the possibility of genetically varying the sex ratio even in XY systems. This general potential is also shown by specific cases (in platyfish and in lemmings) of complex XY systems in which sex ratios far from 1:1 may occur even in large random-mating populations (Bull, 1983:79–80, 94–5).

In addition to variation we might term 'adaptive' (selected for at the level of the whole organism or higher), various other kinds of variation in sex allocation ratios may occur. Hamilton (1967) explores the dynamics of X- and Y-chromosomes that distort sex ratios to maximize their own spread; X-chromosomes of this type, 'sex-ratio chromosomes', are well known from *Drosophila* populations (e.g. Curtsinger, 1991). That '*Drosophila* of sex-ratio research' (May and Seger, 1985), *Nasonia vitripennis*, has yielded a remarkable sex-ratio chromosome which is transmitted via sperm but leads to all-male progenies because it leads to the destruction of the other sperm chromosomes (Nur *et al.*, 1988; Werren, 1991). In addition, microorganism infections are known in a variety of organisms (reviewed by Werren *et al.*, 1986, 1988; Hurst, 1991), including *Drosophila*, which distort the sex ratios of their hosts. Male-haploids lack sex chromosomes, but *Nasonia vitripennis* apparently has the richest assortment of sex ratio distorting symbionts of any animal studied. Werren *et al.* (1986) describe a maternally inherited trait causing progenies to be 97% female and a maternally and contagiously transmitted trait, son-killer, which kills all male larvae feeding on the same host pupa. In some cases, symbionts convert otherwise arrhenotokous species to thelytokous ones (Stouthamer *et al.*, 1990; Zchori-Fein *et al.*, 1992). We would be surprised if social insects showed quite this richness in sex ratio symbionts, but such a possibility cannot be excluded.

Male-haploids have no sex chromosomes but rather an extremely flexible and behavioural system of sex allocation. Hymenoptera also show highly variable sex ratios, with variation in ratios often following predictions very well. Two types of conditional sex determination have been observed, particularly in parasitoid wasps. In the first type, females are able to determine offspring sex in relation to the target of oviposition. For example, in many parasitic hymenopterans, females lay fertilized eggs in larger hosts and unfertilized eggs in smaller hosts (Charnov, 1982:48–66). In the second type, a female can adjust the sex ratio of her brood depending on the number of females ovipositing on the same host. For example, females of *Nasonia vitripennis* vary the sex ratio among their progeny in just this fashion, in accordance with the predictions of theory (Werren, 1983; Orzack and Parker, 1986; Orzack, 1986). Because of this behavioural flexibility, care is needed in the search for evidence of genetic variation for the sex ratio. Nevertheless, a variety of direct and indirect pieces of evidence point to extensive interspecific and intraspecific genetic variation for sex ratio in Hymenoptera.

Intraspecific variation for the sex ratio is evidenced by laboratory and field studies on strains of parasitoids, and by interpopulation differences in several other species. Thus, strains of *Nasonia vitripennis* established from single wild-caught females, as well as lines maintained for long periods in the laboratory, all differed significantly in sex ratio, and there was a significant response to selection for a higher proportion of males (Parker and Orzack, 1985). Parker and Orzack (1985) confirmed that the shift in sex ratio did not result from increased male sterility, the explanation for the response to selection in favour of male production in another parasitoid, *Dahlbominus fuscipennis* (Lee and Wilkes, 1965; Wilkes, 1966). *Nasonia vitripennis* thus provides the best example in Hymenoptera of intraspecific genetic variation for the sex ratio. Luck *et al.* (1993) and Orzack (1993) discuss further the intricacies of sex ratio variation in parasitoids, especially *Nasonia.*

Variation in sex allocation ratios between colonies of ants in the same population appears to be quite marked (Herbers, 1979; Nonacs, 1986a; Elmes, 1987b; Boomsma, 1988), and in many species the nests can be divided into female-producers and male-producers (Pamilo and Rosengren, 1983). Much, probably most, of this variation is facultative, reflecting conditional strategies faced with different colony sizes, social organization of colonies etc. There have been no published reports of data enabling a test for genetic variation in sex ratios in ants, such as data from ant colonies matched for size and condition or, better, between colonies of known genealogical relationship. The only evidence for genetic variation in sex ratios comes from the honey-bee, where the amounts of worker comb and of drone comb produced were shown to have genetic variation (Page *et al.*, 1993).

1.7 Male-haploidy implies greater male mortality

Smith and Shaw (1980) made a challenging point when they noted that the males of Hymenoptera will suffer higher mortality than females because of the expression of deleterious recessive alleles in the haploid sex. If the frequency of recessive lethal mutations, or lethal equivalents (one lethal equivalent results from many mutations causing as much lethality as would one lethal), is U per haploid genome, the expected ratio of male survival to female survival is e^{-3U}, which is less than one. Smith and Shaw estimated this ratio from pooled data of four species of the parasitoid wasp genus *Apanteles* and obtained a maximum likelihood estimate of 0.90 (with 95% confidence limits of 0.838–0.951). If the lowered male survival is due to new recessive mutations, the rate of such mutations is 0.035 per genome per generation. Based on these results, Smith and Shaw considered mutations sufficiently significant to be taken into account by sociobiologists.

As noted by Charnov (1982:49), a mutation effect of the magnitude observed by Smith and Shaw (1980) leads only to a slight female bias. If the numerical sex ratio without mortality effects were 1:1, a 10% reduction in male survival relative to that of females would only reduce the proportion of males from 50% to $(0.9 \times 50)/(0.9 \times 50 + 50) = 47\%$, too small a change to be readily detected in field studies. It would be interesting to obtain further estimates of the relative survival of the two sexes in order to determine whether or not the mutation effect might be of some significance. (Such estimates could be made for species such as the solitary bee *Megachile rotundata*, in which the laying of fertilized and unfertilized eggs can be distinguished and they are left to develop without further attention from their mother (Gerber and Klostermeyer, 1970).) Smith and Shaw (1980) thought that they underestimated male mortality, because early larval stages were not included in their study. Low mortality and high precision of sex ratios in some highly inbred parasitoids (Green *et al.*, 1982) argue against a significant mortality difference between the sexes.

Even if male mortality were higher than thought to be the case, it might not form a significant problem to sex allocation studies of social insects. If mortality occurs in early developmental stages, there has not yet been much investment in the dying larvae, and in many species these can be largely recycled. If mortality is high in later developmental stages, or if a species is a mass provisioner (lays an egg in a cell with sufficient provisions to last the entire larval period), mortality should be taken into account. It would be necessary to estimate the allocation ratio not only on the basis of adults but also taking into account the dead larvae and their provisions.

Much of the genetic variation expressed in honey-bees is limited to females (Crozier, 1976; Kerr, 1976); this limitation reduces the rate of mutations with deleterious effects and increases the male/female survival ratio.

1.8 Importance of sex-linked translocations in termites

Both sexes in termites are diploid, and both sexes are represented in the non-reproducing caste (although in some species there is a sexual bias within soldiers or workers). At first thought, one might therefore expect that the relatedness structure of termite colonies would simply be that of autosomal genes, but this is not strictly true for some species.

Significant deviation from the relatedness structure expected from an autosomal system will occur in those termites in which there have been extensive translocations between sex chromosomes and autosomes (Syren and Luykx, 1977). As discussed by Luykx (1985), such a system can lead to effective sex-linkage for a large part of the genome, for example about 50% in the Florida species *Incisitermes schwarzi* near Miami. Interestingly, similar sex-linked translocations are found in spiders, including at least one highly social species (Rowell, 1987).

Both X- and Y-chromosomes of termites carry active genes; each X-chromosome is homologous to a Y-chromosome, and vice versa. Hence, a relatedness structure of the X-linked or male-haploid type operates within each sex of termites with respect to X- or Y-linked genes. Both parents are equally related to sons and daughters, but sons are more closely related to each other than to their sisters, and the daughters are more closely related to each other than to their brothers. This result follows because all of a male's Y-chromosomes are passed on as a group to sons, and all his X-chromosomes to daughters.

If the proportion of the genome that is sex linked be denoted L, then the relatednesses between the progeny of a termite king and queen are given by

$$\text{relatedness between offspring of same sex} = (1-L)(0.5) + L(0.75)$$

$$\text{relatedness between a son and a daughter} = (1-L)(0.5) + L(0.25)$$

Sex chromosome systems are best developed among the lower termites, and seem to arise readily in termites rather than being ancestral in them (Crozier and Luykx, 1985; Fontana, 1991), with only scattered occurrence elsewhere (Table 1.5).

1.9 Social insects have many potential levels of selection

Fitness in social insects depends on survival and offspring production, as in other organisms. A differentiating feature is that in social insects both components depend on the function of the colony and not on individual performance alone. Therefore, the colony can be an important level of selection in social insect populations. Colony efficiency, high survival, and large reproductive output, should be in the best interests of all colony members. We expect selection to lead to cooperation and coordination of tasks in order to yield that efficiency—hence the colonies have sometimes been referred to as superorganisms (Wheeler 1911; Lumsden, 1982; Moritz and Southwick, 1992). However, the various colony members often differ in their interests concerning the production of sexual offspring: who will reproduce and how are the resources allocated? A colony is simultaneously a scene of both cooperation

Table 1.5. Animal species in which sex-linked translocation heterozygosity is known (from Rowell, 1987).

Species	Common name	$2n$	Ring (R) or chain (C) length
Austragalooides sp.	Leaf hopper	18	C4
Delena cancerides	Huntsman spider	22	C5, C9
Diaptomus castor	Copepod	34	R6
Cyrsylus volkameriae	Flea-beetle	30	C4
Mesocyclops edax	Copepod	14	R12, R14
Nosopsyllus fasciatus	Flea	20–27	C4
Otocryptops sp.	Centipede	25	C5
Isoptera spp.	Termites (24 spp.)	32–63	R4–R18

and conflicts. The conflicts concern reproductive decisions whereas the co-operation focuses on the overall function of the colony. We will later (Section 5.2) return to the question whether there is colony-level selection for sex allocation, and just note here that the sex ratio game is usually considered as a game between individuals.

Two approaches have been used when examining colony-level selection: the phenotypic approach and the allele frequency approach.

The phenotypic approach is based on various 'strategies' (characteristics) and asks what would be the evolutionarily stable strategy given selection at the colony level. The question of colony-level selection, and hence of some form of optimization of colony characteristics, thus enters into discussions of the units of selection (see Sober, 1985; Crozier, 1987a). We will not enter into such discussions here, but note that, for characteristics to be regarded as selected at the colony level, they must pertain to the colony as a whole and not be determined solely by one parent or the other. Possible examples of such characteristics include caste polymorphism and polyethism (division of labour) in general (Oster and Wilson, 1978:21–3) and communication (Hölldobler, 1984; Seeley, 1989). While this approach is indeed tempting and illuminating in the study of caste evolution, there are risks to it if applied widely, because a eusocial insect colony is *not* a single organism, or even a clone, but rather a group of genetically distinct individuals.

The allele frequency approach to colony-level selection uses exact population genetic models to determine the behaviour of genetic variation under selection at this level (Crozier and Consul, 1976). Owen (1986) considers the dynamics of selection at the colony level in both male-haploid and diploid genetic systems. For male-haploids, Owen includes examination of the effects of multiple mating by queens and of the workers producing all or none of the males. A number of stable equilibria are possible under colony-level selection, with the occurrence of multiple mating by queens affecting the number of equilibria, and the production of males by workers affecting the equilibrium gene frequencies.

The basis of these colony-level selection models is that workers of different genotypes also differ functionally. A colony with a mixture of two worker genotypes can be superior to a colony with only a single worker genotype present if a two-genotype nest manages better to carry out the functions of the colony (Oldroyd *et al.*, 1992). We can associate this genotypic diversity with incipient caste polymorphism. The coexistence of functionally specialized individuals in the same colony can have synergistic effects leading to increased survival of the whole colony. The functional differences between the worker genotypes can be based on a variety of characters, for example physiological differences (such as temperature optima, Bernstein, 1976), morphological differences, or behavioural differences. It is particularly interesting and important in this context to note that the task preferences of workers have a genetic component in the honey-bees *Apis mellifera* (Calderone and Page, 1988, 1991, 1992; Frumhoff and Baker, 1988; Robinson and Page, 1988; Robinson *et al.*,

1994) and *A. florea* (Oldroyd *et al.*, 1994b), as well as in the ants *Leptothorax rudis* (Stuart and Page, 1991) and *Formica argentea* (Snyder, 1992).

Colony-level selection is affected by the kind of genetic variation within and between colonies (Wade, 1985). These components depend on the level of variation in the population on the one hand and the colony types on the other. Colony types are affected by multiple mating by queens, number of reproductive queens in a colony, and the spatial organization of the colony (monodomy versus polydomy). Polygyny leads to there being more than one matriline per colony, although these are often but not always related (Section 4.4). Polyandry means that the worker progeny of a queen fall into a number of patrilines; within patrilines workers are related to each other by the familiar 0.75 value in male-haploids and 0.5 in diploids, but between patrilines relatedness is only 0.25 in both genetic systems. We can recognize various levels of organization in social insects which could in some contexts be the levels of selection in that the genetic homogeneity within groups is greater than the similarity between them (Table 1.6).

We can also turn the question around and ask whether colony-level selection has been a driving force in the evolution of polyandry, polygyny, and polydomy. Is genetic variability within a colony, which has a functional dimension, so important that it selects for those modifications of colony types? We will return to this question in Chapter 4.

Table 1.6. Levels of selection in social insects, at or above the level of the individual. Selection may also take place on DNA sequences and gametes (Crozier, 1987a).

Level
Individual
Patriline
Matriline
Nest
Colony
Population

1.10 Summary

The 'social insects' are those colonial insects with the eusocial life pattern, in which some individuals specialize in reproduction and survive for more than one generation of their relatively infertile assistants, the workers. Although some other insects fit the definition of eusociality, we confine ourselves mainly to the 'traditional' social insects, the termites, ants, and the eusocial bees and wasps, because their biology exemplifies adequately the interactions between social life and sex allocation.

A primary feature of most social insects according to the narrow focus we employ is the possession of the male-haploid genetic system, in which males are normally haploid and arise from unfertilized eggs, and females are diploid, arising from fertilized eggs. While the genetic-level mechanism leading to male-haploidy is completely obscure in most Hymenoptera, in many aculeates and in some parasitoids it involves heterozygosity at one or more sex loci: hemizygotes and homozygotes are male, heterozygotes are female. Male-haploidy affects sex allocation chiefly through (a) enabling females to choose the sex of their offspring with greater freedom than is the case with sex determination systems based on sex chromosomes, and (b) creating intrinsic asymmetries of relatedness between male and female family members. Male-haploidy also leads to intrinsically greater mortality from genetic causes in males than in females, but this difference is too small to alter the sex ratio significantly compared with selection stemming from such factors as relatedness asymmetry.

Termites are not male-haploid. There is a strong tendency in many species for the occurrence of translocation complexes tying much of the genome up in the sex chromosome system. These translocation complexes lead to termites being more closely related to siblings of the same than of the opposite sex. It seems more likely that these sex chromosome translocation complexes arise readily in termite evolution than that they have led to the evolution of termite eusociality.

There are many levels at which selection can act on social insects, with the colony level being an unusually important one. The colony is not, however, a unitary entity, and the chief questions about social insect sex allocation can be framed in terms of interactions within colonies. Who reproduces in the colony? How are resources allocated between colony growth and the production of new reproductives? How are resources allocated between male and female reproductive functions? These questions devolve to a central question that can be understood either in melodramatic terms of power, or in terms of the relative strengths of selection: *Who controls the allocation of colony resources*? In particular, the males and females in a colony's brood are often related differently to the workers as against the queen, and this leads to differences in the sex allocation pattern which the workers have been selected to produce from that which the queen is expected to favour. This difference in relatedness patterns is the basis for queen–worker conflict, the central conflict at the heart of hymenopteran societies.

2

Inclusive fitness and sex allocation

An important point in understanding the function and evolution of social insects lies in understanding how the workers can increase their own fitness. In the first chapter we already learnt that workers can play an active role in the evolution of social life, and when doing this they can run into conflicts with the queens, at least in the eusocial Hymenoptera.

In order to study the role played by different parties in the conflicts, we have to quantify their evolutionary successes and this is done using the inclusive fitness. An individual can increase its inclusive fitness either directly by producing its own offspring or indirectly by helping a relative to reproduce. The inclusive fitness of an individual depends on the number of offspring raised, on the genetic relatednesses and the reproductive values of the offspring produced, and on the mating successes of females and males in the population. Because the queens and workers commonly prefer different sex ratios when trying to maximize their respective inclusive fitnesses, we expect conflicts between the castes, and the sex ratios play a central role in our understanding of the function and evolution of social life. We shall in this chapter present the general sex ratio theory and the basic framework of inclusive fitness models used throughout the book.

2.1 The Shaw–Mohler equation for sex allocation

We start by presenting a general sex ratio theory. Consider a population in which there is genetic variation for the sex ratio produced by individual mothers or by pairs of parents. Can we determine the average sex ratio that the population will evolve? In order to do this we have to examine the fitness of an individual as a function of its sex allocation pattern. This fitness can be constructed on the basis of how the individual manages to pass on its genes in the population. If the individual can pass on its genes more successfully by producing daughters, the sex ratio of its brood should be female biased. If it will be more successful producing sons, its brood sex ratio should be male biased. It turns out that a fairly simple equation, the Shaw–Mohler equation

(Shaw and Mohler, 1953; Charnov, 1982:13–18), enables understanding of a great deal of sex allocation theory.

In humans, as in termites but not in hymenopterans, everyone has a father and a mother. For autosomal genes, this means that the two sexes contribute equally *as groups* to succeeding generations. Thus, if a population with N_F females and N_M males has N mating pairs, then the average number of matings per female is N/N_F and that per male is N/N_M. For example, if there are more males than females, then the average mating success of a male will be less than the average mating success of a female. The quantity 'mating success' refers to the number of matings per individual. This number is clearly inversely proportional to the frequency of same-sex individuals in the population. Because fitness can be viewed as a relative rather than an absolute quantity, we can also use related expressions for the mating success. The crucial point is that the ratio of [male mating success] to [female mating success] is N_F/N_M. Whether we define mating successes as N/N_M and N/N_F or as $1/N_M$ and $1/N_F$ will not affect the conclusions as long as they are defined similarly for the two sexes.

If, for the moment, we assume further a lack of complicating factors such as male-haploidy, inbreeding, population subdivision, and differential cost of males versus females, we can derive the fitness, W_i, of any sex ratio phenotype. Let the proportion of males in the population be r and that of females be $1 - r$, and let these proportions among the progeny of a given individual i be r_i and $1 - r_i$. If we further assume that the brood size is constant, say K, for each of the N breeding pairs, we have in the next generation $N_M = rNK$, $N_F = (1 - r)NK$, and the numbers within the brood of our focal individual are $n_M = r_i K$ males and $n_F = (1 - r_i)K$ females. For evaluating fitness of the sex ratio phenotypes, it is not sufficient to count offspring. We also have to pay attention to the mating success of these offspring. If the population sex ratio is not 1:1, a parent producing offspring of the rare sex will have, on average, more grandchildren than a parent who produces just as many offspring, but of the common sex. The reason for this is that the offspring of the rare sex have more matings per individual. It takes two generations until the effects of selection are apparent. Because the number of grandchildren is proportional to the number of matings of the offspring, we can now obtain the fitness of a sex ratio phenotype as

$$W_i = n_F \frac{N}{N_F} + n_M \frac{N}{N_M} = (1 - r_i)K \frac{N}{(1 - r)NK} + r_i K \frac{N}{rNK}$$

$$= \frac{1 - r_i}{1 - r} + \frac{r_i}{r} \qquad (2.1)$$

which is the so-called **Shaw–Mohler equation**. Note that if the sex ratio phenotype in question has a brood size of K_i instead of K, the fitness of that individual will then be multiplied by the ratio K_i/K, but is still proportional to the term $(1 - r_i)/(1 - r) + r_i/r$.

Equation (2.1) is based on the number of grandchildren and, in fact, if the daughters and sons are equally valuable, eqn. (2.1) gives the expected number of genes an individual contributes to future generations in a population of stable size. Thus, it can be used to determine the population sex ratio that forms the **evolutionarily stable strategy** (ESS) (Maynard Smith, 1982:10–27). An ESS is a phenotype 'such that, if all members of a population adopt it, then no mutant' phenotype 'could invade the population under the influence of natural selection'. This definition is the same as Maynard Smith's (1982:10), save that we have used the more general description 'phenotype' instead of 'strategy'. We will later emphasize the point that sons and daughters often are *not* equally good transmitters of their parents' genes, and in such cases W_i must be modified appropriately.

Let us examine eqn. (2.1) more closely. When $r_i = r$, $W_i = 2$, which means that in a population of stable size the phenotype producing the population sex ratio contributes, on average, two genes to future generations, whatever the sex ratio is. In a population with $r = (1 - r) = 0.5$, all sex ratio phenotypes (r_i) contribute on average two genes to future generations. In other words, when the population has an equal sex ratio, no individual can gain an advantage by, nor does it suffer from, producing some other sex ratio among its own progeny. (This principle is violated if the offspring quality (fitness) in one sex depends on the amount of resources that can be used to raise them. For example, the resources available might suffice to produce competitive males, but not high-quality nest-founding females. In that case, parents with few resources would benefit most by producing males and not females.)

To examine the evolution of sex ratios, we can write eqn. (2.1) in the form

$$W_i = \frac{1}{1-r} + r_i \frac{1-2r}{r(1-r)}$$

which is a linear function of the individual sex ratios, r_i, with the sign of the slope depending on $(1 - 2r)$, as shown in Fig. 2.1. The important messages from this treatment are as follows.

1. When $r = 0.5$, all sex ratios have the same fitness of $1/(1 - r) = 2$, as noted above. This means that the ESS is a population phenomenon so that the population can consist of various sex ratios yielding the mean of $r = 0.5$.

2. When $r \neq 0.5$, any rare sex ratio which will bring the population ratio closer to 0.5 has a fitness higher than 2. The phenotype at the opposite extreme to r has the highest fitness.

3. It follows from point 2 that a sex ratio of $r_i = 0.5$ always does better than the population average when $r \neq 0.5$, i.e. it can invade any such population.

The ESS of $r = 0.5$ is quite a general one under the conditions of autosomal inheritance and no population subdivision or inbreeding, but it properly

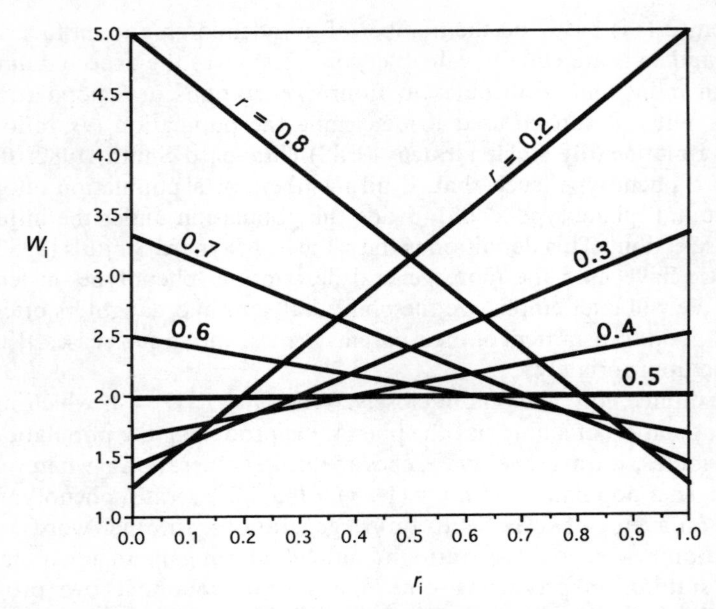

Fig. 2.1 Fitness (W_i) of a sex ratio variant r_i in populations of various overall average sex ratios (r), according to eqn. (2.1). When $r = 0.5$, all sex ratios have the same fitness, but otherwise sex ratios are favoured if they are closer to 0.5 than r or have the opposite bias.

pertains to the investment (amount of resources) allocated to producing each sex, and not to the numerical ratio.

To see that it is the investment and not the numerical ratio that is the chief subject of selection, consider a case where the cost of producing one female is c_F and that of producing one male is c_M, and the total cost of producing a brood is C. Let the proportional investments in females and males respectively be F and M in the population and f and m for the given sex ratio phenotype. Following the same scheme as above, we find that $N_F = FNC/c_F$, $N_M = MNC/c_M$, $n_F = fC/c_F$ and $n_M = mC/c_M$. From this can be obtained the investment equivalent of eqn. (2.1):

$$W_i = \frac{f}{F} + \frac{m}{M}. \tag{2.2}$$

This fitness is of the same form as in eqn. (2.1), except that the proportions of males and females are now counted in units of investment. The investment ratio of 1:1 is stable against invaders, and itself can invade populations fixed for other sex ratios: it is an ESS, but the numerical sex ratio is not. The result is intuitively expected. It is possible to produce more individuals of the cheaper sex, but overproduction of them lowers their mating success and a balance is achieved.

The numerical proportion of males in the population is

$$r = \frac{Mc_F}{Mc_F + Fc_M} \,. \tag{2.3}$$

In particular, when $M = F = 0.5$, $r = c_F/(c_F + c_M)$. For example, for populations with 1:1 investment ratios the numerical proportion r at equilibrium is 0.67 if the cost of a female is twice that of a male, and $r = 0.8$ if $c_F = 4c_M$.

In order for the sex allocation of 1:1 to be an ESS, a number of special conditions are required, such as lack of inbreeding and population subdivision, and equal genetic value of the males and females in the brood to the individual capable of biasing the brood sex ratio. The prevalence of the 1:1 ratio in groups such as higher vertebrates therefore indicates that the conditions for this ratio must have predominated over a very long period, selecting for sex chromosome systems with heterogamety (XY or ZW), although even such a system can be modified to produce other ratios (Karlin and Lessard, 1986:17).

2.2 Genetic values of the offspring

The derivation of the Shaw–Mohler formula in Section 2.1 was based on equalizing the mating successes of males and females. However, mating success is but one component of the evolutionary value of the offspring. The offspring should be valued in terms of how well they transfer into future generations the genes of the individual controlling sex allocation. For this reason, we must also take into account *genetic relatedness,* i.e. the probability with which the offspring carry genes identical with those in the controlling individual, and the *reproductive value* of these offspring (i.e. how many copies of its genes an offspring successfully transmits to future generations). The reproductive value depends partly on sex-specific effects (haploid males are less effective than diploid females as gene carriers) and partly on other fitness components (Hedrick, 1985:164).

Genetic relatedness has attained a special significance in evolutionary studies due to Hamilton's (1963, 1964a) theory of social evolution under **kin selection** (so named by Maynard Smith, 1964). In brief, the theory can be reduced to a single condition for altruistic or donoristic behaviour to evolve, which we can call Hamilton's Principle or Rule, namely

$$\text{cost/benefit} < g_{BA} \,. \tag{2.4}$$

In words, an altruistic trait which is genetically determined increases in frequency if the ratio [cost to the actor]:[benefit to the recipient] is less than the genetic relatedness (g_{BA}) of the beneficiary to the actor. The costs and benefits are measured in terms of decrements and increments to the number of offspring produced and they also include the reproductive values ν. There is a large literature concerning the conditions and formulations under which eqn. (2.4) is valid (e.g. Uyenoyama and Feldman, 1980; Michod, 1982; Grafen,

1985; Queller, 1992). One of the crucial points is obviously how relatedness is defined. Grafen (1985) suggests cutting the Gordian knot by *defining* relatedness so that eqn. (2.4) is true.

It is worthwhile inserting a brief remark about the biological definition of altruism. In biological studies of social behaviour, altruism is defined in terms of *consequences*, whereas in human discourse altruism is defined in terms of *intentions*. The emphasis on consequences rather than intentions stems from the need for experimental practicality: one cannot easily ask animals their intentions (it is also true that one cannot rely on the accuracy of answers of humans either!).

Hamilton's Principle eqn. (2.4) can also be expressed in terms of inclusive fitness, which measures the contribution of an individual to the gene pool. This contribution can be mediated by its own reproduction or by aid it gives to the reproduction of others. The inclusive fitness of an individual I can be defined as (Hamilton, 1964a; Michod, 1982)

$$W_{\mathrm{I}} = W_{\mathrm{I}}^{*} + \sum g_{\mathrm{JI}} \, \Delta \, W_{\mathrm{JI}} \tag{2.5}$$

where W_{I} is the inclusive fitness of individual I, W_{I}^{*} is the fitness of individual I as it would have been without social interactions involving altruism, ΔW_{JI} is the change to the fitness of individual J due to the interaction of individual I with it, and g_{JI} is the relatedness of individual J to I.

Note that each interaction also affects individual I itself, and the recursive fitness effect ΔW_{II} is included in the summation term. Behaviour which maximizes the inclusive fitness of carriers should increase in frequency in the population (Michod and Abugov, 1980).

In social insect colonies, we can write the inclusive fitness in a somewhat different form that counts the total output from a colony. Based on the formulations of Taylor (1988) and Pamilo (1991a), the inclusive fitness of an individual I in a colony producing N offspring with allocation ratio m:f is

$$W_{\mathrm{I}} = N(g_{\mathrm{FI}} \nu_{\mathrm{F}} \frac{f}{F} + g_{\mathrm{MI}} \nu_{\mathrm{M}} \frac{m}{M}) \tag{2.6}$$

where g_{FI} and g_{MI} are the relatednesses of the female and male offspring to individual I, and ν_{F} and ν_{M} are the sex-specific reproductive values.

This formulation is different from that given in eqn (2.5). In eqn. (2.5) the inclusive fitness was counted in terms of increments and decrements caused by individual interactions, whereas eqn. (2.6) counts the total number of offspring after summing all the effects in the process of producing them. In order to maximize its fitness according to eqn. (2.6), an individual can either increase the total number of offspring produced, N, or increase the term in parentheses by biasing the sex ratios in favour of the offspring with the highest genetic value to itself. We will now examine the latter possibility, and denote the term in parentheses V, the inclusive fitness effect of the given sex ratio phenotype:

$$V = g_{\mathrm{FI}} \nu_{\mathrm{F}} \frac{f}{F} + g_{\mathrm{MI}} \nu_{\mathrm{M}} \frac{m}{M}. \tag{2.7}$$

For example, the inclusive fitness functions for a monogynous colony are, for the queen,

$$V_{\mathrm{Q}} = g_{\mathrm{daughter,Q}} \nu_{\mathrm{F}} \frac{f}{F} + g_{\mathrm{son,Q}} \nu_{\mathrm{M}} \frac{m}{M} \tag{2.8a}$$

and for the workers,

$$V_{\mathrm{W}} = g_{\mathrm{sister,W}} \nu_{\mathrm{F}} \frac{f}{F} + g_{\mathrm{brother,W}} \nu_{\mathrm{M}} \frac{m}{M}. \tag{2.8b}$$

In order to maximize V, the individuals should decide how to allocate investments between female and male offspring, i.e. they have to compare the terms

$$g_{\mathrm{FI}} \nu_{\mathrm{F}} \frac{1}{F} \text{ and } g_{\mathrm{MI}} \nu_{\mathrm{M}} \frac{1}{M}.$$

At the population's sex ratio equilibrium, the fitness returns through sons and daughters are equal and it does not matter how the investment is allocated between the sexes. Hence, the equilibrial proportions of males (M^*) and females (F^*) in the population are found by setting

$$g_{\mathrm{FI}} \nu_{\mathrm{F}} \frac{1}{F^*} = g_{\mathrm{MI}} \nu_{\mathrm{M}} \frac{1}{M^*} \tag{2.9}$$

from which we can obtain the equilibrium

$$M^* = \frac{g_{\mathrm{MI}} \nu_{\mathrm{M}}}{g_{\mathrm{MI}} \nu_{\mathrm{M}} + g_{\mathrm{FI}} \nu_{\mathrm{F}}}. \tag{2.10}$$

The fitness as given by eqn. (2.7) is a linear function of the allocation proportion m, as can be shown by rearranging to yield

$$V = \frac{g_{\mathrm{FI}} \nu_{\mathrm{F}}}{F} + m \frac{g_{\mathrm{MI}} \nu_{\mathrm{M}} - (g_{\mathrm{MI}} \nu_{\mathrm{M}} + g_{\mathrm{FI}} \nu_{\mathrm{F}}) M}{M(1-M)}.$$

from which can be verified the expected result that, when $M > M^*$ all $m < M$ are selected for and when $M < M^*$ all $m > M$ are selected for.

Let us illustrate this modification of the Shaw–Mohler equation in terms of sex allocation in eusocial Hymenoptera. Assume first that the sex allocation is controlled by the queens and that there is only one, monandrous, queen in each colony, i.e. the colonies are monogynous. Assume further that all reproduction is by the queen. We will discuss relatednesses and reproductive values in the next two sections, but it can be noted here that in the simple case of our present example the daughters are related to the queen by $g_{\mathrm{FQ}} = 0.5$ and the sons by $g_{\mathrm{MQ}} = 1.0$ and the sex-specific reproductive values of the females and males are $\nu_{\mathrm{F}} = 2$ and $\nu_{\mathrm{M}} = 1$. Therefore, the expected sex ratio from eqn. (2.10) is $M^* = \frac{1}{2}$, namely a 1:1 ratio between females and males. On the other hand, if sex allocation is controlled by the workers as a group, then the relevant relatednesses are $g_{FW} = 0.75$ and $g_{MW} = 0.5$, and it then follows from eqn. (2.10) that the expected sex allocation ratio is given by $M^* = \frac{1}{4}$, so that the sex allocation ratio is expected to be 3:1 in favour of females (Trivers and Hare, 1976).

This example brings out several points. Firstly, males and females may differ in their relatedness to the individuals capable of controlling the sex allocation ratio. Such differences are likely to be common in particular genetic systems due to special aspects of chromosome mechanics, as in the case of the sex chromosome systems of many termites, and under male-haploidy, as in the Hymenoptera. Secondly, in eusocial insects, parents are not necessarily the chief influences on the allocation ratio, and relatedness values should therefore be determined with respect to the individual or individuals controlling sex allocation. Thirdly, genetic relatedness in its simplest form, that used in this book, is a necessary but not sufficient guide to the determination of the sex allocation ratio: reproductive values have to be used as well. We have used here only the sex-specific reproductive values. In reality, the reproductive values within a sex can vary depending on the available resources and the brood sex ratio. This variation can be an important factor directing brood sex ratios and causing variations in sex ratio between broods.

Let us next examine a case where an individual I has a choice between two alternatives: it can either raise its own brood with N_1 offspring and an allocation ratio m_1:f_1, or it can raise N_2 siblings with an allocation ratio m_2:f_2. The individual should now compare the inclusive fitnesses,

$$W_1 = N_1(g_{\text{daughter,I}} v_F \frac{f_1}{F} + g_{\text{son,I}} v_M \frac{m_1}{M}) \tag{2.11a}$$

$$W_2 = N_2(g_{\text{sister,I}} v_F \frac{f_2}{F} + g_{\text{brother,I}} v_M \frac{m_2}{M}), \tag{2.11b}$$

and choose that which is larger. When $W_2 > W_1$, an individual should give up its own reproduction and start helping its relatives. Of course, the individual might instead select a combination of sons and sisters, for example, if that approach gives a higher inclusive fitness.

When the relatednesses and reproductive values differ between the sexes, sex allocation can become an important determinant of social life: selection may lead to both cooperation and conflict over the sex ratio.

2.3 Genetic relatedness

Because genetic relatedness is extremely important in determining the inclusive fitness, in the evolution of cooperative behaviour and also in sexual allocation, we briefly discuss this concept, its definition and some problems in estimating it in natural populations. The determination of genetic relatedness can be complicated because it depends not only on the recent pedigree connections of the individuals concerned but also on population structure (in particular on population subdivision) and on the effects of evolutionary forces such as selection and drift. We can distinguish two completely different meanings of relatedness: absolute relatedness (which refers to the true identity of genes by descent), and relative relatedness (which compares genetic identity to the average similarity between individuals in the population). This distinction can be clarified with an example (see also Jacquard (1975) for a similar discussion of inbreeding). Suppose that an island is successfully colonized by a solitary mated queen, so that some generations later her descendants occupy the island as a number of colonies. The absolute relatedness of the individuals is now high, because they are all descended from the one ancestral queen. But when the population is considered in isolation we can consider the individuals from different nests as being unrelated to each other and estimate the relatedness within colonies in reference to this background. Although absolute relatedness can be important, it is relative relatedness which matters when considering the evolutionary dynamics of a single population. Hence, we use genetic relatedness in the relative sense in this book.

Relatedness is a measure of similarity, and the mean determinant of similarity is pedigree structure. If ancestries are traced back far enough, it is a truism that all copies of a gene will be found to have descended from the same ancestral copy. In order to avoid obtaining this trivial result of absolute relatedness, the pedigree is traced for only several generations and all genes in the population at that point are considered to be unrelated. But similarity can also be affected by selection modifying the genotype frequencies. Two different definitions of relative genetic relatedness can be recognized. First, relatedness has been defined as the expected genetic similarity on the basis of the pedigree, ignoring the effects of selection or drift (Crozier, 1970a). Second, relatedness has been defined as a measure of the genetic similarity of one individual to another by taking into account the effects of selection when measuring similarity (e.g. Uyenoyama and Feldman, 1981). Relatedness during selection has often been defined in terms of the covariances between genotypes and phenotypes (Orlove and Wood, 1978; Michod and Hamilton, 1980; Uyenoyama, 1984; Grafen, 1985; Taylor, 1988), because such a definition applies to Hamilton's rule of social evolution (Hamilton, 1964a).

The first concept, genetic similarity in the absence of selection, is useful when analysing the genetic structure of colonies and populations using genetic markers because the estimates obtained can be easily compared with expectations based on pedigree connections. The latter concept, genetic similarity under selection, is relevant in studies concerning specific characters such as social behaviour and sex allocation which are the very targets of selection. This concept is important in theoretical selection models, but because we generally have no information about the genetic architecture of these characters, the concept is difficult to apply in studies of actual populations. When selection on the locus, or loci, under consideration is weak, then relatedness under selection is very close to that expected for selectively neutral loci. Therefore, for most purposes it is possible to use the concept of genetic relatedness based on the expected similarity of neutral genes as a guideline for examining evolutionary dynamics under selection. This is the approach we use in this book.

We define the relatedness of individual Y to individual X, g_{YX}, as the proportion of Y's genes which are identical by descent to genes in X. If we denote by f_{YX} the probability that two gametes, one produced by Y and one by X, share the same allele through derivation from the same ancestral allele, and by f_{XX} the same probability for two gametes each from X, then

$$g_{YX} = f_{YX}/f_{XX} \, . \qquad (2.12)$$

A male-haploid pedigree and the coefficients of relatedness were shown in Fig. 1.2 and Table 1.3. Clearly, male-haploidy imposes a strong intrinsic bias towards asymmetrical relatednesses between individuals of different sex. This intrinsic bias is expected to have a strong effect on the sex allocation at equilibrium.

Both in theory and, especially, in empirical studies, it is useful to present the coefficient of relatedness with the help of covariances. Relatedness can thus be estimated as a mean value over repeated observations of pairs Y and X, e.g. of daughters and mothers, using genetic markers. We have suggested that the relatedness of Y to X be estimated as the regression coefficient of additive genetic values of Y on X, denoted earlier b_{YX} (Crozier and Pamilo, 1980; Pamilo and Crozier, 1982). We should note that many authors have used a reverse wording, calling the regression of Y on X the relatedness of X to Y. But relatedness is the measure of the genetic worth of one individual to another as a carrier of its genes, say the worth of Y to X, and it is logical to emphasize this directionality when defining the concepts (Taylor, 1988).

In many cases it will be more appropriate to estimate the mean relatedness between colony members using allele frequencies rather than pedigree information. There are methods for estimating relatedness in this way using genotype frequencies (Pamilo and Crozier, 1982; Pamilo, 1984a, 1990a; Queller and Goodnight, 1989); these methods have also been used in analysing DNA fingerprinting data (Reeve *et al.*, 1990, 1992).

A promising technique for detecting genetic variation and estimating relatednesses is 'microsatellite' variation, which involves variation in the number of 2–4 base repeats in the DNA sequence; Moritz *et al.* (1991) were able to distinguish patrilines in honey-bee colonies using a simple repeat, and highly polymorphic microsatellite loci have been found in bees (Estoup *et al.*, 1993), in social wasps (Choudhary *et al.*, 1993; Hughes and Queller, 1993) and in ants (Evans, 1993; Hamaguchi *et al.*, 1993; Gertsch *et al.*, 1995).

Microsatellites give access to highly polymorphic and easily scored loci, but require considerable effort to derive because of the need to construct genetic libraries for the species concerned (but sometimes the primers are applicable to closely related species). The RAPD technique, in which relatively non-specific primers reveal variation among the products from amplification using the polymerase chain reaction (Welsh and McClelland, 1990; Williams *et al.*, 1990) offers a relatively easy route to large amounts of genetic variability, although the results are often hard to analyse (see Crozier, 1993). The first results from the honey-bee and fire ants show good polymorphism and Mendelian inheritance of the markers (partly of dominant and partly of codominant type) suggesting that RAPD markers will be useful in studies of social insects (Hunt and Page, 1992; Fondrk *et al.*, 1993; Shoemaker *et al.*, 1994).

2.4 Sex-specific reproductive values

The concept of reproductive value goes back to Fisher (1930; see also Crow and Kimura, 1970:20–2), as of course does all sex allocation theory. We can define the reproductive value, v, of a breeding individual as its genetic contribution to future generations and measure it as the number of genes contributed by it to the gene pool. It is not enough to count the genes transferred to the offspring, because the gene frequency dynamics through male and female offspring can differ from each other. It is well known that the gene frequencies of sex-linked loci oscillate if there is an initial frequency difference between the two sexes. The same oscillation is expected under male-haploidy. The frequencies approach a final value asymptotically, and we can measure an individual's genetic contribution with the help of this concept of the asymptotic gene pool (Oster *et al.*, 1977). Often it will be sufficient to count contributions among the grandchildren, but the reproductive value can be difficult to estimate, as in the case of age-structured populations.

We measure the expected contribution by ignoring the effects of stochastic changes caused by finite population size. When the population size remains constant, each individual is expected, on average, to replace itself and, similarly, each gene is expected to replace itself. For X-linked loci (females XX and males XY or XO), and under male-haploidy, there are two copies of each gene in females and only one in males. A female passes on her genes through both daughters and sons, but the males can pass on their genes only to daughters, and the relative reproductive values of a female and a male are $v_F = 2$ and $v_M = 1$, respectively (but see Section 4.2.1 for the case of worker reproduction). These are the sex-specific reproductive values and they can vary individually within each sex depending on the condition of the individuals concerned. Note also that the absolute values given to the v are arbitrary, the really important quantity is the ratio v_F/v_M. In this book we use values for the v which refer to the numbers of genes passed on to future generations in a population of stable size.

Taylor (1988) and Pamilo (1991a) have derived ways to calculate general sex-specific reproductive values using slightly different approaches. Taylor derived these values by examining the probability that a gene in the population originated from a female (or alternatively from a male) in an ancestral population. Pamilo derived sex-specific reproductive values by using the probabilities that a female (or a male) succeeds in transmitting copies of her (or his) genes to a new generation. Taylor's approach is to look at where the genes have come from, whereas Pamilo looks at where the genes are going to. The two approaches give identical results that depend on the transmission rules between successive generations. A further alternative would be to use recursion equations (Oster and Wilson, 1978:82). We present here Pamilo's approach.

To derive the sex-specific reproductive values, namely the relative contributions of females and males in the remote future gene pool, we need the probabilities with which a parent transmits a copy of a given gene to an offspring. The probabilities we need are those of a female transmitting a gene to a random daughter (p_{FF}) or son (p_{MF}) and of a male transmitting a gene to a random daughter (p_{FM}) or son (p_{MM}). We can write these in the form of a transition matrix **P**:

$$\mathbf{P} = \begin{pmatrix} p_{FF} & p_{FM} \\ p_{MF} & p_{MM} \end{pmatrix}.$$

When a hymenopteran queen produces all of the offspring, the probabilities are $p_{FF} = p_{MF} = 0.5$, $p_{FM} = 1$ and $p_{MM} = 0$. The transmission probabilities from the parental to the nth offspring generation are given by the nth power of the matrix above ($\mathbf{P}^n$), and the transmission probabilities to the asymptotic gene pool in the remote future are obtained by letting n increase to infinity:

$$\lim \mathbf{P}^n = \mathbf{P}_\infty = \begin{pmatrix} \pi_{FF} & \pi_{FM} \\ \pi_{MF} & \pi_{MM} \end{pmatrix}$$

where π_{FF} is, for example, the probability that a female contributes a given gene-copy to a female in the remote future.[1] For the probabilities given above, i.e. for the case where all reproduction is by the queen, these asymptotic probabilities become

$$\mathbf{P}_\infty = \begin{pmatrix} 2/3 & 2/3 \\ 1/3 & 1/3 \end{pmatrix}.$$

The sex-specific reproductive values of males and females can be obtained from the matrix $\mathbf{P}_\infty$, remembering that females have two genes to contribute and males have one, so that the reproductive value of a male is

$$v_M = \pi_{FM} + \pi_{MM} \tag{2.13a}$$

and that of a female is

$$v_F = 2(\pi_{MF} + \pi_{FF})\ . \tag{2.13b}$$

1 The elements of the matrix can be calculated for male-haploids as follows. Let $c = 4p_{MF} + p_{FM}$ and then $\pi_{FF} = 4p_{MF}/c$, $\pi_{FM} = 2p_{FM}/c$, $\pi_{MF} = 2p_{MF}/c$ and $\pi_{MM} = p_{FM}/c$.

So, when the queen does all the reproducing, the reproductive values are $\nu_M = 1$ and $\nu_F = 2$, as we have already given somewhat more intuitively. These reproductive values tell us how many copies of genes the male and female in a mating pair are expected to contribute on average to the asymptotic gene pool of a population of constant size. Naturally, population size is expected to change often over time, and in that case the expected numbers of copies are altered. In that respect, the reproductive values are only relative, for example Taylor (1988) defines them in such a way that ν_M and ν_F sum to unity. In fact, we need only the ratio ν_F/ν_M, which for hymenopteran pedigrees can be shown (Pamilo, 1991a) to be

$$\nu_F/\nu_M = (1 - p_{MM})/(1 - p_{FF}) = 4p_{MF}/p_{FM} \,. \tag{2.14}$$

With these analytical tools we can determine the sex-specific reproductive values for cases where the matrix **P** differs from the basic form given above (Table 1 of Pamilo, 1991a). We will return to such cases later in Sections 4.2.1 and 5.1, but it can already be remarked that the ratio ν_F/ν_M does not always equal two but depends on reproductive characteristics, such as on male-production by workers.

Several authors have suggested weighting relatedness by the ploidy level (Hamilton, 1972; Crozier, 1979; Pamilo and Crozier, 1982). This concept has been called the *life-for-life* or *weighted* relatedness, and it has been used in several important papers, e.g. by Trivers and Hare (1976:Table 1). The argument for weighted relatedness is that the genetic worth of offspring depends on both genetic relatedness and reproductive value. The ploidy level does not, however, necessarily equal the reproductive value (see Sections 4.2, 5.1); for example the weighted coefficients have to be modified if there is male-production by workers. We agree with Taylor (1988) that clarity is best served by keeping genetic relatedness and reproductive value separate, and that is our approach for the rest of this book.

2.5 Fitness functions can be non-linear

The equilibrium theories presented above and originating from Fisher (1930) are based on the assumption that the genetic return, the evolutionary profit, of an investment is linearly related to the amount of investment. Therefore, at equilibrium, selection favours each individual sex ratio equally, i.e. it does not matter how the investment is divided between male and female offspring. But this is not always the case. It may be that a double investment in daughters more than doubles their success, or that a double investment in sons does not increase their joint success, or increases it only slightly.

We may start by examining the situation in non-social, parasitic hymenopterans, where the female lays one egg per host. The size of the host thus determines the amount of resources available for the developing offspring. It has been known for a long time that the ovipositing females can decide the sex of the offspring on the basis of the size of the host, and that females normally tend to emerge from bigger hosts. Careful studies show that this correlation is not due to different mortality of male and female offspring in hosts of different size, but arises from the decision of the egg-laying female whether to fertilize an egg or not (Charnov, 1982:48–66).

The adaptive significance of this behaviour is that the fitness of a female is more strongly affected by the available resources than is the fitness of a male. Although big males are fitter than small males, the fitness difference between a big and a small female appears to be greater. Therefore, an egg-laying female gains more when fertilizing the egg laid on a big host.

A formal analysis of this situation can be based on an equation derived by MacArthur (1965). If we denote by $\omega_F(.)$ the fitness function of daughters and by $\omega_M(.)$ the fitness function of sons, then at equilibrium

$$\frac{\omega'_F(f)}{\omega_F(f)} = \frac{\omega'_M(1-f)}{\omega_M(1-f)} \tag{2.15}$$

where ω' refers to the derivative of the function ω.

If both $\omega_F(.)$ and $\omega_M(.)$ are linear functions of f (or m), the equation reduces to $1/f = 1/(1-f)$, and the solution is $f = 1/2$. When the functions (one or both of them) are not linear, the investment in male and female functions may differ (Charnov, 1979; Maynard Smith, 1980). Various fitness functions have been examined by Frank (1987a).

We should note here that eqn (2.15) examines the fitnesses of sons and females in terms of the grandchildren they produce to future generations. We have already shown, in eqn. (2.14), that the fitness return to the controlling individual is given by the product of genetic relatedness (g), reproductive value (ν), and the mating success of an offspring. If the total fitness of daughters receiving the amount f of resources used in sexual production follows function $\omega_F(f)$, and the total fitness of sons $\omega_M(m)$, the fitness associated with sex ratio (f, m) of individual I is proportional to the function

$$V_I = g_{FI}\nu_F \frac{\omega_F(f)}{\overline{w}_F} + g_{MI}\nu_M \frac{\omega_M(m)}{\overline{w}_M} \tag{2.16}$$

where $\overline{w}_F$ and $\overline{w}_M$ are the mean fitnesses of females and males in the population. Two types of fitness have been discussed here. V_I is the fitness of a sex allocation type I and measures only the effects of differential allocation. ω refers to the fitness of the offspring and includes fitness components such as survival, mating success, and fecundity. V can be used to examine both individual strategies and population equilibria. To find the population equilibrium, the function is maximized at the point where the derivative ($\mathrm{d}V/\mathrm{d}f = V'$) equals zero. To determine this, we have to assume that the relatednesses g_{FI} and g_{MI} are the same for each individual in the population (say g_F and g_M). The solution, derived by Pamilo (1991a) is

$$\frac{dV}{df} = V' = g_F v_F \frac{\omega'_F(f)}{\overline{w}_F} - g_M v_M \frac{\omega'_M(1-f)}{\overline{w}_M} = 0 \qquad (2.17)$$

and the derivatives (ω') are taken for f. If we give ω_F and ω_M as functions of f and m, we can write the condition for equilibrium as

$$g_{FI} v_F \frac{\omega'_F}{\omega_F} = g_{MI} v_M \frac{\omega'_M}{\omega_M} \qquad (2.18)$$

where ω'_F and ω_F are both evaluated at f, and ω'_M and ω_M are both evaluated at m. It is easily seen that from the point of view of the queen this reduces to eqn. (2.15). From the point of view of the workers the product of g and v may differ between male and female offspring. When the fitness functions are linear, eqn. (2.18) reduces to eqn. (2.9).

We can next ask what is the effect of varying the individual costs of sons and daughters? As before (Section 2.1), let C be the resources expended on a brood. Assume that a colony spends mC resources in producing males and fC in producing females, with individual costs of offspring being c_M and c_F respectively. If a small decrement δc to c_M decreases the survival and mating success of a male only a little but the same amount δc as an increment in a female considerably increases her survival as a nest foundress, producing bigger females and smaller males would certainly be favoured. But that should not affect the sex allocation ratio at the level of a colony or population. It only changes the packaging strategy: instead of mC/c_M big males the nest produces $mC/(c_M-\delta c)$ small males. The fitness functions become important when the fraction δc in mC and fC cannot be exchanged without a change in total fitness. Such a situation can arise, for example, when an offspring requires so large a proportion of parental resources that the allocation problem cannot be solved by altering the package sizes.

The population and individual investment patterns interact. The resources available for sexual production vary among individuals in the population, and there are situations where the allocation of these resources depends on the amount that can be used. Such a situation arises when the physical condition of the offspring correlates with the amount of resources in a sex-dependent way. Such a correlation may lead to variation between colonies in the optimal sex allocation.

2.6 Importance of collateral relatives

The sex ratio may vary markedly during the ontogeny of a brood. It is conventional for studies on vertebrates to distinguish between the *primary* sex ratio (that occurring at fertilization), the *secondary* sex ratio (that at birth), and that among mature adults, the *tertiary* sex ratio (Cavalli-Sforza and Bodmer, 1971:651). Because insect eggs are fertilized while they are being laid, the primary and secondary sex ratios are tightly linked together. Hence, the classification of Cavalli-Sforza and Bodmer is not particularly useful for insects.

Mothers still have great influence over the sex ratio in Hymenoptera due to their power of choosing whether or not to release sperm, and possibly also in termites through differential selection on X- and Y-bearing sperm (although there is no evidence that they do so).

Because the nomenclature derived for vertebrates is inappropriate for insects, we instead distinguish simply the *initial* and *final* sex ratios. The initial sex ratio is that at the time eggs are laid. The final sex ratio is that at the time individuals become adult. The parents determine the initial sex ratio or, rather, the initial sex ratio is determined within the mother's body.

The conversion of the initial to the final sex ratio takes place during the time that the progeny mature into adults. During this period, they are subject to extensive influences of not only their parents but also other relatives, perhaps including very distantly related ones depending on the structure of their colony and population. In mature social insect colonies, the vast numerical preponderance of collateral relatives (workers) compared to parents implies a similar likely predominance of the power of these relatives over the sex ratio of the developing brood (Fig. 2.2).

In termites, which have an XY sex determination system, the father has an obvious influence over the initial sex ratio through variation in the ratio of X-

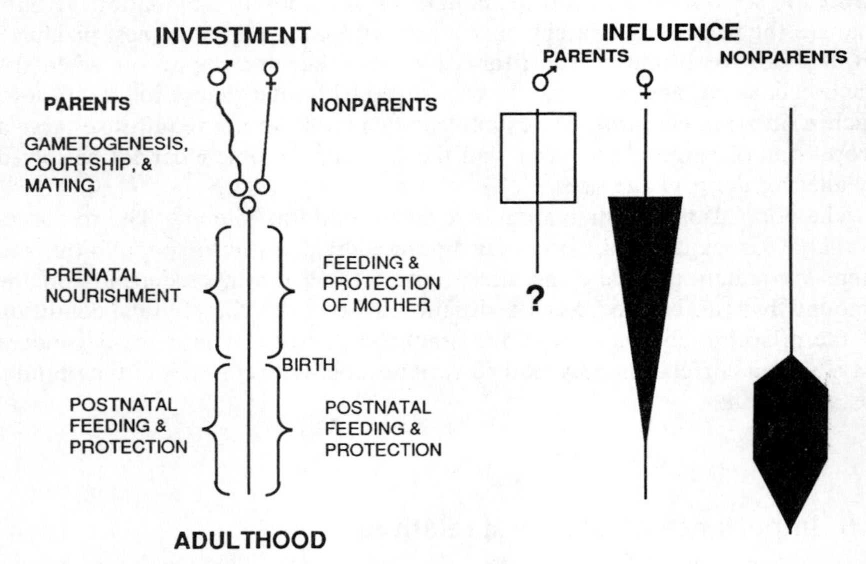

Fig. 2.2 Relative influence of parents and collateral relatives over the sex ratio of a developing brood in a eusocial insect colony. While parents have significant investment (gametogenesis, courtship) before fertilization, that of non-parents begins thereafter. Males in male-diploid species (termites) have a large potential influence on sex ratio via the proportions of the sex chromosomes produced, but probably have no influence on sex ratio in male-haploid species (ants, bees and wasps).

and Y-bearing sperm. This variation is not present in male-haploids, although heritable variation in the proportion of sperm able to fertilize eggs has been reported in the parasitoid *Dahlbominus fuscipennis* (Wilkes and Lee, 1965). All that is open to hymenopteran males is to produce 'muscular sperm', capable of forcing their way out of the spermatheca (female's sperm storage device) and fertilizing passing eggs (Starr, 1984), but this seems rather unlikely given the formidable valvular nature of the spermathecal duct (Dallai, 1972, 1975).

The capacity of workers to modify the sex ratio comes with their role in caring for the developing offspring. They can use their power either by biasing the numerical sex ratio or by biasing the individual costs of females and males without changing the numerical sex ratio. The workers can try to change the proportions of males and females by killing immatures of the sex they do not favour (Aron *et al.*, 1994), or by shifting individual larvae either away from or towards development as reproductives (Brian, 1983:186–97). The payoff for such biasing becomes less as the progeny approach adulthood and the possible saving of investment by affecting them becomes less and less. Even young adults are not safe from molestation, however, because under some circumstances, such as the end of the mating season, it is better to eject or, if possible, to recycle now useless reproductives rather than to care for them further. It is also possible that the workers could influence the proportion of haploid eggs laid by the queen, for example if the queen uses ambient temperature as a guide to the proportion to lay and the workers control the nest temperature, or if fertilized and unfertilized eggs are laid in different places, as in a honey bee hive. Allocation ratios can also be affected by differential feeding of adults or late instar larvae. For example, the females of the ant *Lasius niger* gain 75% of their final weight while remaining in their natal nest as adults prior to the mating flight (Boomsma and Isaaks, 1985).

The relatedness of the male and female progeny to the workers may differ from that to the parents, leading to a conflict of interest as we noted above (Section 2.2) and discuss further below. There are various ways by which the mother can counter, in an evolutionary sense, the vast numerical advantage of the workers. As two examples, she can continually bias the initial sex ratio in a kind of 'race' against the effects of worker biasing (see Section 5.2), and she can also invest more in each egg, thus reducing the payoff for such biasing by the workers.

2.7 Interaction of sex and caste determination

The genetic system of Hymenoptera is male-haploidy: females arise from fertilized eggs and most males, and all functional ones, arise from unfertilized eggs. Females are thus diploid and normal males are haploid. As discussed below, a fraction of the fertilized eggs in many species yield diploid males, but these are effectively sterile because their sperm are diploid, and ineffective in passing on their genes to future generations.

Once an individual has been set on the course of development as a female, it can then be further specified as a reproductive or a non-reproductive. The dichotomy between queens and workers is particularly important. Although genetic influences on the switch between queen and worker have been reported for some species, as in the ant *Harpagoxenus sublaevis* (Buschinger, 1975, 1978; Winter and Buschinger, 1986) and in bees of the genus *Melipona* (Kerr, 1950, 1969), it is generally accepted that in most species environmental influences during development determine the caste into which she will develop. Additionally, behavioural interactions between adults may determine whether or not a female adopts the mated egg-layer role, as in many bees and wasps. There are cases where caste is either determined or biased non-genetically in the egg (Passera, 1984:60–7), but in general the most important of these environmental influences appear to be those acting during larval life (Wilson, 1971:146–56; Oster and Wilson, 1978:140–1; Brian, 1983:206–20).

Both sex and caste determination are therefore linked in the sex allocation of social Hymenoptera (Fig. 2.3). Sex determination sets the initial ratio of diploid to haploid eggs, and caste determination determines the proportion of diploid individuals that actually become reproductive females.

Figure 2.3 brings out the fact that similar considerations to those discussed in the previous paragraph for Hymenoptera apply to the termites, in which both sexes are diploid. In termites, to an extent depending on species, larvae retain the capacity to transform into reproductives for a number of moults. Both the reproductives and the workers may influence whether immatures develop as reproductives or not (Section 5.5).

2.8 Effect of population structure

Subdivision of the population may have a marked effect on the sex ratio if it leads to different levels of competition between related individuals of the same sex. This effect on the sex ratio arises because of the devaluation to the parent of one sex among the progeny relative to the other. Selection then favours decreasing investment in the devalued sex. As the allocation in this sex decreases, leading to a reduction in its numbers, its average reproductive success increases, and an equilibrium is reached when its average reproductive success equals that of the other sex.

The important variables are the relative dispersal rates of the two sexes, and the persistence time of the patches into which the population is subdivided. There are four basic models (Charnov, 1982:67–92; Bulmer, 1986; Table 2.1).

Local resource competition (LRC) (Clark, 1978) arises when males disperse widely but females do not, but rather attempt to settle close to their birthplace and compete with each other and their mother for scarce resources. To simplify, consider that the resources are fixed, and conversion of the resources into further reproductives is independent of the number of daughters produced. In these circumstances, only as many daughters should be produced as are

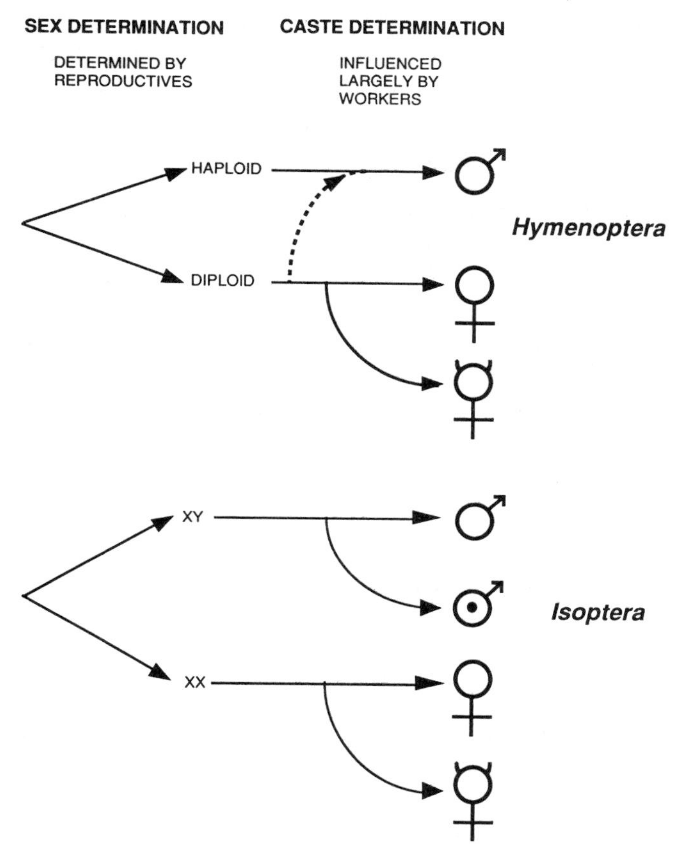

Fig. 2.3 Interconnections between sex and caste determination in determining sex allocation patterns in eusocial insects, with ☿ denoting a female worker and ⚩ a male one. In most if not all social Hymenoptera a fraction of the diploid individuals develop as males, as indicated by the dotted line. Diploid males in Hymenoptera are often sterile (e.g. in the ant *Solenopsis invicta* (Hung and Vinson, 1976)), but in some species can produce diploid sperm capable of fertilization; female progeny resulting from these diploid sperm are triploid and largely sterile because of meiotic abnormalities (e.g. in the bumble-bee *Bombus atratus* (Garofalo, 1973) and the honey-bee (Chaud-Netto, 1975)). Diploid males are therefore best regarded as all being sterile genetically because they fail to pass on their genes to succeeding generations. Female reproductives determine sex in Hymenoptera (by control over fertilization) whereas male reproductives determine sex in Isoptera (termites) via the production of Y- or X-bearing sperm. In both orders, workers are largely responsible for subsequent caste determination, although the effects of pheromones produced by reproductives differ between hymenopterans and termites.

necessary to secure the resources for the next generation. Depending on the precision of reproduction, one daughter might be enough. Sons, on the other hand, each have an equal chance of mating with females in the population at

Table 2.1. The four basic types of population subdivision affecting sex allocation. The symbols + and – refer to the effect of same-sex individuals on each other's fitness.

	+	–
Males	Local mate enhancement (LME)	Local mate competition (LMC)
Females	Local resource enhancement (LRE)	Local resource competition (LRC)

large. This kind of LRC is found, for example, in army ants: colonies divide by fission and daughter colonies may compete for resources, at least initially.

The end result of LRC is to devalue daughters in favour of sons, to an extent depending on the brood size, the disparity in dispersal rates, and the degree of isolation of the habitat patches (if the isolation is not great, then a higher premium might be placed on preventing the daughters of other females from intruding on the patch).

Local mate competition (LMC) (Hamilton, 1967) results from daughters dispersing widely but sons staying together as a group. Females mate when a group of males is encountered, of whatever size it is. Consequently, the number of sons produced is largely irrelevant to their success as a group, and this number is therefore minimized by selection.

Many species showing LMC also show inbreeding; parasitic wasps in which females mate with wingless males before leaving their birthplace have been a much studied example (Charnov, 1982:76–88). But inbreeding is not necessary for LMC to occur: LMC can occur even if sibling mating is rigorously eschewed. However, inbreeding changes the optimal sex ratio. Thus, if inbreeding is allowed, the optimal sex ratio (r^*) for patches with n mothers present is (Charnov, 1982:68)

$$r^* = \frac{n-1}{2n} . \tag{2.19}$$

If inbreeding (here sibling mating) is prevented, then (Charnov, 1982:70)

$$r^* = \frac{n-2}{2n-3} . \tag{2.20}$$

Parasitic wasps such as *Nasonia vitripennis* typically have groups of only one or two females ovipositing on one host, and typically have very low proportions of males (Whiting, 1967; Charnov, 1982:79–87).

Trivers and Hare (1976) reported female-biased sex allocation ($M \approx 0.25$) in many ants whose colonies have one queen each (monogynous). Alexander and Sherman (1977) made a spirited critique of the interpretation of Trivers and Hare of this result, suggesting that the female bias is due to LMC. But it seems unlikely that such ratios could result from LMC because the degree of population subdivision would have to be much more extreme ($n = 2$ or 3) than is generally thought plausible for ants.

The effects of LMC on investment may be illustrated with an example. Let fitness through male function be determined by $w_M = m^a$, with $a < 1$, and let fitness through female function be determined by $w_F = f$, in each case irrespective of M and F. (The relationship $w_M = m^a$ could arise, for example, where a female has one son and competition between males for mates is severe, so that a large male has a disproportionate mating success.) From eqn. (2.15), at equilibrium

$$\frac{a}{m^*} = \frac{1}{f^*}, \text{ so that}$$

$$f^* = \frac{1}{1+a}.$$

The graphs of w_M and w_F are shown in Fig. 2.4.

LMC can be generalized as the **haystack** model (Maynard Smith, 1964; Bulmer and Taylor, 1980; Wilson and Colwell, 1981), in which the patch is not dispersed after one generation but persists for some generations. Thus, a population is founded by some mated females, and persists for a certain number of generations with each generation produced only by inhabitants of that patch. Eventually the patch degenerates (such as by the haystack burning down), and

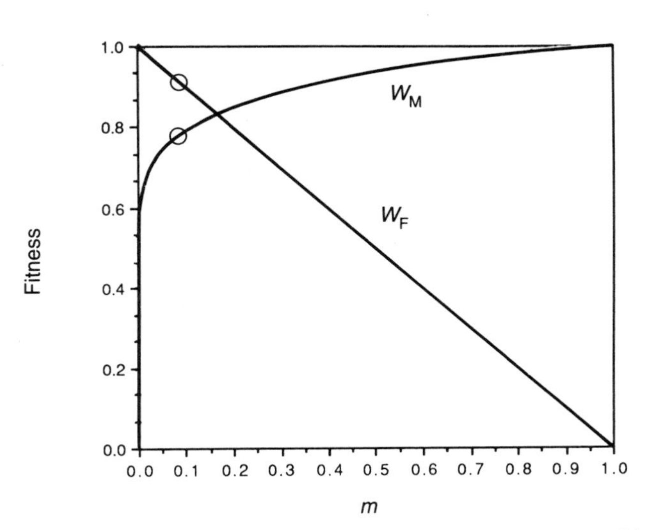

Fig. 2.4 Sex allocation equilibrium in a species with local mate competition. The male and female fitnesses are determined by the functions $W_M = m^a$ and $W_F = f = 1 - m$, respectively. The equilibrium sex allocation ratio is at $\hat{m} = a/(1 + a)$ (see text). The figure shows the male function resulting from $a = 0.1$. The open circles show the position of the equilibrium sex allocation for these conditions ($\hat{m} = 0.0909$).

the mated females of that generation disperse. A variant of this model is for females of the last generation to mate after dispersal.

The haystack model can be formulated in terms of intragroup and intergroup selection (Colwell, 1981). *Within* a group, selection favours 1:1 investment, but groups with higher investments in females produce more dispersers. Colwell also described one-generation LMC as group selection, but was criticized for this by Charlesworth and Toro (1982) who argued that the term (group selection) should be restricted to cases involving partially isolated groups with some permanence (e.g. existing for at least two generations).

Rather than competing with each other, female relatives may enhance each others' reproductive success. Such **local resource enhancement** (LRE) (Schwarz, 1988a) will also lead to differential investment in one sex compared to the other at equilibrium. In the case of the bee *Exoneura bicolor* studied by Schwarz (1988a,b), brood production per female is enhanced in intermediate sized groups compared with larger or smaller groups (Fig. 2.5). Nest foundresses are apparently drawn from individuals derived from the same nest of the preceding season. Schwarz therefore suggests that smaller broods will tend to be female biased, so as to insure descendant colonies of the optimum size or greater, and indeed found that larger broods are more male biased than smaller ones. Many colonies of *E. bicolor* are semisocial (not all females are reproductive), and LRE can therefore be regarded as a phase in the evolution of eusociality (Chapter 3). In *E. bicolor* the non-reproducing bees aid their sisters, and are hence correctly regarded as part of their mothers' investment in female function.

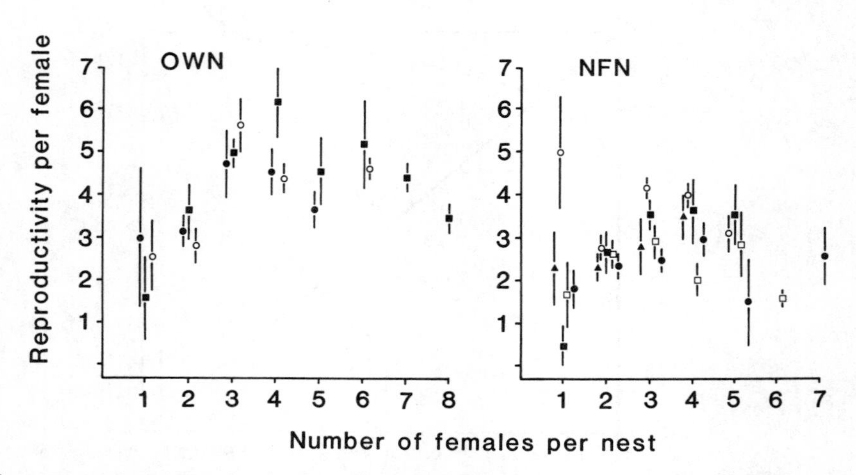

Fig. 2.5 Mean reproductive rate (brood per female) as a function of the number of females per nest in overwintered (OWN) and newly founded nests (NFN) of the primitively social bee *Exoneura bicolor*. Each point denotes the arithmetic mean number of brood per female calculated over all nests for each colony size, with the bars indicating ± mean standard error. Open and filled symbols indicate different dates. Categories with fewer than two colonies are not included, nor are nests parasitized by the cuckoo bee *Inquilina*. Reprinted with permission from *Nature* (Schwarz, 1988a).

If they had been able to aid their mothers instead then they would not be part of female function because they would have then assisted in the rearing of both sexes.

It seems unlikely that LRE is important in many eusocial species, because in eusocial insects the role of the supplementary females reported for *Exoneura* is taken by workers, but LRE has been reported for foundress associations of relatives in *Polistes* (Pickering, 1980). Of course, LRE need not lead to a sex bias, because if the optimum number of collaborating reproductives is small relative to the output of the colony, then optimum groups can be produced without the need for a sex bias. LRE has also been suggested to occur in bird species with helpers-at-the-nest; such populations typically show a sex bias (Emlen *et al.*, 1986; but see also Brown, 1987:81–3).

In the same way as LMC and LRC are differentiated by reference to the sex involved, **local mate enhancement** (LME) in formal terms is identical to LRE except that male relatives cooperate to enhance their reproductive success as a group. Such cases are not known in eusocial insects, but have been reported for vertebrates.

Because the sex allocation pattern expected depends on the occurrence of LMC, LRC, LRE, or LME, one cannot specify an 'intrinsic' sex allocation pattern unless the pattern of population structure is specified. The description of any particular sex ratio as 'biased' reflects expectations based on human patterns of population subdivision; a social insect might look at things differently had it the cognitive capacity to do so (Crozier and Pamilo, 1993).

2.9 Summary

It is perhaps surprising that so much of sex allocation theory can be approached using one simple equation, the Shaw–Mohler equation. Consideration of this equation shows that it is the investment, and not the numerical ratio, that is modified by selection. Another major result is that at equilibrium under selection there should be equal expenditure on the production of the two sexes among the progeny.

Further insights come from relaxing the assumptions of the Shaw–Mohler equation. The first such assumption concerns equal relatedness of the two sexes in the brood to an individual controlling the sex allocation pattern. Social insects are characterized by the great importance of non-parents in determining such allocation patterns. Social Hymenoptera are characterized by the often strong asymmetry of relatedness between male as against female juveniles and the workers. These relatedness values vary with life pattern. Inclusion of relatedness values yields the 'modified Shaw-Mohler equation', fundamental to understanding social insect sex allocation.

A further assumption of the Shaw–Mohler equation is that the sons as a group have equal reproductive value to the daughters. But reproductive value also varies with genetic system, so that, for example, the reproductive value of hymenopteran females is twice that of males in the important case of all reproduction being done by the queens.

Application of the Shaw–Mohler equation assumes linearity of the reproductive payoffs obtained by shifting investment from one sex to another. This assumption is violated for various population structures involving greater competition between the members of one sex than the other, and for the case where there is greater cooperation between reproductive members of one sex. Under such conditions, the equilibrium sex-allocation may see greater expenditure on one sex than on the other. The effect of population structure is such that if it is not specified there is no 'intrinsic' sex allocation pattern.

3

The evolution of eusociality in insects

Hamilton (1964a,b) formalized the modern approach to the evolution of social behaviour with his notion of inclusive fitness, and also noted that the exceptionally high relatedness among sisters might lead to a predisposition of the Hymenoptera to evolve eusociality because females could benefit more by raising sisters than own daughters. A decade later Trivers and Hare (1976) combined Hamilton's kin selection theory with Fisher's (1930) sex ratio theory and pointed out that hymenopteran workers benefit from the high sister relatedness only if they can bias the sex ratios in favour of females. These insights give sex allocation theory the central place in considering the evolution of eusociality. In the years since Hamilton's seminal papers, opinion has varied greatly over whether male-haploidy was, after all, important in the evolution of eusociality. In this chapter we follow Pamilo (1991c) and evaluate the current status of views on male-haploidy as a predisposing agent to the evolution of eusociality in the light of sex allocation theory. We apply here the inclusive fitness formulation developed in the preceding chapter. The main aims in this approach are to see whether there is a difference between diploid and male-haploid genetic systems in how readily they allow the evolution of a worker caste, and to examine how different features of insect life cycles might affect such evolution.

3.1 A framework based on inclusive fitness

It was first suggested by Hamilton (1963, 1964b) that hymenopteran females might prefer raising sisters (with relatedness of $g = 0.75$) instead of daughters (with $g = 0.50$). That would open the way to the evolution of eusociality. If a female replaces its own daughter with a sister, the relative benefit is $0.75/0.5 = 3/2$. Therefore, the alternatives of raising two sisters or three daughters give the same genetic payoff to a female.

Extending Hamilton's pioneering work, Trivers and Hare (1976) noted that female workers may not, after all, benefit from the high relatedness among sisters if they are simultaneously replacing sons ($g = 1$) with less related brothers ($g = 0.5$). When weighting relatedness with the reproductive values and summing up over the sexes, the result is the same in both cases: $2(3/4) + 1(1/2) = 2(1/2) + 1(1) = 2$. This means that the fitness of a worker equals that of contemporary reproductive females, provided that they raise the same number of offspring. Trivers and Hare suggested that the incipient female workers would benefit from unequal relatednesses only if they could make the brood sex ratio female biased.

We can analyse the situation using the fitness function derived in eqn. (2.6), namely

$$W_{\mathrm{I}} = N(g_{\mathrm{FI}} v_{\mathrm{F}} \frac{f}{F} + g_{\mathrm{MI}} v_{M} \frac{m}{M}) = NV_{\mathrm{I}} \tag{3.1}$$

where V_{I} is the inclusive fitness effect for genotype I.

Assuming that the population investment ratio is $M = F = 0.5$, and that the females mate only once, the fitnesses can be calculated both for females producing their own offspring and for females raising their siblings. For reproductive females, the fitness is $V = 2$ independently of the brood sex ratio (as already shown, Section 2.1). For worker females, the fitness is 2 when the brood sex ratio is $m = f = 0.5$, but becomes $V = 3$ if $f = 1$, and $V = 1$ if $f = 0$. The workers raising a female-biased brood sex ratio have the highest fitness in the population, and we could expect such a behaviour pattern to increase in frequency. When that happens, the whole population sex ratio will become female biased. The contemporary reproductive females will in such a situation benefit by producing only sons in order to counteract the female bias. If we plot the fitnesses as a function of the population sex ratio (Fig. 3.1), we note that the population should reach an equilibrium where there are two

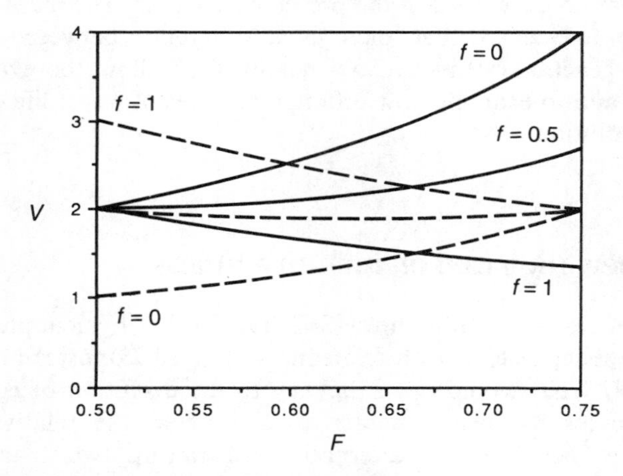

Fig. 3.1 Inclusive fitness function, V, from eqn. (3.1) for workers (dotted lines) and solitary females (solid lines) in the same population, as functions of the population investment in females, F. The three lines in each case are for brood sex ratios of $f = 0$, 0.5, or 1. (Based on Fig. 1 of Pamilo, 1991c.)

female types, workers raising sisters and solitary females producing sons (Charlesworth, 1978; Craig, 1979). The population sex ratio at this equilibrium would be female biased. This result, that the worker behaviour would not so easily become fixed in the population, suggested that male-haploidy has not been that important in biasing many hymenopteran lineages to eusociality (Andersson, 1984; Crozier, 1982).

When examining eqn. (3.1) we implicitly assumed that the numbers of offspring raised per one worker female and one reproductive female are same. In that case the function (3.1) reflects the fitnesses of these females. But the numbers are likely to differ from each other. Let a reproductive female raise N_S offspring, of which $f_S N_S$ are females with relatedness to her of $g_{daughter}$ and $(1 - f_S)N_S$ are males with relatedness to her of g_{son}. Let these same parameters be N_W, f_W, g_{sister} and $g_{brother}$ for a worker female. When a female adopts the solitary role her inclusive fitness is

$$W_S = N_S(g_{daughter} v_F f_S / F + g_{son} v_M m_S / M)$$

and when she behaves as a worker it is

$$W_W = N_W(g_{sister} v_F f_W / F + g_{brother} v_M m_W / M) .$$

The net effect is given as a difference between these two values. When the gain is greater than the loss, the worker behaviour is expected to spread in the population, and the condition for this becomes

$$E = \frac{N_W}{N_S} > \frac{g_{daughter} v_F f_S / F + g_{son} v_M m_S / M}{g_{sister} v_F f_W / F + g_{brother} v_M m_W / M} = E^* \qquad (3.2)$$

where f_W and m_W are the investment proportions raised by workers in the maternal colony.

We call the ratio $E = N_W/N_S$ the **worker efficiency**, and the threshold condition given by inequality (3.2) is the critical efficiency threshold E^* which allows worker behaviour to evolve. It should be remarked here that the condition obtained from eqn. (3.2) is not always true because the real threshold depends on the specific genetic mechanism determining worker behaviour (dominance of alleles, number of loci etc.). Equation (3.2) can, however, be used as a useful heuristic guide, because the inclusive fitness approach generally yields the same result as does a one-locus model with small allelic effects (Taylor, 1989).

In order to understand the conditions favouring the evolution of eusociality, it is necessary to know the life cycle of the non-eusocial ancestor and to specify how the step crossing the boundary to eusociality may have taken place. Three major classes of hypotheses have been proposed in the earlier literature, each referring to a specific selection process and behavioural pattern. The hypotheses can be grouped as follows (e.g. Charlesworth, 1978; Crozier, 1979): kin selection (Hamilton, 1964a,b), parental manipulation (Alexander, 1974; Michener and Brothers, 1974), and mutualism (Lin and Michener, 1972).

These hypotheses view the evolution from the viewpoint of the non-reproductive helpers, the reproductive individuals, and both, respectively. In that sense, they remind us of the hypotheses on the evolution of polygyny as presented in Section 4.4. Mutualism refers to a situation where all parties benefit in terms of direct reproductive success. By itself mutualism cannot (by definition) lead to eusociality, a situation in which many individuals lose personal reproductive success. But mutualism could establish the conditions (communal life) which could lead to eusociality, given other factors such as an appropriate relatedness structure among the colony members.

Because there are only two basic parties of interest, reproductive and non-reproductive individuals, we reduce the hypotheses to the following two questions.

1. Do the non-reproductive individuals benefit by raising offspring of other individuals and giving up their own reproduction?

2. Do the reproductive individuals benefit by forcing (letting) other individuals to raise the offspring and by preventing these helpers from reproducing?

Benefit is measured by the change of gene frequencies. If the frequency of any gene of an individual is expected to increase in the population due to the specific behaviour chosen by that individual, we say that the individual benefited from the behaviour. If we make a hypothesis that some individuals 'want' to become non-reproductive workers, the answer to the first question must be 'yes'. If we make a hypothesis that the parents want to manipulate other individuals to become helpers, we must have a positive answer to the second question. It is possible that there is conflict between the manipulating parents and the would-be workers, or it may be that they both prefer the same solution. Our aim here is to examine what are the thresholds of worker efficiency required for eusociality to evolve and then to examine how these thresholds can most easily be met in nature.

3.2 Male-haploid models

3.2.1 *Three basic life cycles present alternatives to potential workers*

The hypotheses presented above can be discussed in connection with various insect life cycles. The two main alternatives commonly discussed in this context are the matrifilial (extending from subsocial) and family-group (semisocial) associations (cf. Table 1.1). We will use simple life cycles associated with these alternatives, schematically shown in Fig. 3.2, when developing models of social evolution.

In a subsocial colony, the mother continues taking care of her developing offspring, but there is neither overlap of adult generations nor reproductive division of labour. If the mother lives long enough to overlap with the offspring generation, the offspring have an opportunity to help the mother and such a

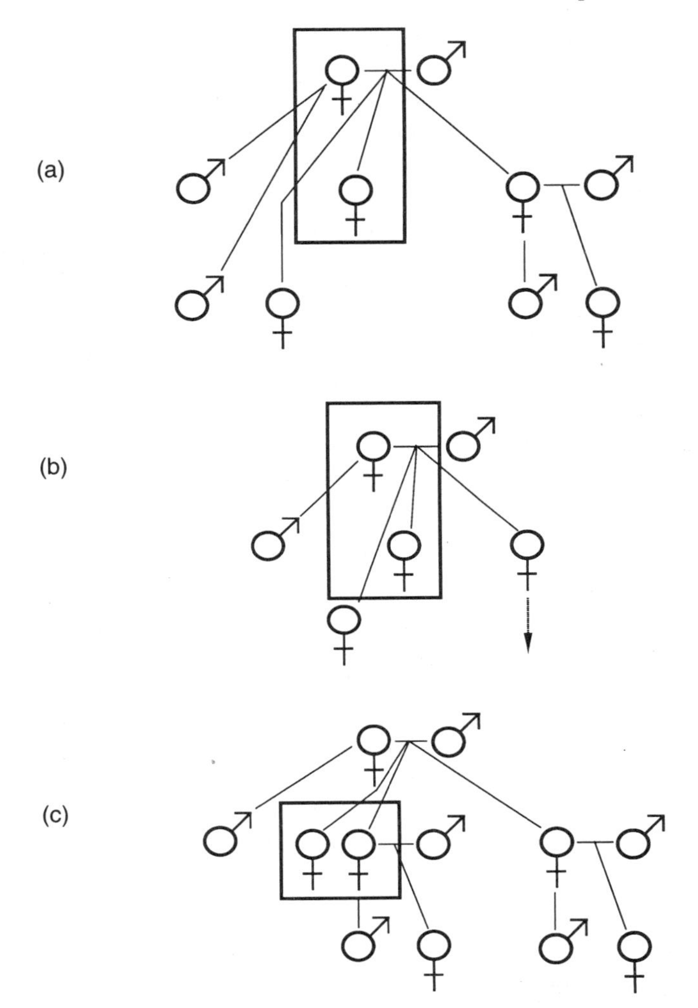

Fig. 3.2 The basic candidate life cycles considered in models of the evolution of hymenopteran eusociality. Boxed individuals are nest mates and cooperate. (a) Bivoltine model in which offspring of the first females to emerge may either assist their mother in the production of the second brood (i.e. act as workers) or leave and produce their own offspring. (b) Univoltine model in which the first offspring females may either act as workers or leave to diapause and reproduce themselves next year. (c) Semisocial model in which the first offspring females either assist their sisters or leave to reproduce themselves. (Based on Fig. 2 of Pamilo, 1991c.)

matrifilial colony can become primitively eusocial. Overwintered females reproduce in spring and the offspring emerge as a summer generation. These offspring then mate and produce a second generation, an autumn generation, if the life cycle is bivoltine (Fig. 3.2(a)). The females of the autumn generation start the cycle again next spring after having overwintered. Let us now assume that a new behavioural type arises in the population, and some of the summer females start to behave as workers. Instead of mating and producing their own offspring, they raise the offspring produced by the old mother. These offspring will emerge as part of the autumn generation. Yanega (1988) suggested that in halictine bees the original life cycle may have been univoltine (Fig. 3.2(b)). The summer females normally mate and overwinter without any reproductive activity until next spring. Alternatively, instead of overwintering, they could become helpers in their mother's nest, raising further siblings which will enter the reproductive population next spring.

A semisocial colony consists of individuals belonging to the same generation, e.g. of sisters. In this case, help is directed to such collateral individuals, some of them reproducing and some raising the offspring. The life cycle presented in Fig. 3.2(c) has three options open to each of these sisters: (i) they may be solitary and produce their own brood; or they may form a semisocial association, in which (ii) one female reproduces and (iii) a sister behaves as a worker. Such associations may later expand into more diverse family groups including siblings, cousins etc. (West-Eberhard, 1978).

We will not review the biology of primitively social insects, however interesting and important to our topic that would be (for such a review, see Eickwort, 1981). Instead, we aim to examine whether male-haploidy might affect the conditions for eusociality to evolve and what might be the role of sex ratio biasing in this evolution. As pointed out above, the once highly praised role of male-haploidy has met with strong scepticism recently. We will first examine in very general terms the subsocial and semisocial alternatives in male-haploid species (i.e. Hymenoptera), then compare the results with diploid models, and finally introduce several additional aspects important to our theme.

3.2.2 *Matrifilial associations: bivoltine life cycle*

Let us assume that the expected number of offspring for a single reproductive summer female in the life cycle shown in Fig. 3.2(a) is N_S and that for a single worker is N_W. The ratio N_W/N_S defines the worker efficiency in terms of offspring numbers. It is clear, of course, that if this efficiency is very high, worker behaviour can evolve without any obstacles. If the efficiency is extremely low, worker behaviour does not evolve, whatever the relatednesses and sex ratios. Our analysis is based on examining the critical worker efficiency thresholds which allow worker behaviour to evolve. It should be noted that the offspring numbers N_S and N_W include both mortality and fertility effects. In other words, if the mortality of worker females is, say, 50% and each surviving worker raises two offspring, the mean number of offspring per original workers is one. Similarly, if the successful reproductive females have ten offspring each, but the risks when dispersing and establishing the nests raise the mortality to 90%, the mean number of offspring per original reproductive female is also one.

We can now examine the two alternatives of the summer females from the viewpoint of these females themselves and from that of the mother. Assuming that the population sex allocation ratio is 1:1 (as expected in non-eusocial, outbred Hymenoptera), we can apply eqn. (3.2) to calculate the inclusive fitness effects depending on how the summer females behave. These effects are calculated from the viewpoints of both the mother (called queen) and of the daughters (either workers or solitary) by setting $F = M = 0.5$, $\nu_F = 2$ and $\nu_M = 1$. The threshold worker efficiency, E^*, can then be calculated by dividing the fitness effect gained by a solitary female by that gained by a worker female. If this ratio equals one, the summer females should prefer worker behaviour whenever they can raise more offspring as workers than as solitary females. If the threshold ratio is less than one, worker behaviour is favoured even though the workers could not raise as many offspring as the contemporary solitary females do. Inserting the values of genetic relatedness in the above formula, we obtain the following worker efficiency thresholds:

from the viewpoint of a non-laying worker

$$E^* = \frac{2}{1 + 2f},$$

a laying worker

$$E^* = \frac{2}{2 + f}, \qquad (3.3)$$

a non-laying worker raising nephews and sisters

$$E^* = \frac{2}{1.5 + 1.5f}.$$

The first two of these are also presented as graphs (Fig. 3.3). It is easily seen that the preferred behaviour for a summer female is to behave as a laying worker; she would then be making the best of both worlds, raising her own sons and highly related sisters. If the brood sex ratio is female biased, worker behaviour is favourable even though the queen does all the egg-laying (this was already seen in Fig. 3.1). If the sisters lay some of the haploid eggs, the fitness will be intermediate between these two. Because the colonies of primitively eusocial insects are small, the contribution of a single laying worker to male production must be significant.

What about the mother, would she happily continue egg laying and let her daughters behave as workers? First, she must live long enough to give the daughters this opportunity. Second, she should do that only if the contribution to her own inclusive fitness from a helping daughter exceeds that from a solitary daughter. The worker efficiency thresholds for maternal manipulation, from the viewpoint of the mother will be

daughter is a non-laying worker $\quad 1/2$,

daughter is a laying worker $\quad \dfrac{1}{1 + f}. \qquad (3.4)$

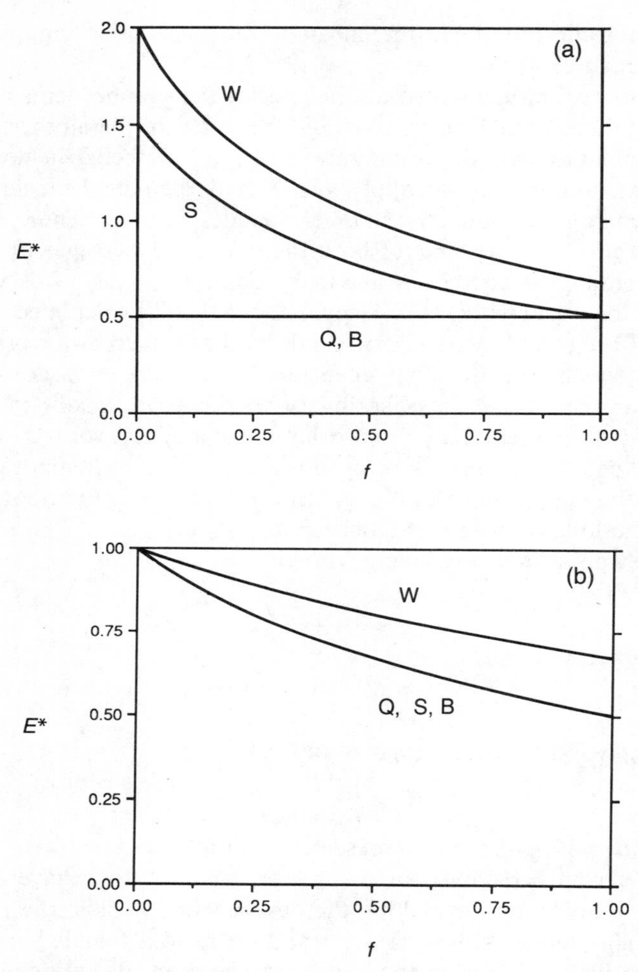

Fig. 3.3 Worker efficiency thresholds (E^*) for selection for worker behaviour in a subsocial, male-haploid species for (a) a potential non-laying worker and (b) a potential laying worker from the viewpoints of the worker herself (W, eqn. (3.3)), of the mother (Q, eqn. (3.4)), of a sister (S, eqn. (3.5)) and of a brother (B, eqn. (3.6)). Values above each curve favour the evolution of worker behaviour from the point of view of the individual concerned. (Based on Fig. 3 of Pamilo, 1991c.)

These are also presented as graphs in Fig. 3.3, and it is easy to see that the thresholds are always smaller than one (unless $f = 0$). Whenever the summer females can raise nearly as many offspring whether behaving as helpers or as solitary reproductives, the mothers should manipulate them to become workers — even though the workers would produce the male eggs.

Parental — more precisely maternal — manipulation has been discussed as a possible alternative to worker altruism (Alexander, 1974; Michener and

Brothers, 1974; Stubblefield and Charnov, 1986). It is clear from our presentation that the critical threshold of worker efficiency from the mother's point of view is $E^* = 0.5$ from eqn. (3.4), whereas from the daughters' point of view it is $E^* = 2/(1 + 2f)$ from eqn. (3.3), which ranges from 0.67 to 2, and equals 1 when $f = 0.5$. From these values we easily see that the interests of the mother and daughter either agree or disagree depending on the real worker efficiency as follows:

$E > \frac{2}{1+2f}$ both agree that daughters become workers,

$\frac{2}{1+2f} > E > 0.5$ there is a conflict of interest,

$0.5 > E$ both agree that daughters should not become workers.

The above conditions concern the early origin of eusociality. When worker behaviour spreads in the population, the population sex ratio may change (if the workers prefer biasing it) and the reproductive values of males and females will change (if workers participate in male production). For these reasons, it is important to follow further changes in the population.

It has been shown that if the early workers were able to bias the brood sex ratio, they would make it female biased and then evolution readily leads to the polymorphic equilibrium described above (Craig, 1980; Aoki and Moody, 1981; Pamilo, 1987). The situation is different if the population sex ratio does not become highly female biased but remains close to the 1:1 ratio. This would be the case if the brood sex ratios are controlled not by the workers but by the mother queens. If there is some initial association between worker behaviour and female-biased brood sex ratio, worker behaviour is favoured, but while it increases in frequency in the population, the worker-raised sex ratio should come closer to 1:1. In such a case worker behaviour can become fixed even if the workers can raise less offspring than contemporary solitary females do, i.e. when $E < 1$ (Pamilo, 1987).

Maternal influence is certainly essential in the evolution of eusociality through a subsocial route, because the mother has to continue egg laying. But it is interesting also to examine the potential interests of collateral relatives, brothers and sisters. It may happen in primitively social species that the old mother dies before all the first-generation offspring have emerged, but she may have left eggs and developing larvae behind her. The siblings, however, can coexist in the nest for some time before dispersing, and during this time they can try to manipulate each other to start raising the youngsters left by the mother (Stubblefield and Charnov, 1986). (Of course, in those species in which the mother has provided a complete store of provisions for her progeny, there is little incentive for such manipulation.) We can proceed in the same way as above and calculate the efficiency threshold of sister workers from the viewpoint of sisters and brothers:

from the viewpoint of a female whose sister is

a non-laying worker $$E^* = \frac{0.75}{0.5 + f},$$

a laying worker $$E^* = \frac{1}{1 + f}; \qquad (3.5)$$

from a viewpoint of a male whose sister is

a non-laying worker $$E^* = \frac{1}{2},$$

a laying worker $$E^* = \frac{1}{1 + f}. \qquad (3.6)$$

The thresholds are shown also in Fig. 3.3. The males would be selected to favour their sisters entering the worker role even though the efficiency would be low. And even though a summer female herself might prefer being a solitary reproductive, she might prefer manipulating the sisters to become workers in the mother's nest. The results indicate that when there is mother–daughter conflict over the future role of the daughters, the mother could expect some help from her other offspring to resolve the conflict in her favour.

3.2.3 *Matrifilial associations: univoltine life cycle*

The worker efficiency in a univoltine species (Fig. 3.2(b)) cannot be determined in the same way as above, because the offspring raised by the females taking different behavioural roles belong to different generations. What a worker female has to do is to replace herself by siblings. Assuming that the population sex ratio is $F = M = 0.5$, the number of siblings (N_W) raised by a worker must satisfy the inequality

$$N_W(f \times 2 \times \frac{3}{4} + m \times 1 \times \frac{1}{2}) > 2$$

which can be written as

$$N_W(0.5 + f) > 2 \qquad (3.7)$$

for the worker behaviour to be advantageous. The threshold includes, for example, the cases where a worker raises one sister and one brother, or four brothers, or an average of 1.33 sisters.

3.2.4 *Semisocial life cycles*

As for the mother–daughter associations considered above, we can examine the worker efficiency thresholds when the help is given to sisters (Fig. 3.2(c)).

Such groups of same-generation females with reproductive division of labour are called semisocial (Table 1.1). The interested parties are the potential workers and the reproductive females who are helped. For outsiders (the old mother, other sisters, and brothers) it does not matter which of the females does reproduce as long as the total reproductive output does not decrease: the threshold for them is 1. The critical efficiency thresholds for the sister–sister associations are as follows:

from the viewpoint of a worker whose sisters

produce all offspring $E^* = \frac{4}{3}$,

produce females while the worker produces sons

$$E^* = \frac{4}{3+m}; \tag{3.8}$$

from the viewpoint of a reproductive sister who

produces all the offspring $E^* = \frac{3}{4}$,

produces females but lets a sister produce males

$$E^* = \frac{3}{3+f}.$$

the thresholds for workers are always greater than 1 and for the reproductive sisters less than 1.

3.3 Why are there no male workers in Hymenoptera?

Sociality in the Hymenoptera is based almost exclusively on females. The males play very little part in any social activity, and they show no systematic reproductive division of labour. The lack of male workers, a contrast between hymenopterans and termites, has been generally attributed to two major factors: genetic factors and behavioural or anatomical factors. We will here review the ideas concerning genetic factors.

As already noted by Hamilton (1964a,b), sisters and brothers are of equal genetic value to a male (the product of relatedness and sex-specific reproductive value is $gv = 0.5$). His daughters are of a higher value ($gv = 1$) but his contribution in the genome of the sons produced by his mate is nil. Let a male have two choices: to mate and father a brood of size N_P, a proportion f_P of which are his daughters, or to become a worker and raise N_W siblings with proportions f_W of sisters and $1 - f_W$ of brothers. If the population sex ratio is 1:1, the threshold condition for worker behaviour being favoured in males is

$$E = \frac{N_W}{N_P} > 2f_P \cdot$$

Because the expectation is $f_P = 0.5$, the threshold reduces into $N_W/N_p > 1$ (Charlesworth, 1978; Charnov, 1978; Craig, 1982). This threshold is the same as in diploid organisms and in hymenopteran females when there are no sex ratio biases. Therefore it has been often asked: Why did hymenopteran males never develop into workers?

We will here examine the thresholds for male workers in somewhat closer detail. We know that in many hymenopteran lineages, females evolved a division of castes. The question is, what should the males do in this situation. We can further note that there may be good reasons to think that the evolution of female workers was followed by sex ratio conflicts. For males to become workers, the efficiency N_W/N_P must satisfy the inequality

$$E > \frac{2f_P(1-F)}{F + f_W(1-2F)} \cdot \tag{3.9}$$

If $F = 0.5$, eqn. (3.9) reduces to $E > 2f_P$, and with $f_P = 0.5$, to $E > 1$. If the female workers, which we assume to have preceded the males as workers, have biased the population sex ratio, as might be expected, this threshold drops below $2f_P$. In Fig. 3.4 we show combinations of the sex ratios F, f_W and f_P which lead to worker efficiency thresholds of 1 or 0.5. When the population sex ratio becomes female biased, the efficiency threshold decreases. This is easy to understand, because in such a case the mating success of the brothers increases considerably while that of the daughters decreases simultaneously. When the population sex ratio is biased by female workers, we expect both F and $f_W > 0.5$, on average. The contemporary solitary females should try to compensate for this bias by producing males, i.e. $f_P < 0.5$. This situation is most favourable for male workers to evolve.

The conclusion is therefore clear, that there has been no genetic bias against the evolution of male workers in the Hymenoptera. It therefore seems most likely that males lack the preadaptations necessary to be efficient workers (Alexander, 1974; West-Eberhard, 1975; Starr, 1985; Kukuk *et al.*, 1989; Kerr, 1990). Hence, female workers are more rapidly evolved, and this leads to female-biased sex ratios in the broods so that males are essentially eliminated before they have a chance to evolve worker-like traits.

We should further note that when female workers evolve, there will not be many solitary females for males to mate with. As a result, the expected N_P decreases because it includes mating success, and it becomes much easier for the worker efficiency N_W/N_P to reach the critical threshold. We can therefore now repeat with good reason the question: Why did the hymenopteran males not evolve into workers? Of course we know what happened to them: their inefficiency eliminated them from those broods which are not reproductive.

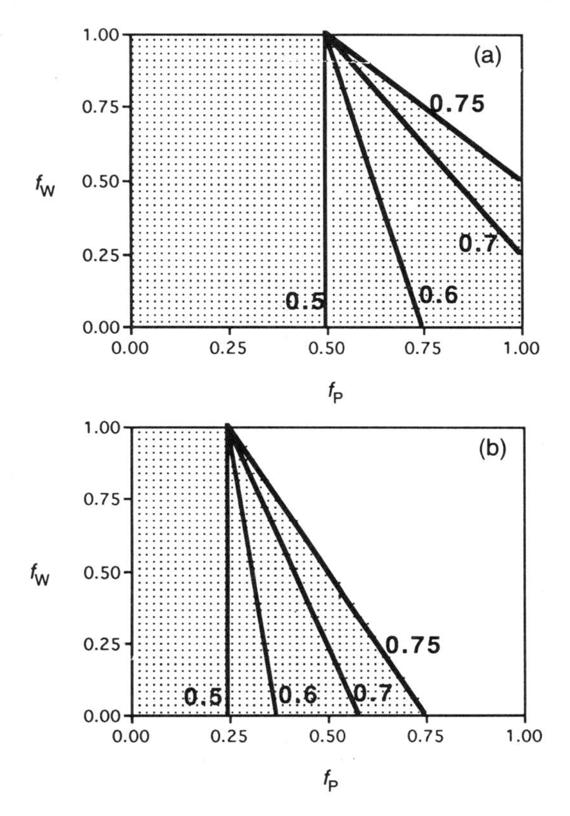

Fig. 3.4 Thresholds for the evolution of male workers in Hymenoptera already possessing female workers, according to eqn. (3.9). Male worker behaviour is selected for combinations of f_P and f_W falling to the left of the lines shown for $F = 0.5$, 0.6, 0.7 or 0.75 if (a) $E > 1$, or (b) $E > 0.5$ (the grey area). Note that the line $F = 0.75$ represents the likely sex allocation ratio under (female) worker control. (Based on Fig. 4 of Pamilo, 1991c.)

3.4 Diploid models

The thresholds can be calculated for diploid organisms in the same way. Because the genotype frequencies should be the same in both sexes (except perhaps if the locus is associated with some exceptional chromosomal arrangements, see Section 1.8), there is no inherent asymmetry between male and female offspring and the sexes need not be considered separately. The efficiency thresholds are obtained in a very straightforward manner when we note that the relatedness of both siblings (assuming single mating) and offspring is 0.5, and that of nieces, nephews and grandchildren is 0.25. The thresholds are in a bivoltine matrifilial association

for a worker $E^* = 1$,
for a parent $E^* = 0.5$,
for a sibling $E^* = 0.5$.

The thresholds for the potential worker and the parents are the same as those in a male-haploid organism with no sex ratio bias. But the offspring of a diploid organism should associate more strongly with the parents in forcing siblings to become workers in the case when there is a conflict (i.e. when $1 > E > 0.5$).

In a semisocial sister group, the thresholds differ from those of a male-haploid species and the area of conflicting interests between the reproductive and non-reproductive individuals is larger. The thresholds are

for a worker $E^* = 2$,

for a reproductive individual $E^* = 0.5$,

for the parents $E^* = 1$.

The semisocial route to eusociality is therefore relatively much less likely to be followed in species with both sexes diploid as against in male-haploid species.

3.5 The effects of multiple mating

If the reproductive females have mated with more than one male, the workers may end up raising half-siblings instead of full-siblings (subsocial model) or offspring of half-siblings (semisocial model). Because of the lower relatednesses of brood to worker in such a situation, the threshold efficiencies for worker behaviour to evolve are consequently affected.

When the contributions of the various males differ from each other, we have to take this variation also into account. In such a case we can use the effective number of matings, or effective promiscuity (Starr, 1979, 1984), which is

$$k_E = \frac{1}{\Sigma^k \gamma_i^2} \tag{3.10}$$

where, k is the number of males mating with the female, γ_i is the proportion of the female's daughters fathered by the ith male.

In the subsocial model with an effective polyandry of k_E and the population sex ratio of $F = M = 0.5$, the threshold will be

$$\frac{2k_E}{k_E + 2f}$$ for a non-laying worker, and

$$\frac{2k_E}{k_E(2 - f) + 2f}$$ for a laying worker.

These thresholds increase with k_E, and we can conclude that multiple mating makes worker altruism less likely to evolve. On the other hand, multiple mating does not affect the mother's attitude, the threshold efficiency remains unaltered at the value of 0.5 in the maternal manipulation model. Similarly, the interests of brothers in a possible mother–daughter conflict are not affected by polyandry, and the male offspring should support the mother in encouraging the sisters to become workers (see Fig. 3.2). The female offspring, however, are affected and should encourage their sisters to behave as workers only if the efficiency exceeds the following thresholds:

$$\frac{3k_E}{2k_E + 4f} \quad \text{for a non-laying worker, and}$$

$$\frac{3k_E}{(k_E + 2)(1 + f)} \quad \text{for a laying worker.}$$

These are increasing functions of k_E, and they are intermediate between the thresholds calculated above from the viewpoints of the workers and queens. Both thresholds may easily be larger than 1.0. When $k_E = 2$, the two thresholds are equal. With $k_E < 2$, sisters should prefer worker-reproduction and with $k_E > 2$ they should prefer the queen to produce all of the males (see also Section 4.2; Woyciechowski and Łomnicki, 1987; Ratnieks, 1988; Pamilo, 1991b).

3.6 Essential differences between male-haploid and diploid populations

We will next try to examine what kind of factors might explain the taxonomic distribution of eusociality. These factors can either affect the required worker efficiency thresholds (as derived above), or they can modify the actual efficiency of workers to meet these thresholds. The necessary condition for worker behaviour being adaptive is $E > E^*$. An inequality has two sides, and we cannot understand the situation completely by examining only one side. We can, of course, examine the factors affecting one side; that is what we have done for E^*, and we continue this expedition here. In this section we focus on factors which characteristically separate male-haploid and diploid populations, and in the next (Section 3.7) we consider factors which do not depend on the sex determination mechanism.

As seen from the above derivations (Sections 3.2, 3.4), the thresholds are similar in the two types of populations (male-haploid and diploid) unless worker behaviour is associated with a female-biased brood sex ratio in male-haploidy. We should therefore examine whether such an association is plausible in primitively social hymenopterans. Another factor characteristic of male-haploidy is the ability of unmated females to lay haploid eggs developing into males. Worker egg-laying was shown to affect the thresholds (Section 3.2).

A third potential factor which we have not yet considered is inbreeding. Inbreeding alters the values of genetic relatedness and it may also affect the mating successes in the two sexes (local mate competition, LMC). As these consequences of inbreeding can be different in male-haploid and diploid populations, the attractiveness of worker behaviour may also be differently affected by inbreeding.

3.6.1 *Sex ratio biases*

It is evident that an association between worker behaviour and female-biased brood sex ratio lowers the worker efficiency threshold (e.g. Fig. 3.3) and favours the evolution of eusociality. Such an association could be created by active manipulation by workers, active manipulation by the egg-layer being helped, or other non-random associations of worker behaviour and sex ratio.

Theoretical studies clearly show that if the workers can manipulate the sex ratio in the brood they raise, they will do so (Section 2.2). One possibile way to achieve this is selective feeding or elimination of the developing larvae; another would be to manipulate the egg-layer to fertilize her eggs. In the first option, the crucial question is whether the early workers could distinguish male and female larvae. It would also be essential to distinguish them at an early stage in order to avoid the heavy expenses of eliminating big larvae. Selection would favour the ability to distinguish between haploid and diploid eggs. Given the initial situation of no eusociality, it is unlikely that the first potential workers had this crucial ability (Crozier, 1977a), except in the case of spatial cues such as when male and female eggs are differentially distributed beween cells in a burrow (Evans, 1977). We can also note that a laying worker replacing a random egg laid by her mother with her own haploid egg will not improve her fitness (unless the mother is multiply inseminated, the brood sex ratio is male biased, or the population sex ratio is strongly female biased). Selection favours a laying worker being able to recognize the ploidy of eggs and to replace only the haploid ones with her own.

If we assume that the workers were able to bias the sex ratio from the very beginning of eusociality, we also face the problem that a female-biased population sex ratio will finally stop the spread of worker behaviour. For this reason, the hypothesis of eusociality evolving with laying workers has gained support (e.g. Iwasa, 1981).

The hypothesis that early workers manipulated the egg-layer to bias the sex ratio is attractive, because it circumvents the problem of distinguishing the larval sex. There is, of course, the possibility that the egg-layer benefits from such a bias without any intentional manipulation by the workers. As emphasized by Frank and Crespi (1989), helpers in a nest will bring in additional resources and the level of resources tends to correlate with female bias in investment. If the investment ratio in the population is 1:1, each brood sex ratio is equally advantageous to the egg-layers (see Section 2.1) and they should not resist any changes within a colony.

The third cause for an association between worker behaviour and female-biased brood is a non-random association caused by factors other than adaptive manipulation. The possibilities giving rise to such an association include the following.

1. Because some hymenopteran females normally remain unmated and produce only males, the workers (which raise offspring of a mated female) will be, on average, associated with a brood sex ratio more female biased than the population average (Iwasa, 1981; Godfray and Grafen, 1988). The coupling of worker behaviour and sex ratio due to unmatedness may be a weak one, but it is a factor which is known to exist.

2. In a matrifilial association workers will be raising offspring of a late brood of the mother and this brood can, because of protandry, be more female-biased than an average brood. Protandry has been documented in many species (Section 6.7), and although not universal, seems to be rather common. The simulation models of Pamilo (1987) assumed that late broods, raised by workers, can be female biased. Although the simulations assumed a deterministic genetic determination of sex ratios, they can also be considered applicable to the case of protandry.

3. Seger (1983) pointed out important differences between those bivoltine life cycles which include either diapausing of both sexes (as larvae) and those in which there is overwintering only of mated females. According to characteristic representatives of the models, he termed these the sphecid and halictid models, respectively. The importance of these models lies in the feature that some males can outlive their own generation and enter the mating pool of the next generation. This feature will devalue the new males in that succeeding generation and predicts female-biased sex ratios in it. Interestingly, the biases are predicted to depart in opposite directions in the two models. In the sphecid model, the first brood produced will be female biased and the second brood male biased. In the halictid model, the first brood will be male biased and the second brood female biased. Because the potential workers born in the first brood raise offspring belonging to the second generation, the workers are raising female-biased broods in the halictid model. Although each female, solitary or worker, produces the same biased sex ratio in the second generation, the mating successes of the two sexes can be equal because of the surviving old males, and the inclusive fitness effects depend on the relatednesses and reproductive values. Therefore workers would benefit from the female-biased sex ratio (high relatedness) and we can predict that the halictid model is more favourable than the sphecid model for the evolution of worker behaviour.

4. A non-random association could also be created if both brood sex ratio and the appearance of first workers correlate with a third factor, e.g. with a habitat or a nest site or size (Grafen, 1986; Stubblefield and Charnov, 1986).

We conclude that the genetically determined thresholds for the evolution of worker behaviour differ between male-haploid and fully diploid species, and that there are a number of factors which can on occasion yield the female-biased sex ratios which favour the evolution of eusociality in male-haploid populations.

Gadagkar (1991b) compiled estimates of genetic relatedness in eusocial hymenopteran species and found that in most cases relatedness between sexual brood and workers is so low that the workers would benefit from producing and raising their own offspring (see Table 4.7). He took this as a test of the haplodiploidy hypothesis and concluded that male-haploidy is insufficient by itself to promote the origin of eusociality or to maintain eusociality. There is no doubt that ecological and behavioural determinants have a major role in the evolution of eusociality (Strassmann and Queller, 1989; Pamilo, 1991c; Section 3.7), but it seems also still true that the threshold for worker behaviour to evolve can be lower under male-haploidy than under diploidy.

3.6.2 *Inbreeding*

The offspring of an inbreeding couple are more related to each other than those of an outbreeding couple. Therefore, it has been suggested that inbred offspring might more easily become workers raising these highly related siblings. The problem of theoretical treatments of inbreeding is that there are many possible models. We will here discuss only three alternatives to derive some insight on the effects of inbreeding. The cases inspected are regular inbreeding, one-generation inbreeding in matrifilial associations, and one-generation inbreeding in semisocial sister groups.

By regular inbreeding we mean a system of mating where average inbreeding remains the same in each generation. Although such inbreeding may alter conditions for sociality by increased between-group variance and reduced within-group variance, it leads to increased relatedness not only among siblings but also between offspring and parents. For this reason, individuals do not necessarily gain from high sibling-relatedness because they obtain the same benefit from the high offspring-relatedness. In male-haploid species inbreeding tends to even out the relatedness asymmetries among siblings: the relatednesses of brothers and sisters become closer to each other and the female workers do not benefit from biased sex ratios as much as in outbred populations. It therefore seems that inbreeding does not much affect the threshold efficiency for worker behaviour in diploid populations but may make them slightly more unfavourable in male-haploid species. On the other hand, the outcome is also affected by changes in the mating success of the two sexes (inbreeding leading to LMC) which will devalue the males to some extent. This will give more weight to sisters and the workers may benefit from biased sex ratios even though the relatedness asymmetry is weaker. The question needs more modelling of concrete situations.

We turn next to a model in which matrifilial colonies are founded by inbreeding pairs (sibling mating) but this inbreeding ceases in the next generation. When the life cycle is bivoltine, there will be an alternation of outbred and inbred generations. The offspring of the inbreeding couple show elevated relatedness; that among sisters is 0.8 and that of brothers to sisters is 0.6. Because the next generation is assumed to be outbred, the offspring of a female are related to her by 0.5 (daughters) or by 1 (sons). The efficiency threshold now becomes $E^* = 2/(1.2 + 2f)$ instead of the $2/(1 + 2f)$ from eqn. (3.3). Inbreeding thus slightly lowers the threshold. In a diploid species the relatedness among inbred siblings in this model would be 0.6, which leads to an efficiency threshold $E^* = 1/1.2$ instead of the value 1 from eqn. (3.3). It is interesting to note that this threshold is lower than that of male-haploid populations without any sex ratio bias ($f = 0.5$).

A similar result is obtained if the offspring of an inbreeding couple form a semisocial colony where some siblings reproduce and the others raise nieces and nephews. The relatedness of them to a female worker in a male-haploid colony would be 0.5 (nieces) and 0.8 (nephews). These relatednesses decrease the efficiency threshold from 4/3 (eqn. (3.8)) to $E^* = 5/(4 + f)$. In a diploid species the relatednesses of nieces and nephews under this inbreeding model would be 0.5 and the efficiency threshold drops from 2 (eqn. (3.9)) to $E^* = 1$. We see again that the threshold in diploid species has become smaller than that in a male-haploid one.

The above models of a bivoltine life cycle assumed that the colonies are started in spring by siblings and the first generation offspring will be inbred but themselves outbreed. It may be more plausible to assume that the colonies in spring are founded by outbreeding parents but that the first offspring generation may inbreed and produce an inbred second generation. The inbreeding female of a male-haploid colony would produce daughters related to her by 3/4 and sons related by 1. These are exactly the same as the relatednesses to a laying worker, whereas the relatedness of males to a non-laying worker are 1/2. It therefore seems that the alternatives of being a laying worker or inbreeding with your brother might be equally advantageous alternatives. Some differences do, however, exist. A reproductive female might more easily than a laying worker manipulate the brood sex ratio and get a higher benefit from the unequal values of sons and daughters. On the other hand, an inbreeding female prevents her brother from mating with other females (assuming single mating by males). If the brother would have mated with a non-related female, that female would have produced daughters related by 1/4 to the female we are concerned about. Therefore, if outbreeding is possible in this generation, selection might favour worker behaviour over sibling-mating.

In diploid species, the offspring resulting from sibling-mating are related to the parents by 3/4, whereas the siblings raised by a worker are related only by 1/2. An inbreeding couple would thus raise N_P offspring related to them by 3/4. If both of them would have become workers, they would have raised offspring related by 1/2 and we can assume that the total number of these offspring would have been N_P. Sibling-mating would therefore be advantageous to them, a result different from what we might expect in male-haploid species.

If the colony is a semisocial association where the egg-laying female has mated with her brother, the relatedness in a male-haploid species would be daughter to mother 0.75, son to mother 1, daughter to worker (aunt) 0.625, and son to worker (aunt) 0.75. Your own, inbred, offspring will be more highly related to you than those of your sisters by a ratio of about 4:3. In diploid species the offspring would be related to the parents by 0.75 and to the aunts by 0.5, a ratio of 3:2. The ratio is higher in diploids.

The conclusion is again that the risks of dispersal favouring staying at the natal nest site can more easily select for worker behaviour in male-haploid organisms and for inbreeding in diploid organisms.

3.6.3 *Other genetic models*

Male-haploidy does not only bias genetic relatednesses within families but affects also the dynamics of allele frequency changes in the population. Effects on the mutation load (Section 1.7) and effective population size (Section 4.2) are briefly discussed elsewhere. Some recent genetic models have suggested that such differences between the two genetic systems could partly explain why eusociality has evolved more frequently in male-haploid than in diploid species.

An evolutionary change depends both on a directional pressure (e.g. selection) and on stochastic factors (drift, influenced by the effective population size), and their joint effects can be examined with diffusion equations (Crow and Kimura, 1970:Chapter 8). The lack of heterozygous males in the Hymenoptera alters the effects of drift and it also makes directional selection stronger. Based on these factors, Reeve (1993) suggested a *protected invasion hypothesis*: dominant alleles for maternal care are more resistant to loss by drift in male-haploid than in diploid populations. The same applies to alloparental care in which females raise siblings instead of own offspring. The hypothesis also predicts that both sexes could equally well evolve to workers in a diploid species, whereas there is a bias toward female workers in male-haploids. Reeve's model is for a finite population and it would still be desirable to examine its effects as a function of the population size.

Deleterious alleles in Hymenoptera often have female limited expression (Section 1.7). Saito (1994) suggested that if such alleles affect fecundity, females that are homozygous for recessive deleterious alleles might be biased toward worker behaviour. As deleterious alleles are otherwise selected effectively against in male-haploids, populations could tolerate some inbreeding and produce many females with low fecundity This model still waits for a quantitative evaluation. The author also notes that the model cannot explain the existence of suicidal behaviour in clonal aphids and polyembryonic wasps (Section 1.2).

Another model aiming to explain the differences between male-haploid and diploid species is a kin selection model of Yamamura (1993). The model indicates that whereas kin selection leads more easily to worker behaviour in male-haploids, self-sacrificing soldier behaviour evolves most readily in asexually reproducing species. Diploid sexual populations are predicted to be intermediate. The results of this model therefore agree with the existence of workers in the Hymenoptera and of suicidal individuals in clonal organisms, while the diploid termites have commonly both workers and soldiers.

3.7 Increased efficiency favours the evolution of worker behaviour

So far we have focused on the thresholds, E^*, set by genetics. We should now move to examine the other side of the inequality condition (3.2), i.e. the factors which determine the real efficiency of the workers. The two worlds, that of genetic dynamics and that of potential worker efficiency, are not completely independent of each other. The efficiency can be largely determined by the context where the worker meets the choice: help or reproduce. For example, a worker helping a mother to raise siblings may have much higher efficiency than does a worker associating with a sister to raise nieces and nephews. So, the threshold efficiency required, E^*, and the actual efficiency E may covary.

The important question is: How do various factors affect the efficiency, N_W/N_P, which factors will tend to increase it and which decrease? The literature on primitively social insects is vast and offers many concrete examples which are relevant to our topic. Instead of a detailed review, however, we prefer classifying the factors in broader categories which are then discussed briefly with only a limited number of examples. The factors discussed here are as follows: care of young, the number of young, dispersal risks of non-reproductive individuals, advantages and disadvantages of staying, synergistic effects in offspring care, and asymmetry of reproductive capacities.

3.7.1 *Care of young is an important prerequisite*

For N_W to be of any significant value, some type of parental or sibling care must predate the transfer from reproductive to non-reproductive behaviour. It is often suggested that subsociality, (extended parental care) is a major prerequisite of eusociality (Alexander *et al.*, 1991). Although it is true that eusociality seems to have arisen in lineages where ancestral species probably (and in some non-eusocial relatives certainly) showed parental care, this may not be the only way. In aphids and termites the care is directed to siblings (or other relatives) developing in the same cohort. Such behaviour does not require parental care and subsociality. The major requirement is a coexistence of individuals which can help and those which can receive the help. Coexistence of different age classes can give such an opportunity; the adults could then help the developing young as is the case in hymenopteran societies. But the developing individuals of the same cohort can also provide help to each other if they are able to perform such behaviours. This is more easily accomplished in hemimetabolous than in holometabolous insects, because the nymphs of the previous group can more easily show different activities than the helpless larvae of the latter.

In species with parental care of larvae, it also becomes important whether care is given by one sex or by both sexes. When shifting to a worker role, an individual trades N_P of its own offspring for N_W relatives. The value of N_W depends entirely on the worker's own activity, whereas that of N_P depends on the activities of both the individual itself and its mate. If the contribution of one reproductive individual is to raise N_S offspring, we expect $N_P > N_S$. This means that a worker should be efficient enough to replace the parental efforts of two individuals (Charnov, 1981; Pamilo, 1987). Assuming that $N_P = 2N_S$, the thresholds for both worker altruism and parental manipulation are doubled compared to the situation were $N_P = N_S$.

3.7.2 *Workers should be allowed to use their full capacity to raise young*

Not only should workers be prepared for brood care, but there should also be enough brood to be taken care of. If colony life does not provide as many young for the workers to raise as they have capacity for, N_W will be lowered and worker behaviour may not be adaptive.

The number of young could be below a worker's capacity due to limitations in the mother's egg-laying ability, longevity or nest construction. It should be noted that the first workers in the nest can markedly increase the potential productivity of the mother and she may not be able to meet the demands. It has also been shown in several species that the productivity of a colony does not increase linearly with the size of the worker force (Michener, 1974:246–7). Above a certain threshold, each new worker may add a diminishing amount to the total productivity; in other words, N_W is a diminishing function of the colony size (but see Oster and Wilson, 1978:48; Schwarz, 1988a).

If workers cannot use their whole working capacity because their mother dies at an early age or does not lay enough eggs, we expect worker egg-laying. If unmated (in Hymenoptera), they will produce sons, but at the dawn of eusociality we may expect that a certain proportion of female workers were mated and could then become replacement reproductives (Kukuk *et al.*, 1989).

3.7.3 *When dispersal risks are high, it may pay to stay home*

The offspring numbers N_W and N_P used to calculate the worker efficiency ($E = N_W/N_P$) cannot be simply estimated from the productivities of successful colonies. They should be estimated per individual choosing a particular behavioural pattern. For example, if the mortality of workers is 50% and each surviving worker raises two offspring, the mean number of offspring per worker is $N_W = 1$. Similarly, if the successful reproductive females produce ten offspring each, but the risks when dispersing and establishing the nest raise the mortality to 90%, the mean number of offspring per reproductive female is $N_P = 0.1 \times 10 = 1$.

We therefore expect that staying can be favoured in conditions where dispersal is risky. Such conditions include limitation of nest sites, high predation outside the nest (e.g. during mating), high costs of constructing a nest and high mortality of young and small nests. But it is relevant to ask whether avoidance of dispersal should necessarily lead to helping; it could also lead to inbreeding. The examples worked out in the previous section (3.6.2) showed that dispersal risks could more easily select for worker behaviour in male-haploid populations but for inbreeding in diploid populations (see also Pamilo, 1984b). Risks due to inbreeding depression might act against sibling-mating, but that does not remove the difference of thresholds.

Conditions making staying advantageous are the same which make dispersing costly. When good nest sites are limited or hard to occupy (nest construction being expensive), staying will become adaptive. Its attractiveness can be further enhanced by possible synergistic effects, whereas diminishing returns per additional workers in big groups make staying the less attractive alternative. These conditions are largely the same as those favouring polygyny in already eusocial species (Section4.4). Because of the limitations mentioned above, it seems that worker behaviour might more easily evolve in relatively small groups. A worker may also have the opportunity to become a replacement reproductive if the previous egg-layer dies.

Against these considerations, there can also be disadvantages in staying at home if an old nest site is vulnerable to diseases, parasites or predators because of either age or size effects (Shykoff and Schmid-Hempel, 1991a,b,c).

3.7.4 *Synergistic effects*

There are good arguments suggesting that the workers might not benefit as much as they could otherwise because the nest size is limited (physically or in terms of the mother's egg-laying capacity) and the additional effect of an extra worker is a diminishing function of the number of workers. But this need not be universally true and the exceptions form interesting cases as they provide opportunities for worker behaviour to be adaptive. The efficiency per worker can be increased by synergistic effects between the colony members. Such interactions can take place either between workers or between the egg-layers and workers.

Schwarz (1988a) described an interesting phenomenon of local resource enhancement (see Section 2.8) in an Australian allodapine bee. His observations indicate that the productivity of nests increases with group size in such a manner that the actual productivity per single colony member increases up to a threshold. Such synergism would favour group formation. It does not, however, explain why some group members would become non-reproductive. Yet it shows that N_W can be high once such groups are formed.

Another, and possibly very important, case of synergism can occur between the egg-layers and workers. This is especially so in matrifilial associations because they have a suitable age structure. If the offspring are emerging at different times, the first offspring are accompanied by developing younger siblings. If these developing larvae benefit from provisioning, protection, or other care which the mother alone cannot provide, the older siblings can start taking care of them (Queller, 1989; but see Nonacs, 1991). If the younger siblings would have suffered considerably without that care (e.g. if the mother had died), a small investment by the helpers can benefit them a lot, i.e. N_W can be high.

3.7.5 *Asymmetry in reproductive capacities*

The factors discussed above can be classified as extrinsic, ecological factors affecting the number of young which can be produced by solitary individuals (N_P) or by workers (N_W). These factors concern each individual similarly when they face the choice in the same situation. But it is also possible that individuals have intrinsic differences which will bias their choice.

West-Eberhard (1978) proposed an intuitive hypothesis verified quantitatively by Craig (1983) that variance of fertility can promote worker behaviour. If some females are less fertile than others (subfertile), they could become workers because their personal reproductive loss (N_P) is minimal. This hypothesis requires that offspring production is limited by fertility but the same females still have the capacity to raise a larger brood than they can produce. This does not require complete independence between fertility and working capacity. For example, if working capacity increases linearly with female size, but offspring production follows a sigmoid function, we can expect that big females produce more offspring than they can raise and small females can raise more brood than they are able to produce. Strassmann and Queller (1989) and Queller and Strassmann (1989) did not find much support in favour of this subfertility hypothesis in *Polistes* wasps, but positive evidence is provided by Gibo (1974) from *Polistes fuscatus* and Packer (1986b) from *Halictus ligatus*. Roisin (1994) has further developed the idea in order to explain the origin of sociality in termites.

There is also an asymmetry between the mother and offspring in matrifilial associations in that the mother is already known to be productive. She can then manipulate the offspring to become workers. Once the asymmetry exists, selection can strengthen it. Iwasa (1981) postulated that N_W/N_P increases in evolution because workers become better workers and queens become better queens. This agrees with the concept that eusociality proceeds in small steps (Trivers and Hare, 1976; Charlesworth, 1978).

Queller (1989) and Gadagkar (1990, 1991a) have pursued further models invoking an advantage of worker behaviour to individuals of poor personal reproductive capacity, or reduced survivorship. These models rest on the benefits, called by Gadagkar (1990) 'assured fitness returns', of workers being able to contribute to the next generation through the inclusive fitness effect even if they do not survive until the brood mature. By contrast, a lone foundress must survive until her offspring mature if these offspring are to survive. The workers can start their assistance on eclosion, whereas foundresses have to become reproductively mature before they can lay eggs and begin to care for their young. Following the arguments outlined above from West-Eberhard (1978) and Craig (1983), these factors are best regarded as stabilizing eusociality once it has arisen. Initially, all individuals would be selected for acting as foundresses. Once reproductive asymmetries allowed, via kin selection, superior inclusive fitness for some individuals by acting as workers than as foundresses, then caste biasing by foundresses would enhance their reproduction through parental manipulation. Shykoff and Schmid-Hempel (1991c) note that this effect may be enhanced in eusocial species through the effects of disease in delaying reproductive maturity of individuals thus increasingly predisposed to act as workers.

3.8 Summary

Historically, views on the factors leading to eusociality were grouped under three hypotheses: **kin selection**, in which the workers gained in inclusive fitness, **parental manipulation**, in which queens gain through manipulating some offspring to aid them in spite of a loss of inclusive fitness for the offspring, and **mutualism**, involving mutual increases in personal fitness. Because there are only two parties to the interaction, and because mutualism cannot by definition lead by itself to eusociality, we can reduce the three hypotheses to two questions concerning the inclusive fitness benefits to potential workers and potential queens. The task is then to specify the conditions under which genes favouring altruistic behaviour will increase in frequency and give rise to eusociality.

Considerations of the evolution of eusociality are intimately associated with questions of life cycle. Although univoltine life cycles could yield eusociality under some conditions, the bivoltine life cycle under which a worker can still produce sons gives rise to the most lenient conditions for evolving eusociality. Multiple mating reduces the likelihood of eusociality evolving through its effects at the worker level, although the advantage to the mother remains the same. Male haploidy favours the initial evolution of eusociality through the ability of workers to bias the sex ratio and the production of sons by unmated females maintaining an advantage for female production.

Although there are genetic factors promoting worker performance in male-haploids, other conditions are necessary as well to lead to worker performance values capable of leading to eusociality. Similarly, although there are genetic models seeking to explain the absence of male workers in eusocial

Hymenoptera, basic biological factors such as a poor capacity for parental care are more likely.

Given the profusion of male-haploid eusocial groups, the evolution of the male-diploid termites has been suggested to be an enigma. Inbreeding has been suggested several times to have been the engine driving the evolution of eusociality in the ancestors of termites. However, although inbreeding is expected to be favoured under some conditions given the termite life-pattern, it does not itself appear to promote the evolution of eusociality through any genetic bias. Other ecological or life-pattern factors need to be sought to provide a satisfactory explanation for the evolution of termites.

Once termite eusociality has arisen, workers may either be permanent and sterile or a transient life cycle stage, as discussed earlier. Higashi *et al.* (1991) develop a cost-benefit inclusive fitness model, and find that true (sterile) workers are much more liable to evolve when the species does not consume its nest as food but rather forages for food away from the nest.

Many hymenopterans are not eusocial, and there are relatives of termites that possess many similarities to them in life cycle but lack eusociality. Rather than assume that these other species are on the way to eusociality, it is reasonable to assert that time has been sufficient and therefore to ask the question 'what has prevented the evolution of eusocial behavior?' in those cases. These extant species should be seen as ones possessing features *preventing* the evolution of eusociality.

4

Evolution of colony characteristics

Colony life with a reproductive division of labour is the hallmark of the eusocial insects. In earlier chapters we have presented simple family models in which the colony is headed by one reproductive female, and in most cases we have assumed that she has mated with a single male. This gives the simplest possible colonial structure that results in the typically high sister-relatednesses in hymenopteran species. Even though some social insect colonies do have such a simple familial composition, many (probably most) species have genetically more complex colonies. In many species females mate with several males and a colony can have more than one reproductive female. Either several to many queens reproduce in a single colony or workers can compete with the queen and lay haploid eggs that develop into males. Colonies can also be subdivided into several physically separate nests that exchange individuals. Such complex colonial structures pose new problems to evolutionary theory: what is the functional unit and what is the unit of selection (cf. Table 1.6)? When a colony has many reproductives in it and the worker force can be divided into different kin groups (patrilines and matrilines), we can also ask whether individuals can distinguish between various levels of kinship and direct their behaviours in favour of closely related nest-mates. We can also wonder how much the genetic heterogeneity of a colony can increase before its social harmony breaks down. These are the questions we deal with in this chapter.

Variation in colony structure affects both the relatedness structure of the colony and also the reproductive values of males and females. What brings about differences in colony structure? How do differences in relatedness and reproductive value interact in affecting the behaviour and general biology of eusocial insects? As will be seen, the data bearing on these points are quite voluminous, but tend to be incomplete for most species. The way ahead then must lie in obtaining more complete data about fewer species, and also experimentally manipulating colonies to assess the costs and benefits of various strategies. Therefore we can seldom come to definitive conclusions about the relative strengths of factors identified by theory as likely to be important, and hence we here discuss these factors to stimulate the necessary field and laboratory studies.

4.1 The single family is the simplest colony type

The simplest possible structure for a social insect colony comprises a single, once-mated female and her worker progeny, with the queen producing all the reproductive individuals for the next generation. Such colonies are indeed widespread in the social insects and characteristic of many species (Table 4.1).

Not all monogynous, monandrous social Hymenoptera fit this simplest model, because the workers may produce some or all of the males (Section 4.2). One species in which we can be certain that the males come only from the queen-laid eggs is the fire ant *Solenopsis invicta*, because in this genus the ovaries of workers are totally degenerate (Wilson, 1971:320). Allozyme data also show that in this species queens mate only once (Ross and Fletcher, 1985b). The monogynous form of *S. invicta* therefore has the simple colony structure discussed in this section.

The pedigree type for the social hymenopteran colony structure discussed in this section is shown in Fig. 4.1. The relatedness values between various relatives were given in Table 1.3.

It is clear that the simple colony structure shown in Fig. 4.1, while widespread, is far from the general rule and may even be in the minority (Table 4.2). We therefore need to explore in some detail the consequences of departing from this starting point. We would also like to remark that it is difficult to use the information in our tables to draw evolutionary conclusions because the levels of polyandry and polygyny (Table 4.1) are rarely estimated quantitatively, and because the associations in Table 4.2 are not corrected for

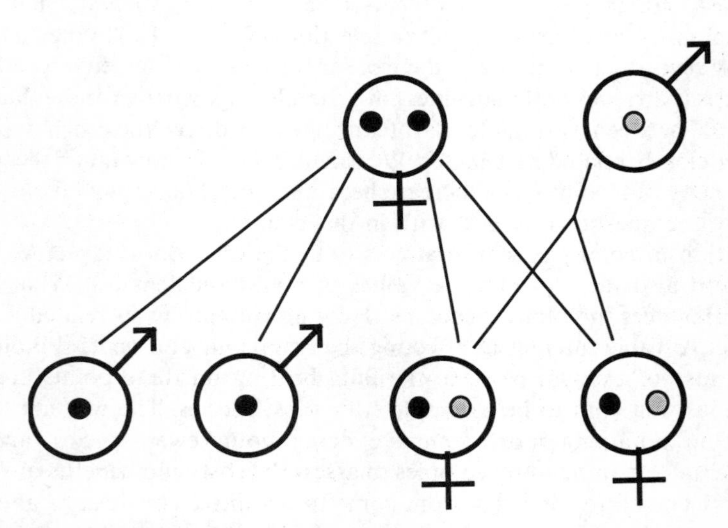

Fig. 4.1 A simple male-haploid pedigree, corresponding to the situation in eusocial species in which all offspring arise from queen-laid eggs, showing the fate of alleles passed on by the mother and father. In species with such pedigrees, females have twice the reproductive value of males.

Table 4.1. Number of matings by females and number of egg-layers per colony in eusocial Hymenoptera. The ants are also grouped into subfamilies. The information on mating rates comes primarily from Table 1 of Page (1986). Unless otherwise mentioned, all other information also comes from the references listed by Page. The letters in the column 'Notes' denote, following Page, the means by which the number of matings were inferred: A allozyme or DNA markers, D dissection, O observation, V visible phenotypic marker. Other symbols are as follows: M monogynous, P polygynous, G production of fertilized eggs is by gamergates (mated workers), U populations are unicolonial (Wilson (1971:457), lacking colony boundaries). Reproductives in termites are either primary (colony founding) or secondary (additional reproductives added in established colonies); with rare exceptions primary reproductives are exclusively monogamous (Nalepa and Jones, 1991).

Family *Species*	Number of matings, range [mean]	Number of queens	Notes	References
BEES				
Apidae				
Apis				
andreniformis	< 4	M	D	Moritz and Southwick (1992:205)[1]
cerana	14–30	M	D	
dorsata	> 1	M	D	Moritz and Southwick (1992:205)[1]
florea	< 4	M	D	Moritz and Southwick (1992:205)[1]
mellifera	7–20	M	V,D,O,A	Seeley (1985:21), Estoup *et al.* (1994)[12]
Bombus[3]				
atratus	1	M	D	Sakagami (1976)
bifarius	1.3	M	O	Foster (1992)10
b. nearcticus	1	M	O	
californicus	1.3	M	O	Foster (1992)[10]
frigidus	1.14	M	O	Foster (1992)[10]
hortorum	1	M	D	Sakagami (1976)
humilis	1	M	D	Sakagami (1976)
huntii	1–3	M	O	Hobbs (1967)
hypnorum	1–4	M	A,D	Röseler (1973), Estoup *et al.* (1995)
lapidarius	1	M	D	Sakagami (1976)
lucorum	1	M	A,D	Estoup *et al.* (1995)
melanopygus	1	M	V	Owen and Plowright (1980, 1982)[1]
morio	1	M	D	Sakagami (1976)
pascuorum	1	M	D	Sakagami (1976)
pratorum	1	M	A,D	Sakagami (1976), Estoup *et al.* (1995)
rufocinctus	1.06	M	O	Foster (1992)[10]
sylvarum	1	M	D	Sakagami (1976)

Table 4.1 *Continued*

Family *Species*	Number of matings, range [mean]	Number of queens	Notes	References
terrestris	(>)1	M	A,D,O	Röseler (1973), Van Honk and Hogeweg (1981), Estoup *et al.* (1995)
variabilis	1	M	D	
Melipona				
marginata	1	M	O,V	
quadrifasciata	1	M	D,O	Kerr (1975)[2]
quinquefasciata	1	M	O	Kerr (1975)[2]
subnitida	1	M	A	Contel and Kerr (1976)[1]
Plebeia				
droryana	1	M	A	Machado *et al.* (1984)[1]
Halictidae				
Augochlorella striata	1	M	A	Mueller *et al.* (1994)
Halictus ligatus	≈ 2	M	A	Richards *et al.* (1995)
Lasioglossum				
malachurum	> 1	M	O	Packer and Knerer (1985)
marginatum	> 1	M	O	
rohweri	1–3	?	O	
zephyrum	> 1	M?	DO[11]	
WASPS				
Vespidae				
Polistes				
metricus	> 1	P	A	
variatus	> 1	P	A	
versicolor	1	?	D	
Ropalidia				
marginata	> 1	M,P	A	Muralidharan et al. (1986)[1]
Vespa				
crabro	> 1	M	O	
Vespula				
atropilosa	> 1	M	D,O	
germanica	1	M	O	
maculifrons	2–7	M,P	O,A	Ross and Visscher (1983)[5], Ross (1986)[1]
pensylvanica	1	M	O	
squamosa	2–7	M	A	Ross (1986)[1]
ANTS				
Formicidae				
Dolichoderinae				
Conomyrma				
insana	> 1	M	A	Berkelhamer (1984)1
Iridomyrmex				
humilis	1	P(U)	D,O	Bartels (1985)[1]
purpureus	1	P	A	

Table 4.1 *Continued*

Family *Species*	Number of matings, range [mean]	Number of queens	Notes	References
Tapinoma				
minutum	1	P[8]	A	Herbers (1991)
Ecitoniane				
Eciton				
burchelli	1–5[9]	M	O	
Formicinae				
Brachymyrmex				
depilis	2–3	?	O	
Cataglyphis				
cursor	> 1	M	O	Lenoir *et al.* (1988)
Colobopsis nipponicus	1	M	A	Hasegawa (1994)
Formica				
aquilonia	1–6 [1.9]	P	A	Pamilo (1993)
argentea	1	M,P	A	Bennett (1986)
bradleyi	1–3	M	O	Halverson *et al.* (1976)
dakotensis	(>)1	?	O	Talbot (1971)
exsecta	> 1 [< 2]	M&P[7]	A	Pamilo (1991d)
montana	> 1	?	O	
opaciventris	(>)1	P	O	Scherba (1961)
pergandei	(>)1	?	O	Kannowski and Johnson (1969)
pressilabris	1–2 [1.3]	P	A	Pamilo (1982c)[1]
rufa	> 1	M&P[8]	O	Rosengren and Pamilo (1986)[2]
sanguinea	> 1 [1.5–2]	P	A	Pamilo (1981, 1982c)[1]
subintegra	1–4	?	O	
subpolita	1–4	?	O	O'Neill (1994)
transkaucasica	1–2 [1.1–1.2]	P	A	Pamilo (1982c)
truncorum	> 1 [1.3]	P	O,A	Rosengren *et al.* (1985)[1], Sundström (1989)
yessensis	> 1	P	O	Higashi (1976)[2]
Lasius				
alienus	1–2	M	O	Bartels (1985)[1]
niger	> 1 [1.2, 1.9]	M	A,O	Van der Have *et al.* (1988)[1,6]
Polyergus				
lucidus	(1)–6	?	O	Marlin (1971)
Prenolepis				
imparis	> 1	M	O	
Myrmicinae				
Acromyrmex				
landolti	> 1	?	O	
Anergates				
atratulus	> 1		O	Forel (1928:407)

Table 4.1 *Continued*

Family *Species*	Number of matings, range [mean]	Number of queens	Notes	References
Aphaenogaster				
rudis	1	M	A	
Atta				
laevigata	3	M	D	Corso and Serzedello (1981)
sexdens	3–8	M	D	
texana	> 1	P	D	
Cardiocondyla				
wroughtoni	> 1	P	O	Kinomura and Yamauchi (1987), Yamauchi *et al.* (1991)
Carebara				
vidua	> 1	M	O	Lepage and Darlington (1984)
Harpagoxenus				
canadensis	1	M	O	Buschinger and Alloway (1978)[1]
sublaevis	1	P	O	
Leptothorax				
acervorum	1	P	A	Lipski *et al.* (1992)
gredleri	1	M	A	Lipski *et al.* (1992)
longispinosus	1	P	A	Herbers (1986)[1]
muscorum	1	P	A	Lipski *et al.* (1992)
Monomorium				
pharaonis	1	P(U)	O	Petersen and Buschinger (1971)[1]
Mycocepurus				
goeldii	4	?	O	
Myrmica				
americana	1	P(U)	O	Kannowski and Kannowski (1957)[4], Ayre (1971)[2]
punctiventris	1	M	A	Snyder and Herbers (1991)[1]
rubra	1–7 [2.1][10]	P	O	Woyciechowski (1990a)
ruginodis	1–3 [1.2]	M&P[8]	A	Seppä (1994a)
Pheidole				
sitarches	1	?	O	
Pogonomyrmex				
badius	2–4	?M	O	
barbatus	> 1	?M	O	Hölldobler (1976)
californicus	1–6 [2.9]	M	O	Mintzer (1982a)
desertorum	> 1	?M	O	Hölldobler (1976)
maricopa	> 1	?M	O	Hölldobler (1976)
occidentalis	> 1	M	O	
rugosus	1–6 [3.1]	M	O	Hölldobler (1976)

Table 4.1 *Continued*

Family *Species*	Number of matings, range [mean]	Number of queens	Notes	References
Solenopsis				
invicta (M)	1	M	A	
invicta (P)	1	P	A	
geminata	1	M	A	Ross *et al.* (1988)
richteri	1	M	A	Ross *et al.* (1988)
Ponerinae				
Rhytidoponera				
chalybaea	1	M&P(G)	A	
confusa	1	M&P(G)	A	

[1]The authority for both the mating frequency and queen number result.
[2]Queen number authority.
[3]*Bombus* colonies are typically monogynus, but *B. atratus* colonies begin with 10–20 queens who successively eliminate each other over 3–6 months of combats; should all queens be thus killed, a mated worker takes over the production of fertilized eggs (Michener, 1974:322–3).
[4]Mating frequency authority.
[5]*Vespula maculifrons* is monogynous in northern North America, but Ross and Visscher (1983) record a polygynous colony from Florida. Jeanne (1980) notes that vespine colonies are monogynous save for exceptions found in the southern United States and various warm climates to which they have been introduced.
[6]The proportion of females mating twice varied between populations, from 0.25 to 0.92. The authors note that the harmonic means for the numbers of matings are 1.12 and 1.63. Other populations had intermediate values.
[7]Some populations are monogynous, others polygynous.
[8]Most nests are monogynous.
[9]Mates repeatedly but only once per year (Franks and Hölldobler, 1987).
[10]Based on laboratory experiments.
[11]Allozyme data (Kukuk, 1989; see also Crozier *et al.*, 1987) fail to separate multiple mating from polygyny as the cause of mean relatedness between workers in the first brood of the year being lower than 0.75; second-brood females have relatednesses very close to 0.75.
[12]Estoup *et al.* (1994) estimated by using microsatellites that the numbers of patrilines ranging from 7 to 20 correspond to the effective numbers of matings (eq. 4.15) ranging from 6.6 to 17.9.

Table 4.2. The distribution of queen mating frequency and queen number per colony from the data presented in Table 4.1. Values of 0.5 arise due to counting polymorphic populations half in both relevant columns.

	Monogynous	Polygynous	Queen number unknown
WASPS			
Monandrous	2	–	1
Polyandrous	4	3	–
ANTS			
Monandrous	9	11	3
Polyandrous	18.5	9.5	6

phylogenetic correlations. The tables serve here mainly to emphasize the diversity of colonial structures in social insects.

4.2 Worker reproduction

4.2.1 *Production of males by workers yields a more complicated life cycle*

In social bees and wasps it appears that all females have potentially functional ovaries. In primitively eusocial species such as the bee *Lasioglossum (Dialictus) zephyrum*, in which caste determination does not involve morphological differentiation, there are sometimes subordinate mated individuals (hence workers) which may lay both haploid and diploid eggs on the death of the queen (Michener, 1974:322–3, 1990). Very often in such species one or more workers will mate and become replacement queens on the death of the queen. Workers seldom if ever lay diploid eggs in queen-right colonies of primitively eusocial insects, but in many species they produce some or most of the haploid (male-producing) eggs. In ants, Oster and Wilson (1978:102) record 29 of 35 genera and subgenera surveyed as having workers with ovaries (but do not list the taxa surveyed). It therefore seems likely, from this indirect evidence, that in most species of social Hymenoptera some or all of the males can arise from worker-laid eggs. Direct evidence that this is so is relatively sketchy. Bourke (1988) and Choe (1988) have reviewed the information on worker reproduction in ants. Based largely on their compilation of data, a list of known worker reproduction is presented in Table 4.3. The frequency of worker-produced males has been rarely estimated.

In most cases, workers reproduce only (or at least mainly) in the absence of the queen, i.e. in orphaned colonies. Only in a very few cases do the workers contribute significantly to male production in queen-right colonies. Interesting exceptions are further formed by thelytokous species, where the workers and sometimes queens produce also female offspring parthenogenetically (Table 4.4).

Table 4.3. Eusocial Hymenoptera in which male production by unmated workers has been documented. Much of the information on ants comes from the reviews by Bourke (1988) and by Choe (1988). Asterisks denote papers not seen by us. We have excluded from this list cases of reproduction by mated workers (*gamergates*, as in many ponerines) and by ergatoid (worker-like) queens (as in the ant *Technomyrmex* sp., Terron, 1972). The column 'Queen-right colonies?' records whether worker male-production has been recorded in the presence in the same nest of a mated egg-laying queen; 'Yes' within parentheses indicates that worker reproduction in queen-right colonies is rare. The column 'P or M' records whether the queens are polyandrous or monandrous. Worker egg-laying has been documented in many more species, but the development of these eggs into males has not been verified.

Taxa	Queen-right colonies?	P or M	References
BEES			
Apidae			
Apis mellifera	(Yes)[6]	P	Page and Erickson (1988)
Bombus ardens	Yes[8]		Katayama (1988)
B. atratus	Yes (80%)		Sakagami (1976)
B. ignitus			Katayama (1971)
B. melanopygus	Yes (39%)	M	Owen and Plowright (1982)
B. nevadensis			Hobbs (1965)
B. pascuorum			Cumber (1949)
B. terrestris	Yes (≤ 82%)	P	van Honk *et al.* (1981)
B. terricola			Owen *et al.* (1980)
Melipona subnitida	Yes (39%)	M	Contel and Kerr (1976)
Plebeia droryana	Yes (16%)	M	Machado *et al.* (1984)
P. shrottkyi	Yes (most)	M	Machado *et al.* (1984)
Trigona postica	Yes		Beig (1972)
Halictidae			
Augochlorella striata	Yes[10]	M	Mueller (1991), Mueller *et al.* (1994)
Halictus ligatus	Yes		Packer (1986a)
Lasioglossum malachurum[7]	Yes (most)		Packer and Knerer (1985)
L. marginatum	Yes (most)		Plateaux-Quénu (1959)
L. nigripes	Yes (most)		Knerer and Plateaux-Quénu (1970)
L. zephyrum	Yes (≤ 15%)		Kukuk and May (1991)
WASPS			
Vespidae			
Polistes chinensis	Yes		Miyano (1986)
P. gallicus			Pratte *et al.* (1984)
P. jadwigae			Miyano (1991)
P. metricus		P	Metcalf (1980)
P. snelleni			Suzuki (1985)
Vespa orientalis			Ishay *et al.* (1965)
Vespula 'spp.'	Yes		Spradbery (1973:226)
V. flavopilosa	(Yes)		Ross (1985)
V. germanica	(Yes)[9]	M	Montagner (1966), Ross (1985)

Table 4.3 *Continued*

Taxa	Queen-right colonies?	P or M	References
V. maculifrons	(Yes)		Ross (1985)
V. squamosa	No		Ross (1986)
V. vidua	(Yes)	P	Ross (1985)
V. vulgaris	(Yes)[9]		Montagner (1966), Ross (1985)
Dolichovespula silvestris	Yes		Montagner (1966)
ANTS			
Formicidae			
Dolichoderinae			
Dolichoderus quadripunctatus			Torossian (1967a,b, 1968a,b)
Iridomyrmex purpureus		M	Hölldobler and Wilson (1990)
Dorylinae			
Dorylus sp.[1]			Raignier (1972)
Formicinae			
Camponotus aethiops			Dartigues and Passera (1979)
C. herculeanus pictus			Fielde (1905)
C. vagus	No		Brian (1983:155), citing *Benois (1969)
Cataglyphis cursor			Cagniant (1979, 1982, 1988)
Formica argentea			Fielde (1905)
F. canadensis			Hung (1973)
F. cinerea			Lubbock (1885)
F. exsecta		P	Pamilo and Rosengren (1983)
F. fusca			Lubbock (1885)
F. pallidefulva			Fielde (1905)
F. pergandei		P	Hung (1973)
F. polyctena	(Yes)		Ehrhardt (1962)
F. sanguinea	(Yes)	P	Pamilo (1982c)
Lasius niger	Yes	M/P	van der Have et al. (1988)
Oecophylla longinoda			Hölldobler and Wilson (1983)
O. smaragdina			Crozier (1970b), Hölldobler and Wilson (1983)
Plagiolepis pygmaea			Passera (1966)
Polyergus breviceps			Hung (1973)
P. rufescens			Lubbock (1885); Goetsch and Käthner (1937)
Myrmeciinae			
Myrmecia gulosa			Freeland (1958)
M. nigrocincta			Haskins and Haskins (1950)
M. piliventris			Haskins and Haskins (1950)
Myrmicinae			
Acromyrmex rugosus			Fowler (1982)
Aphaenogaster cockerelli			Hölldobler and Bartz (1985)

Table 4.3 *Continued*

Taxa	Queen-right colonies?	P or M	References
A. fulva			Fielde (1905)
A. rudis	No[2]	M	Crozier (1974)
A. senilis			Ledoux and Dargagnon (1973)
A. subterranea			Bruniquel (1972)
Apterostigma dentigerum			Forsyth (1981)
Crematogaster impressa			Delage-Darchen (1974)
Epimyrma ravouxi	?Yes (25–50%)		Winter and Buschinger (1983)
Harpagoxenus americanus	?Yes	M	Buschinger and Alloway (1977)
H. canadensis	Yes	M	Buschinger and Alloway (1978)
H. sublaevis	?[3]		Bourke *et al.* (1988)
Leptothorax[4] *allardycei*	Yes		Cole (1981, 1986)
L. ambiguus[5]			Alloway *et al.* (1982)
L. curvispinosus[5]		M	Alloway *et al.* (1982)
L. longispinosus[5]			Alloway *et al.* (1982)
L. nylanderi	Yes		Plateaux (1981)
L. recedens	Yes		Dejean and Passera (1974)
L. tuberum			Bier (1954)
L. unifasciatus			Bourke (1988)
Messor capitatus			Delage (1968)
Myrmica rubra	Yes	P	Smeeton (1981)
M. ruginodis			Brian (1953), Mamsch and Bier (1966)
M. sabuleti			Brian (1983:203)
M. schencki emeryana			Talbot (1945a)
M. sulcinodis			Elmes (1974)
M. 'near tahoensis'	Yes		Evans (1993)
Zacryptocerus varians			Wilson (1976)
Ponerinae			
Diacamma australe			Peeters and Higashi (1989)
Hypoponera eduardi			*Le Masne (1953,a,b)
Neoponera obscuricornis	Yes[5]		Fresneau (1984)
Odontomachus haematodes	Yes		Colombel (1972)
Plectroctena conjugata			Peeters and Crewe (1988)
P. mandibularis			Peeters and Crewe (1988)

[1]Queenless worker groups laid eggs hatching into male larvae, but these larvae failed to yield adults.
[2]No queen-right colony was found to have worker-produced males in an allozyme analysis, but the number of appropriate colonies was small.
[3]Bourke *et al.* (1988) note that their genetic data 'could neither confirm nor deny' worker male-production in the presence of a queen in this species.
[4]A. Buschinger (personal communication) reports the impression based on maintaining many leptothoracine colonies in the laboratory that male production by queenless worker groups is the rule in this group.
[5]Viable eggs were laid by workers and inferred to be male.

[6]Workers lay male-producing eggs at a very low rate in the presence of the queen, but produce large numbers of males in queenless colonies. It is possible that some species for which worker male-production in the presence of a queen has not been recorded also produce males in queen-right colonies, but that this is not so readily detectable in them as in honey-bees.
[7]Populations in France and Spain had most males derived from worker-laid eggs, but not those of England.
[8]Males are produced by unfertilized worker-like queens.
[9]Ross (1985) considers Montagner's (1966) observations of frequent worker-reproduction as a likely experimental error.
[10] Worker reproduction in parasocial colonies.

Table 4.4. Reported cases of thelytoky in eusocial Hymenoptera. It is likely that thelytoky occurs sporadically in a great many species (Crozier, 1975:15–18, 67–70), therefore where it has been reported to be a significant part of the life cycle this is indicated under 'Major?' below.

Taxa	Major?	References
Apidae		
Apis mellifera	Yes[1]	Verma and Ruttner (1983)
Formicidae		
Formicinae		
Cataglyphis cursor	Yes[2]	Cagniant (1982), Lenoir and Cagniant (1986)
Formica polyctena	No	Otto (1960)
Lasius flavus	No	Leutert (1963, 1965), Bier (1952)
L. niger	No	Crawley (1912)
Oecophylla longinoda	Yes[3]	Ledoux (1950, 1954), Way (1954), Vanderplank (1960), Hölldobler and Wilson (1983).
O. smaragdina	Yes[4]	Bhattacharya (1943), Crozier (1970b), Hölldobler and Wilson (1983).
Myrmicinae		
Aphaenogaster lamellidens	No	Haskins and Enzmann (1945)
A. rudis	No	Haskins and Enzmann (1945)
Atta cephalotes	?	Tanner (1892)[5]
Crematogaster auberti	Yes[6]	Soulié (1960)
C. impressa	?	Delage-Darchen (1974)
C. scutellaris	Yes[6]	Soulié (1960)
C. skounensis	Yes[6]	Soulié (1960)
C. vandeli	Yes[6]	Soulié (1960)
Messor aciculatum	?	Imai (1966)
Pristomyrmex pungens	Yes[7]	Itow *et al.* (1984)
Solenopsis invicta	No	Tschinkel and Howard (1978)

[1]Thelytoky occurs sporadically in all subspecies of honey-bee, but is of major importance only in *A. m. capensis*, which can be reared for several female generations without mating. Crosses between this and other subspecies show that the propensity to thelytoky is inherited as a single-gene trait (Ruttner, 1988:48). Thelytokously laying workers of *A. m. capensis* differ from unmated bees of other races which lay preferentially in drone cells (Page and Erickson, 1988), and are treated as queens by queenless groups of workers from other subspecies.
[2]Queen-right colonies do not show worker egg-laying, but in the absence of the queen, worker production of males, queens and workers is prolific and thelytokous. Lenoir and Cagniant (1986) suggest that the function of thelytoky in this species is to replace lost queens, as suggested also for *Apis mellifera capensis*.
[3]Ledoux (1950, 1954) reported a major role for thelytoky in the life cycle of *Oecophylla longinoda*, with queens giving rise only to workers and males, and workers laying eggs of two sizes, the larger of which yield males and the smaller queens and workers. However, the subsequent researchers listed could not confirm the ability of worker ants of this genus to reproduce reliably via thelytoky. Given the more recent and well-substantiated findings on *Cataglyphis*, however, another examination of *Oecophylla* is desirable; the examination of natural colonies using allozyme markers and a minor modification of protocols used for testing for worker male-production (Crozier, 1974) would probably be decisive.
[4]As did Ledoux for *O. longinoda*, Bhattacharya (1943) reported a large component of thelytoky for the Asian species of the genus. Bhattacharya also reported that workers laid eggs of two sizes, but in opposition to Ledoux reported that the smaller worker-laid eggs yield males. The subsequent researchers listed could not confirm the ability of queenless worker groups to reproduce thelytokously, but an examination of natural colonies using genetic markers would be a good idea.
[5]As discussed by Wheeler (1903).
[6]Soulié reported that queens yield workers and males, and workers queens and males, so that the castes alternate each generation. Confirmation using genetic evidence from natural populations is desirable.
[7]All reproduction is thelytokous, males are extremely rare and functionless, and queens of the form seen in congeneric species are unknown.

A colony structure in which all males arise from worker-laid eggs is shown in Fig. 4.2. Attention to the fate of the genes of the original queen and her mate will show that the reproductive values of the two sexes are the same, contrary to the situation where all reproduction is performed by the queen in which case females have twice the reproductive value of males. In other words, when workers produce all the males, the queens and males contribute equal numbers of genes to remote future generations.

Worker reproduction has several important evolutionary consequences. For example, it affects the dynamics of genetic change in the population. Owen (1985, 1986) has made an extensive examination of the effects of male production by workers on the formal population genetics of eusocial Hymenoptera. He finds that the proportion, ψ, of males that is produced by the workers enters into the conditions for stable allele frequency equilibrium under selection.

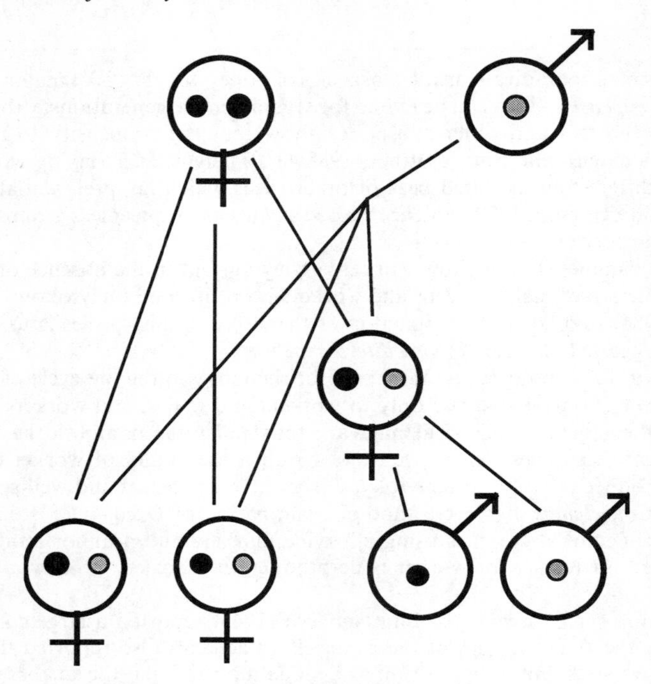

Fig. 4.2 A male-haploid pedigree of a eusocial species in which the queen (topmost female in the figure) produces all of the fertilized eggs and her worker progeny (denoted by the female symbol in the centre of the figure) produce all of the males. The male and female symbols in the bottom row denote the reproductives produced by the colony. In such species queens and males have equal reproductive values.

It may seem an apparent paradox that worker reproduction slightly decreases the effective population size, N_e, despite there being more breeding individuals (Crozier, 1979)! One way to define the effective population size is to consider it as the number of breeding individuals in a population of randomly mating individuals losing genetic variation through random genetic drift at the same rate as the study population. Unless there is an infinite number of laying workers, there will be a greater drift of male allele frequencies each generation when male production is by the workers than when the queen is responsible for all reproduction, because with worker reproduction there is an additional generation in which drift can occur. The drift in the queens will be the same whatever the mode of male production.

Male production by workers will thus, by its effects on effective population size, lead to lower levels of polymorphism of neutral genes, although it is unlikely that the effect is large.

For sex allocation theory, the more important consequences of worker reproduction concern the reproductive values of males and females and the genetic relatednesses within colonies.

Inspection of Figs. 4.1 and 4.2 allows us to track the fate of the genes of the original breeding pair in the next generation. If $\psi = 0$, i.e. all reproduction is by the queen, the number of maternal genes in the offspring is twice that of the paternal genes, and the reproductive values are $\nu_F = 2\nu_M$. At the other extreme, $\psi = 1$, there will be, on average, as many maternal as paternal genes in the offspring and hence $\nu_F = \nu_M$. The effects on the product of relatedness and reproductive value (and hence on the expected sex allocation patterns) of these pedigree changes are shown in Table 4.5.

In general, when a proportion ψ of all the males produced in queen-right colonies are sons of laying workers we can calculate the sex-specific reproductive values using the method presented in Chapter 2. The transition matrix of Section 2.4 now takes the form

$$\mathbf{P} = \begin{pmatrix} 0.5 & 1 \\ \dfrac{(1-\psi/2)}{2} & \dfrac{\psi}{2} \end{pmatrix}.$$

Inserting the transition probabilities in eqn. (2.14) we obtain the ratio of reproductive values

$$\nu_F/\nu_M = 2 - \psi. \tag{4.1}$$

Table 4.5. The products of relatedness and reproductive value ($g_{YX}\nu_Y$) for male-haploids under two extreme pedigree types. The top set of values ($\nu_M = \nu_F$) represents the situation when all males arise from worker-laid eggs, and the bottom set ($\nu_F = 2\nu_M = 2$) that when all males arise from queen-laid eggs. The preferred sex allocation ratio is given by the ratio of the values of the individuals concerned from the viewpoint of the controlling individual. For example, in a monogynous colony with all males arising from worker-laid eggs, the preferred sex allocation of a worker is given by the ratio sisters/nephews, which is 1:1. In a monogynous colony with all males arising from queen-laid eggs, the preferred sex allocation of a worker is given by the ratio sisters/brothers, which can be determined from the table to be 3:1. See Sections 2.4 and 4.2 for further explanation.

X	Y								
	Mother	Father	Daughter	Son	Sister	Brother	Aunt	Niece	Nephew
$\nu_M = \nu_F$									
Female	1/2	1	1/2	1	3/4	1/2	3/8	3/8	3/4
Male	1/2	0	1/2	0	1/4	1/2	3/8	1/8	1/4
$\nu_F = 2\nu_M = 2$									
Female	1	1	1	1	3/2	1/2	3/4	3/4	3/4
Male	1	0	1	0	1/2	1/2	3/4	1/4	1/4

Noting that, according to our definition, $\nu_F + \nu_M = 3$, we can solve the two reproductive values separately:

$$\nu_F = 2 - \frac{\psi}{3 - \psi} \tag{4.2a}$$

$$\nu_M = 1 + \frac{\psi}{3 - \psi} . \tag{4.2b}$$

When the workers produce males, it can be seen that the reproductive value of males increases because they can now father males in the next generation; they do so indirectly, through the haploid eggs of their worker offspring. Consequently, the reproductive value of females must necessarily decrease proportionately.

Figure 4.3 gives the pedigree structure for a colony in which males arise from both queen-laid (QM) and worker-laid (WM) eggs. A worker-laid male in such a colony will be related to his own mother as WM is to W1, and to other workers as WM is to W2. The resulting relatedness values are given in Table 4.6.

In the absence of workers being able to distinguish their own sons from those of the others (queen and other workers), and to direct aid towards them

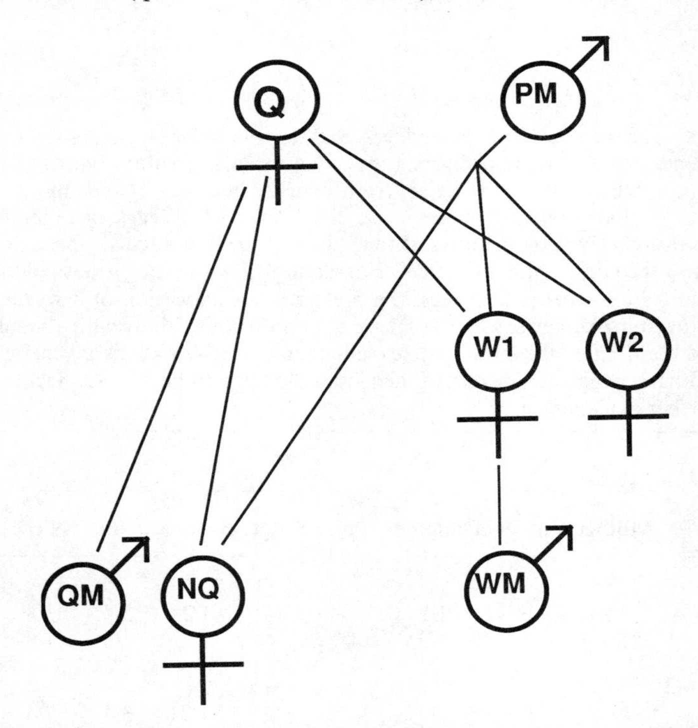

Fig. 4.3 Pedigree structure of colonies in a species in which males arise from both queen-laid (QM) and worker-laid (WM) eggs. Other designations are Q queen, PM queen's mate (paternal male), W1 mother of WM, W2 sister to W1, NQ new queen.

Table 4.6. Relatedness (g_{YX}) values between new reproductives and the female members of colony described by Fig. 4.3. Relatedness is from individual Y to individual X.

X	Y				
	W1	W2	NQ	QM	WM
Q	1/2	1/2	1/2	1	1/2
W1	1	3/4	3/4	1/2	1
W2	3/4	1	3/4	1/2	3/4
PM	1/2	1/2	1/2	0	1/2

preferentially, the mean relatedness of males to workers will be a weighted average involving ψ and the size and composition of the worker force. If there are N_{LW} laying workers contributing equally to the males of the next generation, the mean relatedness of males to such a worker is

$$g_{MW} = (1 - \psi)(1/2) + \frac{\psi[1 + (N_{LW} - 1)(3/4)]}{N_{LW}}. \qquad (4.3)$$

In those species with large colonies, N_{LW} is probably large, so that

$$g_{MW} = 0.5 + \frac{\psi}{4}. \qquad (4.4)$$

This is also the relatedness of males to non-laying workers. Similarly, the relatedness of males to the queen is given by

$$g_{MQ} = 1 - \frac{\psi}{2}. \qquad (4.5)$$

In those eusocial Hymenoptera with small colonies, the size of the worker pool might be a significant factor, in which case the life-history stages of individual workers need to be considered. As the worker progresses from immature to mature and egg-laying, and then to senescent non-laying, the average relatedness of the males to her will decrease (but only if the worker lives through several generations of males).

4.2.2 *Evolution of worker reproduction*

While the essence of eusociality is reproductive division of labour, the production of haploid eggs by hymenopteran workers occurs in many species (Table 4.3), but not in all. What are the selective forces acting on this system?

It is evident from the above analyses that the genetic value of the sons to the queen and to the workers depends on who produces them in the colony. From the queen's point of view, male production by workers dilutes her impact in the gene pool of sons by one half. We therefore expect that the queen should prevent workers from reproducing. For the workers, the situation is different.

It is always best if the worker can produce her own son, related to her by one. The second best option, under the simple family model of Fig. 4.3, is to allow sister-workers to produce sons with relatedness to her of 0.75, because the relatedness of the queen's sons to the workers is only 0.5. This means that there is, at least potentially, a conflict over male production between the queen and workers (Hamilton, 1964b; Trivers and Hare, 1976).

Multiple mating alters this picture. Because some of the worker nest-mates are now half-sisters, the relatedness of their sons to other workers is, on average, less than 0.75. In fact, this average may drop below 0.5. This changing relatedness depending on polyandry leads to varying expectations concerning worker reproduction. Polygyny can have similar effects.

Let us first examine the case of single-queen colonies, where the worker force is divided into k equally large patrilines because of multiple mating by the queen. The average relatedness of worker-produced males to a randomly chosen worker individual is now

$$\frac{1}{k}\frac{3}{4}+\frac{k-1}{k}\frac{1}{4}=\frac{k+2}{4k} \tag{4.6}$$

if we neglect the worker's own reproduction. From the worker's point of view, male production by other workers is advantageous only if this relatedness is greater than that of the queen-produced males (Starr, 1984; Woyciechowski and Łomnicki, 1987; Ratnieks, 1988). Thus, workers should allow other workers to produce males only when

$$\frac{k+2}{4k}>\frac{1}{2}$$

which reduces to

$$k \leq 2\,. \tag{4.7}$$

This means that workers should favour worker reproduction only when the number of matings is less than 2, and they should prevent other workers from producing males when the number of matings, or rather the effective number of matings (Sections 3.5, 4.3), is greater than 2. If the workers control who produces the males, we would expect that in some colonies all males are produced by the queen and in some colonies all males are produced by the workers. Only when $k = 2$ will workers be indifferent. This conclusion assumes that the workers can somehow recognize how many times the queen has mated, i.e. they can recognize the genetic heterogeneity among the worker nest-mates. If they cannot do this, and if the mean number of effective matings varies around 2, we might expect some uncertainty and intermediate values of ψ.

The above analysis depends on the assumption that workers can control each other's reproduction. As noted by Hamilton (1972), West-Eberhard (1975) and Craig (1983), workers will be selected for to lay eggs, which will yield males because of arrhenotoky. Cole (1986) and Ratnieks (1988) examined the invasion of worker reproduction and the invasion of the suppression of worker reproduction ('worker policing' of Ratnieks) by combining the genetic effects of polyandry and assuming a cost to the colony from the invading behaviour. Such a cost might be caused by a lower efficiency due to many workers being involved either in reproduction or in controlling other workers' reproduction instead of in other duties benefiting the colony. Both Cole and Ratnieks used an explicit genetic one-locus model with large effects. By 'large effects' we mean that a mutant genotype has a well-defined effect both in the frequency of worker reproduction and in the colony efficiency. Pamilo (1991b) generalized these two models combining the invasion and the loss of worker reproduction into a single model, and derived the results using both single-locus and general inclusive fitness models.

The inclusive fitness model for the evolution of worker reproduction can be based on the general fitness function for a worker I

$$W_{\mathrm{I}} = N(g_{\mathrm{FI}}\nu_{\mathrm{F}}\frac{f}{F} + g_{\mathrm{MI}}\nu_{\mathrm{M}}\frac{m}{M}) \tag{4.8}$$

as discussed earlier (eqn. (2.6)). N refers to the number of offspring produced and measures the overall ecological success of the colony. Changes in the source of the males, either from the queen or from the workers, affect the relatednesses within the parentheses and may also affect colony success. The question is: how large can the cost of worker reproduction to the colony be and still allow it to persist? We denote by N_ψ the number of offspring produced by a colony where the proportion ψ of male eggs is laid by workers. A simple assumption is that N_ψ is locally a linear function of worker egg-laying, i.e.

$$N_{\psi+\delta} = N_\psi\,(1 - \delta c) \tag{4.9}$$

where δ is a small change in the initial amount of worker reproduction and c is a constant defining the cost of worker reproduction. This constant $c > 0$ if worker reproduction imposes a cost to the colony, and $c < 0$ if worker reproduction increases colony reproductive potential (e.g. when the suppression of worker reproduction imposes an overall cost to the colony).

Let us next look at the relevant relatednesses from the workers' point of view. The relatedness of brothers to workers remains $g_{\mathrm{BW}} = 0.5$ independently of the number of matings, but those of sisters (g_{SW}) and nephews (g_{NW}) will be affected (under our notation $g_{\mathrm{SW}} = g_{\mathrm{NW}}$). When the queen has mated with k unrelated males, these relatednesses are $g_{\mathrm{SW}} = g_{\mathrm{NW}} = 0.25 + 1/(2k)$. Assuming that each colony produces the same sex ratio, i.e. $m_i/M = f_i/F = 1$, the inclusive fitness for a colony can be written as

$$W_i = N_\psi(g_{\mathrm{SW}}\nu_{\mathrm{F}} + [(1-\psi)g_{\mathrm{BW}} + \psi g_{\mathrm{NW}}]\nu_{\mathrm{M}})\,. \tag{4.10}$$

Strictly speaking, this is the inclusive fitness of a non-laying worker, but it also approximates the inclusive fitness of the laying workers if their number is so large that we can neglect each worker's own sons.

The evolution of worker reproduction can be examined by letting the proportion of worker-produced males change slightly from its initial value ψ_0 and calculating the effect of this change on inclusive fitness. Combining eqns. (4.9) and (4.10) and noting that $v_F/v_M = 2-\psi_0$, Pamilo (1991b) showed that the critical threshold of the cost associated with worker reproduction is

$$c_{crit} = \frac{2-k}{4(1+k) - 2k\psi_0} . \qquad (4.11)$$

The amount of worker reproduction increases when the cost $c < c_{crit}$ and reduces when $c > c_{crit}$. In particular, when initially $\psi_0 = 0$, worker reproduction can invade provided $c < (2-k)/(4+4k)$, and when $\psi_0 = 1$, it is lost when $c > (2-k)/(4+2k)$. This result means that, with a single mating, worker reproduction could invade a population even if it causes a cost of 12.5%. Worker policing can invade a population when the cost is less than 10% if $k = 3$ and when the cost is less than 16.7% if $k = 4$.

The single-locus models give exactly the same result as the above inclusive fitness model, provided that the locus has only small effects on worker reproduction (Pamilo, 1991b).

The results of a single-locus model are more favourable for the invasion of worker reproduction if the locus has large effects, i.e. $\delta = 1$, and if only the workers carrying a mutant allele start laying eggs (Cole, 1986; Pamilo, 1991b). The evolution of worker reproduction is made still easier by behavioural dominance, a condition in which the presence of a mutant genotype leads to the whole colony expressing the mutant phenotype (Charnov, 1978; Craig, 1980; Pamilo, 1982b; Bulmer, 1983a).

Ratnieks (1988) examined the reduction in worker reproduction. This reduction can arise either from workers simply ceasing to lay eggs, or from some workers controlling egg-laying by actively preventing their fellows from reproducing. Ratnieks studied the latter case and called it worker policing. He also assumed that, while the controlling workers reduce the reproduction of other workers in the colony, they do not give up their own reproduction. When examining the invasion of an allele in an initially monomorphic population, we need to consider only two types of colonies because it is unlikely that we will find more than one copy of the allele in the parental genomes. If worker reproduction does not affect the colony's overall success, the necessary condition for worker policing to evolve is that the proportion of worker-produced males is smaller in colonies headed by a heterozygous queen (inseminated by non-mutant males) than in colonies headed by homozygous queens inseminated by a mutant male (and also by non-mutant males if the queen is polyandrous). The evolution of worker policing also depends on how efficient the controlling workers are in suppressing egg-laying of their nest-mates (Ratnieks, 1988; Pamilo, 1991b; Ratnieks and Reeve, 1992).

The factor driving worker control with increasing degree of polyandry is the decreasing relatedness of worker-produced sons to a randomly chosen worker in the colony. A similar effect also occurs for polygyny. Analysing an inclusive fitness model, Pamilo (1991b) found that the critical conditions for the evolution of worker reproduction are given by eqn. (4.7), if the number of mates, k, is replaced by the term $\{1 + (n - 1)g_{QQ}\}$, where n is the effective number of coexisting queens and g_{QQ} is the genetic relatedness among these queens. If there are no costs to the colony, the condition for worker reproduction to evolve can be written as (Pamilo, 1991b)

$$n < 1 + \frac{1}{g_{QQ}}. \tag{4.12}$$

The above results, concerning both polyandry and polygyny, are based on the assumption that the number of workers is relatively large. Otherwise one should take into account the relatedness to oneself of one's own sons. The situation will also change if the workers can recognize kin groups within the colony and direct their behaviour preferentially.

How well do the data agree with predictions? As noted by Nonacs (1993), comparative data do not demonstrate any dependence of worker reproduction on polyandry, which may reflect on their reliability. But there is a good case in point in the honey-bee. Worker honey-bees have functional ovaries, and produce males on the death of the queen. In queen-right colonies, worker-laid males occur at very low rates (Page and Erickson, 1988). Worker-laid eggs are generally detected and destroyed by other workers (Ratnieks and Visscher, 1989), but some genotypes escape detection (Oldroyd *et al.* 1994a).

Nonacs (1993) suggests that some cases where the predictions of this section are not met may represent instances of deception, for example of a once-mated queen leading workers to believe that she has mated several times, or that her male-producing eggs are in fact worker laid. It is hard to see how this could arise unless the population is variable for the trait in question. However, some instances may reflect inefficiencies of selection. In ants of the *Rhytidoponera impressa* group, for example, colonies are either queen-right or queenless, with the queen role filled by 4–15 mated workers, gamergates (Ward, 1981, 1983a,b). Ward (1981) suggests that queenless colonies arise from other queenless colonies by fission, but the high relatedness between their gamergates suggests that they were originally queen-right but that their queens died, and that they do not last much longer than one generation. Unmated workers with developed ovaries are rare even in queenless colonies (less than 6%), and it is possible, given evidence from other *Rhytidoponera* species (Peeters, 1987a), that they do not lay any eggs. Given that there are at least four gamergates in most queenless colonies, then reproduction by the unmated workers should be suppressed according to eqn. (4.12). Male production in populations of the *Rhytidoponera impressa* group is disproportionately carried out by the queenless colonies, so that it is plausible to suggest that it is selection in these colonies which determines worker reproductive behaviour overall.

Based on the above inclusive fitness models, Pamilo (1991b) found that worker reproduction is expected to evolve when there is both monogyny and monandry, although it can also be favoured in polygynous species when the number and relatedness of the coexisting queens are low enough. In contrast, Trivers and Hare (1976) suggested that worker reproduction should occur more readily under polygyny. The reason for this suggestion is that worker egg-laying is likely to cause a confrontation between the laying workers and the queen or queens. Because workers in a monogynous colony cannot harm their single queen without seriously endangering the whole colony, the queen is expected to win the conflict. On the other hand, it is more important in monogynous than in polygynous colonies for the workers to retain the ability to lay eggs, because they are more likely to become orphaned than are workers of polygynous colonies. Orphaned workers of monogynous colonies which can lay haploid eggs are thus able to extend the reproductive life of their colony for a while by producing males (Bourke, 1988; Franks *et al.*, 1990).

Unfortunately, the data to test these predicted trends are not impressive. Bourke (1988) surveyed data pertaining to 37 ant species with worker reproduction. Seven (24%) of the species with monogynous colonies and five (62%) of the species with polygynous colonies have reproductive workers in the presence of the queen. The difference is statistically significant and indicates that workers of monogynous colonies are less likely to produce males in queen-right colonies. However, Pamilo (1991b) notes that the observations may be biased in the observed direction because worker reproduction is most easily documented in orphaned colonies producing males, and these colonies are likely to have been monogynous. We still lack reliable quantitative estimates of worker reproduction, as well as of the effective levels of polyandry and polygyny.

Although more ant species are known with worker-produced males than all other eusocial Hymenoptera combined (Table 4.3), this appears to be a consequence of the large number of ant species; the general impression is that worker-produced males are more common in social wasps and bumble-bees than in ants. Single-mating by queens is the rule in most bumble-bees (Sakagami, 1986) but multiple-mating is widespread in vespids (Table 4.1). As in many ants, in bumble-bees and in vespids worker reproduction commonly occurs in the event of the senescence or death of the mother queen.

An interesting phenomenon in some parasocial and advanced eusocial species, but absent in species with primitive eusociality, is the laying of trophic eggs, sterile eggs made to be eaten. In many species, a substantial proportion of the nutrient flow within colonies is due to trophic eggs (reviewed by Crespi, 1992b). Crespi (1992b) remarks that the diversity of trophic egg production patterns in the stingless bees (one of the two advanced eusocial groups to show trophic eggs) defies a general explanation for their occurrence, although queen–worker conflict is probably important. Crespi (1992b) concurs with West-Eberhard (1981) that in ants workers laying trophic eggs do so in order to retain their physiological capacity to produce fertile, male-producing eggs. West-Eberhard's (1981) hypothesis does not conflict with there being adaptive advantages to trophic egg production, but does imply that in the absence of an opportunity to lay fertile eggs there should be no trophic egg production.

There is some negative support for West-Eberhard's (1981) idea: as noted by Crespi (1992b) no species is known in which workers lay only trophic eggs and cannot produce fertile eggs.

4.3 Multiple mating

4.3.1 *Polyandry is widespread*

Multiple mating by both sexes can take place, and if it leads to multiple inseminations, can affect the genetic relationships between colony members. Multiple mating by males is very important in sex allocation when there is local mate competition (Section 2.8), but males are not always capable of inseminating more than one female (Crozier, 1977a; Nonacs, 1986a). Single mating is certainly the case for honey-bee males, because these lose their genitalia after mating (Michener, 1974:361–3). Because multiple mating by a male does not affect the genetic diversity within a colony (unless coexisting queens are inseminated by the same male), we do not discuss it here but concentrate on examining multiple mating by females. As documented in Table 4.1, multiple mating by females is widespread in social Hymenoptera.

Figure 4.4 illustrates some salient connections in a colony with one multiply-mated queen. It is clear that the colony's workers are divided into a number of different *patrilines*. Workers in the same patriline are full sisters and are related by $g_{SS} = 0.75$, but workers in different patrilines (W1 and W2 in the figure) are maternal halfsiblings and are related by

$$g_{PP} = \frac{1}{4} + \frac{1}{2} g_{MM} , \tag{4.13}$$

where g_{PP} is the relatedness between workers in different patrilines and g_{MM} is the relatedness between the parental males (Section 2.3). Where these males are unrelated, $g_{MM} = 0$ and $g_{PP} = 1/4$.

If all of the, say, k males father the same proportion of the offspring, we can easily calculate the resulting average genetic relatedness among the workers as

$$g_{WW} = [0.75 + (k-1)g_{PP}]/k . \tag{4.14}$$

When the contributions of the various males differ from each other, we have to take this variation into account. In such a case we can use the effective number of matings, or effective promiscuity (Starr, 1979, 1984), which was already introduced in eqn. (3.10) and defined as

$$k_E = \frac{1}{\sum^{k} \gamma_i^2} \tag{4.15}$$

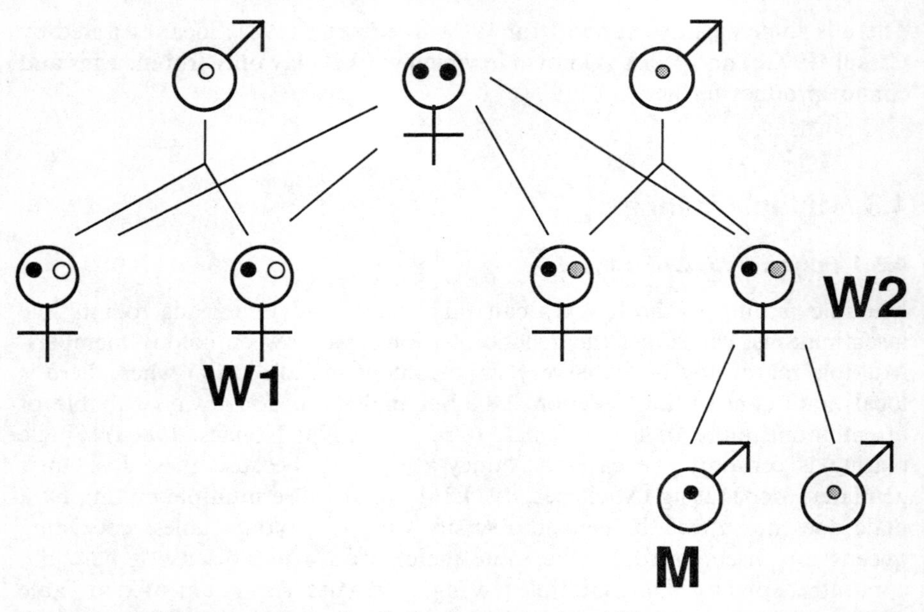

Fig. 4.4 Pedigree structure of colonies in species in which queens mate more than once. Workers W1 and W2 belong to different *patrilines,* and are related by $g = 0.25$. A male progeny, M, of W2 is related to W1 by $g = 0.25$.

where, k is the number of males mating with the female, and γ_i is the proportion of the female's daughters fathered by the ith male.

The average relatedness between workers is now obtained by replacing k with k_E:

$$g_{\mathrm{WW}} = \frac{0.75 + (k_{\mathrm{E}} - 1)g_{\mathrm{PP}}}{k_{\mathrm{E}}}$$

$$= \frac{2 + k_{\mathrm{E}}(2g_{\mathrm{MM}} + 1) - 2g_{\mathrm{MM}}}{4k_{\mathrm{E}}} \tag{4.16}$$

The average relatedness of males to workers in a species with large colonies and with all males produced from worker-laid eggs is also given by eqn. (4.16), which reduces to $(2 + k_{\mathrm{E}})/(4k_{\mathrm{E}})$ if the paternal males are unrelated. The mean relatedness of worker-produced males to workers varies with k_{E}, in contrast to that of the queen-produced males, which is always 0.5 in monogynous colonies. The estimate of g_{WW}, when calculated by sampling many colonies, is affected by both the mean and the variance of the number of matings and of the contribution of each male (Pamilo, 1993; Ross, 1993).

When colonies are small, the approximations of eqn. (4.16) may be inaccurate for calculating the relatedness of male offspring to workers, g_{MW}, because the relatedness of a worker's own sons to her should be taken into account, as should the number of workers in each patriline. The appropriate equation, assuming that all males arise from worker-laid eggs and that every worker reproduces, is

$$g_{MW} = \frac{1}{N_W}(1) + \frac{3}{4}\left(\frac{N_W}{k_E} - 1\right)\frac{1}{N_W} + g_{PP}\left(N_W - \frac{N_W}{k_E}\right)\frac{1}{N_W}$$

$$= \frac{1}{N_W} + \left(\frac{3}{4}\right)\left(\frac{N_W - k_E}{N_W k_E}\right) + g_{PP}\left(\frac{N_W k_E - N_W}{N_W k_E}\right)$$

$$= \frac{1}{4}(1 + 2g_{MM}) + \frac{(1 - g_{MM})}{2k_E} + \frac{1}{4N_W} \tag{4.17}$$

where N_W is the total number of workers and N_W/k_E is the mean number of workers per patriline. If the number of workers is large, the last term is negligible, and if the number of effective matings is large, as it is in honey-bees, the middle term also diminishes. Otherwise, eqn. (4.17) should be used for male–worker relatedness in eusocial insects with small colonies.

If we assume the number of workers to be large and $k_E > 2$, then the average relatedness of worker-produced males to workers is smaller than that of queen-produced males, provided that the paternal males are not close relatives. This situation contrasts with that where the queen mates only once, and where worker-produced males are *more* related to the workers than are queen-produced ones. As discussed earlier (Section 4.2), Woyciechowski and Łomnicki (1987) and Ratnieks (1988) regard this difference as providing a plausible explanation for the lack of worker egg-laying in queen-right colonies: it is to each worker's genetic advantage to prevent egg-laying by other workers.

Polyandry has been commonly inferred from observations of multiple mating in field or laboratory (Page, 1986). Multiple mating need not necessarily lead to multiple insemination, and if it does the contributions of different mates cannot be easily estimated without an actual genetic analysis of the offspring. A genotypic mother–offspring analysis can be used for estimating the number of mates in polyandrous species, e.g. using allozyme variation (Pamilo, 1982c, 1993; Ross, 1986; Sundström, 1989; Seppä, 1994a). (Laidlaw and Page (1984) used phenotypic markers in the progeny of artificially inseminated honey-bee queens to infer that sperm use of each male is appreciable and that therefore in natural populations of honey-bees the relatedness between workers must be close to 0.25.) The power of this method is limited by the amount of genetic variation in the population. If there is a high probability that two unrelated males carry an identical multi-locus genotype (at the loci used in the analysis), it becomes difficult to distinguish all the patrilines in the offspring. For this reason, genetic markers should be sought with the highest possible levels of genetic variability. The advent of DNA sequence variants with high levels of variability is therefore welcome. One class of variants is that of minisatellites, in which there is variation in the number of tandem repeated sequences (50–100 bases long) (reviewed by Burke, 1989). Blanchetot (1991, 1992) examined such 'DNA fingerprinting' variation in the honeybee *Apis mellifera* and in a solitary bee *Megachile rotundata*, finding moderate levels of variation which could be used to distinguish patrilines within a single honey-bee colony. A moderate level of variation was also detected in the ant *Camponotus floridanus* (Heinze *et al.*, 1994a). Other highly variable genetic markers will be useful in studies of polyandry (see Section 2.3).

4.3.2 *Why are some species polyandrous?*

We have seen that there is considerable interspecific variation in how many times females of eusocial Hymenoptera mate. The number of matings of the queen in turn affects the preferred options of the workers in terms of worker male-production and finally the sex allocation pattern of the colony. It is therefore worth asking about possible reasons for favouring monandry in some species and polyandry in others.

There are numerous hypotheses about the evolution of polyandry, but this area was given considerable coherence and direction by Cole's (1983) observation that species with multiply mating queens tend to have larger colonies than species with singly mated queens. Cole (1983; see also West-Eberhard, 1975; Trivers and Hare, 1976) suggested that this trend indicates that the larger colony size imposes a need for more sperm, and that females therefore mate more than once to acquire this sperm.

The problem with Cole's (1983) sperm-need hypothesis is that it begs the question as to why larger males have not evolved to provide more sperm (Crozier and Page, 1985; Hölldobler and Bartz, 1985). In fact, at least for bees, such male size evolution has occurred: Crozier and Page (1985) found a close correspondence between the average numbers of sperm produced by males and stored by females; wasps showed no trend and the ant case needs to be examined when sufficient data become available. The case of the honey-bee *Apis mellifera* is particularly instructive: Moritz and Southwick (1992:194) remark that polyandry in the honey-bee was unsuspected for so long *because* 'single drones do have sufficient semen to inseminate the queen'.

Crozier and Page (1985) examined twelve hypotheses for the evolution of polyandry, in the light of the observed association with colony size (Cole, 1983), finding only three plausible and capable of directing future work. These hypotheses can be stated as follows:

1. Caste determination has at least a partial genetic basis, so that factors maintaining genetic variation within colonies are required to enable expression of the full range of castes,

2. Colonies require genetic diversity to ensure that they have workers able to function under a wide array of environmental conditions; this hypothesis thus includes resistance to pathogens, argued by Hamilton (1987) and

Sherman *et al.* (1988) as the causal agent for eusocial hymenopteran polyandry,

3. In the cases where sex-determination is controlled by heterozygosity at one or more sex loci, then the load of diploid males will select for monandry under some life-history conditions and polyandry under others.

Crozier and Page (1985) considered that hypothesis 3 is the most likely. Pamilo *et al.* (1994) have reassessed their analysis extending the findings. The treatment below draws on these analyses.

We use hypothesis 3 as the entré to our analysis.

The driving force of this model is the effect on colony growth of the production of diploid males. Diploid males take the place of diploid females, and represent a severe cost to the colony because they neither work nor contribute to the next reproductive generation if they survive to adulthood.

Multiple mating does not reduce the proportion of diploid males in the population, but rather, in comparison with single mating, it leads to many females producing some diploid males rather than a few producing half of their diploid offspring as males. The effect of multiple mating is therefore on the variance of diploid male production between mated females. The distribution of diploid males between colonies also alters the mean fitnesses of strains that are either single mating or multiple mating. It is convenient to restrict the latter to mating twice.

We assume just one locus determines sex, as is the case in the honey-bee. Let there be n alleles at the sex locus. If a queen mates once there is a probability of $2/n$ under Hardy–Weinberg assumptions of her mating with a male sharing an allele with her at the sex locus, and in that case 50% of her diploid progeny will be diploid males. If she mates twice, and one of the males shares an allele at the sex locus with her, then 25% of her diploid progeny will be male. In a strain that mates twice, the expected proportions of times the queen mates with no males carrying the same sex allele, with one, and with two are

$$\begin{aligned}&\text{with no males with common alleles } \left(\frac{n-2}{n}\right)^2\\ &\text{with one male with a common allele } 2\left(\frac{2}{n}\right)\left(\frac{n-2}{n}\right)\\ &\text{with both males with common alleles } \left(\frac{2}{n}\right)^2.\end{aligned} \tag{4.18}$$

The occurrence of diploid males in a colony will slow its growth and increase its mortality relative to colonies lacking diploid males to an extent dependent on the proportion of diploids which are male. It is reasonable to assume that colonies grow logistically, and hence the basic logistic growth equation (see Crow and Kimura 1970:22–5; Roughgarden, 1979:299–306) can be modified for our purposes:

$$N(t,D) = \frac{N_{max}}{1 + [(N_{max} - N_0)/N_0]e^{-(1-D)rt}} \quad (4.19)$$

where N_{max} is the maximum possible size attainable by the colony, N_0 is the initial colony size at time zero, D is the proportion of diploids which are male (0, 0.25, or 0.5), and rt is the product of a growth constant (r) and the time elapsed since the beginning of the colony growth cycle (t).

The dynamics of the evolution of the degree of polyandry can now be approached by comparing the fitness, in a monogynous species, of a strain whose queens mate once with one that mates twice, making this comparison for different timings of reproduction during the colony growth curve (Fig. 4.5).

Adding the effects of mortality (Page, 1980; Pamilo *et al.*, 1994) reinforces (with reservations) the message from Fig. 4.5: reproduction during the exponential phase of colony growth imposes selection for monandry and reproduction during the asymptotic phase imposes selection for polyandry. Increasing the numbers of alleles or of loci reduces the strength of this selection by reducing the differences between the single- and double-mating strains in terms of the proportions of queens with $D = 0$ and $D = 0.5$.

What predictions do the three hypotheses make by which they might be distinguished? All three effects might indeed occur: the test would be of which of these is the most important in determining the biology of mating behaviour.

Hypothesis 1 predicts a relationship between polyandry and the complexity of the caste system, or the number of roles.

The predictions of hypothesis 2 depend on how the presence of resistant workers affects the resistance of the colony as a whole. The effect is strengthened if there is behavioural dominance: the occurrence of resistant workers preserves the colony as a whole. There is evidence for individual variation in either infectiousness or susceptibility in the interactions between the bumble-bee *Bombus terrestris* and its trypanosome parasite *Crithidia bombi* (Shykoff and Schmid-Hempel, 1991a,b). These authors also showed that the transmission of parasites to a social group is at least associated with kinship, related individuals preferentially infecting each other. These results demonstrate the possibility of negative frequency-dependent selection by parasites on common host phenotypes. Genetically diverse colonies might therefore have higher resistance to infection. If there is behavioural dominance for resistance, selection by pathogen resistance is unambiguously in favour of polyandry.

The tentative conclusion is that pathogen selection probably does select for increased genetic diversity. If so, then polygyny will have the same effect as polyandry. Hence, polygynous species will have reduced selection for polyandry, whereas monogynous species will have relatively strong selection for polyandry.

Hypothesis 3 predicts an association between passage through the selection period of the exponential phase of colony growth and the occurrence of monandry or polyandry. In polygynous species the effects of diploid male production by one queen are diluted by the normal progeny of the others.

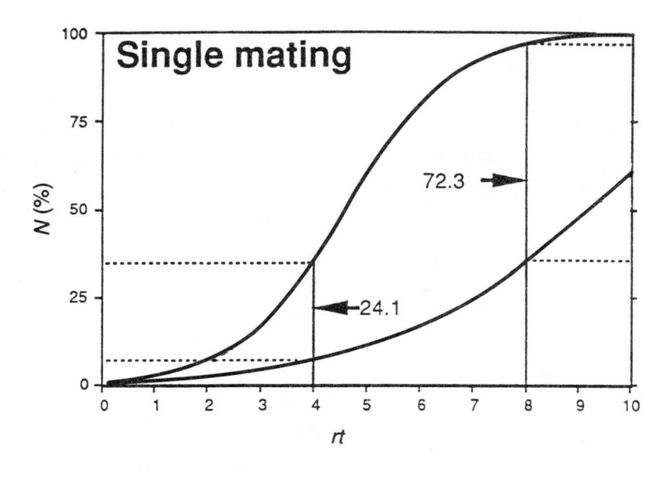

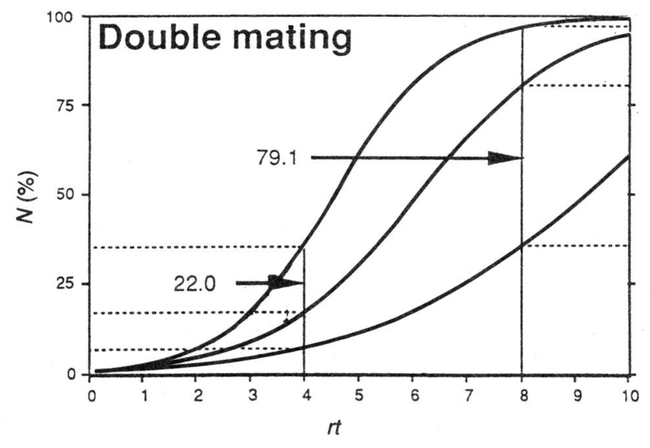

Fig. 4.5 Changes in the size (numbers of workers) of single-versus twice-mating strains of eusocial Hymenoptera under the demographic model, assuming five equally frequent sex alleles. Values of the growth curve parameter *rt* are shown along the *x*-axis and on the *y*-axis is shown the size ($N\%$) in number of workers of colonies during colony growth as standardized to N_{max}. For a single-mating strain there are only two kinds of colony, one with no diploid males and one with 50% of the diploids being male. For a double-mating strain there is an additional colony type (resulting from the queen mating with one male with a sex allele the same as hers and with one male whose allele is different from both of hers); this intermediate type produces 25% of the diploids as males. According to the demographic model discussed by Crozier and Page (1985), the single-mating strain has a larger average size than the double-mating strain early in colony growth (24.1 units versus 22.0 in this example) whereas it has a smaller one later (72.3 versus 79.1). Because honey-bees swarm late in colony growth (high *rt*) multiple-mating is favoured in them, whereas in species reproducing early in colony growth (low *rt*) single-mating is expected to be favoured. (Based on Fig. 2 of Pamilo *et al.*, 1994.)

Hence, as for hypothesis 2, monogynous species will show most clearly the associations of hypothesis 3. For example, species whose colonies reproduce by colony fission are liable to lack essentially the exponential phase of colony growth and be polyandrous, whereas those in which colonies are founded by queens without the aid of workers are liable to endure a long exponential phase and be monandrous. Honey-bees provide an example of a species in which selection is predominantly during the asymptotic phase, and they are polyandrous. Monogyne *Solenopsis invicta* fire ants provide an example of a species with very strong selection during the exponential phase (Ross and Fletcher, 1986): incipient colonies with any diploid males are at a severe disadvantage to those lacking any, and are eliminated under natural conditions (Ross *et al.*, 1988). Naturally, however, two examples do not make a convincing demonstration!

The data (e.g. Table 4.3) are not yet sufficiently numerous or well couched to distinguish between these hypotheses. However, the conceptual framework now appears to be sufficiently developed for future work to make this distinction.

The above hypotheses referred to the effects of genetic diversity. As presented in Section 4.3.1, polyandry also affects genetic relatedness between colony members, and can therefore also affect the optimal behaviour of individuals, particularly with respect to patterns of optimal resource allocation. The evolution of polyandry as a response to queen–worker conflict was proposed by Starr (1984) in the context of conflict over male production and by Moritz (1985) in the context of sex ratio conflicts. Ratnieks (1990) and Crozier and Page (1985) did not, however, consider these conflicts as plausible explanations for the evolution of polyandry, whereas Woyciechowski and Łomnicki (1987) regarded worker reproduction as a likely factor in selection for polyandry. Analysing the conflict models by connecting the inclusive fitness models that we have already become familiar with earlier in this book with the split sex ratio hypothesis of Boomsma and Grafen (1990, 1991), Pamilo (1991b) showed that under certain situations polyandry of the queen can be beneficial both to her and to her worker offspring. This requires that the workers can estimate the level of genotypic diversity in the colony and bias the colony sex ratio accordingly — colonies with multiply inseminated queens should specialize in producing males. This scenario is still very speculative and requires both theoretical development and a formulation that makes it empirically testable (Pamilo, 1991b; Queller, 1993).

4.4 Many species have polygynous colonies

4.4.1 *The occurrence and effects of polygyny*

Many species of eusocial Hymenoptera have one queen or functional mated egg-layer per colony (**monogyny**), but in others there is more than one (**polygyny**). The differences between the two states reflect significant behavioural differences, as shown by the fact that in many monogynous species colonies are founded by several queens (pleometrosis) but all but one are eliminated, either by each other or by workers, once workers have appeared and colony growth begins. Polygyny is common in ants (Buschinger, 1974; Hölldobler and Wilson, 1977; Rosengren and Pamilo, 1983; Rissing and Pollock, 1988), quite common in wasps (Jeanne, 1980; Spradbery, 1986; Itô, 1987b), reasonably common in termites (Thorne, 1985; Roisin, 1987) and rare in bees (Michener, 1974:113; Packer and Knerer, 1985).

There is a wide range of social structures with respect to queen biology (Hölldobler, 1962; Baroni-Urbani, 1968; Buschinger, 1968, 1970, 1979; Wilson, 1974a,b; Buschinger *et al.*, 1980). While some species have mechanisms enforcing the presence of only one queen in a colony, in others unmated queens acting as workers may be retained (pseudopolygyny) or additional mated queens may be present but fail to reproduce (functional monogyny).

A form of polygyny is **oligogyny** (Hölldobler, 1962), in which multiple fertile queens are mutually intolerant of each other's close proximity and inhabit separate parts of the same nest (for which the term *paragyny* was proposed by Pamilo (1991b)). The compartmentation of the nest by the mutually intolerant queens naturally limits the number of queens which can coexist — hence the term oligogyny. Perhaps because of the detailed observation needed to document it, oligogyny has been shown for only a few species (two *Camponotus*, Hölldobler, 1962; *Lasius flavus*, Hölldobler and Wilson, 1990:218; *Iridomyrmex purpureus*, Hölldobler and Carlin, 1985), although suspected by Wilson (1971:332) and by Hölldobler and Wilson (1990:209) to occur in many more.

The number of queens present in colonies often varies within a species in that some colonies are polygynous but others are truly monogynous (Elmes, 1987a; Seppä, 1994a). Such variation, whether it exists among colonies of the same population or characterizes different conspecific populations, can provide important information concerning the evolution of polygyny. It may be possible to correlate the degree of polygyny with the characteristics of the environment or of the colony itself (e.g. its size and age) and to try to find the determinants favouring monogyny or polygyny. Few such studies have been done.

The difference in the level of polygyny can be quantitative (difference in the mean number of queens) or qualitative (a difference between monogyny and polygyny). Such variation in the number of queens exists commonly among colonies but also among populations. Quantitative differences have been reported between different populations of the ant *Formica sanguinea* (Pamilo, 1981) and between populations as well as between years in *Myrmica* ants (Elmes, 1987a). The workers regularly kill a large fraction of queens in the colonies of the Argentine ant, *Iridomyrmex humilis* (recently renamed *Linepithema humile* by Shattuck (1992)), thus causing fluctuations both in the number of queens and in the relatedness among the offspring (Markin, 1970; Keller *et al.*, 1989). The colonies of the ants of the *Rhytidoponera impressa* group provide an example of a qualitative difference in colony types. The colonies can be classified into two types: those headed by one morphologically differentiated queen and those headed by several gamergates (Ward, 1983a). The ants *Solenopsis invicta* (Ross and Fletcher, 1985b), *Formica exsecta* (Pamilo and Rosengren, 1984) and *F. truncorum* (Rosengren *et al.*, 1985, 1986; Sundström, 1989) also have two colony types which usually can be easily identified: monogynous colonies and highly polygynous colonies which often form multinest associations.

The presence of multiple queens in a nest does not necessarily indicate shared egg-laying, because nests can be functionally monogynous, one queen dominating the others (Buschinger, 1968). But multi-queen associations have been generally shown to be truly polygynous. Such associations often include uninseminated as well as inseminated and reproductive queens. For example, although all of 159 queens dissected of *Formica transkaucasica* were found to be inseminated, 24 of 115 *F. aquilonia* and 56 of 224 *F. polyctena* queens lacked sperm in the spermatheca (Pamilo, 1982a), as was also the case for seven of 23 queens from a single nest of *Vespula maculifrons* (Ross and Visscher, 1983). These proportions vary significantly between species. There is a strong association between ovarian status and the presence of sperm in the spermatheca: of the 21 *V. maculifrons* queens which were non-senescent, all six with developed ovaries were mated, but this was the case for only one of the 15 queens with undeveloped ovaries ($P < 0.0001$, Fisher's Exact Test).

Functional polygyny is also clearly demonstrated by the genetic heterogeneity of the offspring, as shown by the relatednesses among worker nest-mates (Table 4.7).

Relatedness between individuals in functionally polygynous colonies depends on the number of mated egg-layers, the relatedness between them, the relative size of their contributions and the effective number of times each has mated. For example, when two females share the queen role in *Polistes metricus*, the α-female is responsible for about 80% of the reproductives produced (Metcalf and Whitt, 1977a).

By analogy with the index of effective promiscuity (eqn. (4.15)), we define an index of effective polygyny in the same way:

$$Q_E = \frac{1}{\Sigma^n q_i{}^2} \tag{4.20}$$

where n is the number of queens producing eggs, and q_i is the proportion of all offspring produced by the ith queen.

The application of the concept of effective polygyny is complicated by the possibility that the queens may contribute differentially to the pool of workers, reproductive males, and females produced by the colony. The effective polygynies may differ depending on which group of offspring is being censused. From the evolutionary point of view polygyny is functional only when it concerns the production of sexuals. But we should note also that polygyny in worker production leads to colonies being composed of more than one *matriline*. The relatedness between workers in a polygynous colony is therefore a weighted average of the relatedness between sisters and nest-mates from other

Table 4.7. Estimates of relatedness in social insect populations, based on pedigree (g), regression (b), or other (o) methods applied to allozyme (A), DNA micro- or minisatellites (D), or external morphological (E) markers (see also Table 4.8). Confidence intervals (95% unless otherwise footnoted) are shown (e.g. 0.04 < 0.50 < 0.65) where these can be determined; in some other cases the ranges of several estimates are given (e.g. 0.04–0.65). Relatedness is given with the standard directionality where intergroup comparisons are made. Use of these values to predict sex allocation patterns requires knowledge of the reproductive values of the two sexes (Section 2.4, 4.2).

Species/classes	Estimate	Method	Notes and references
BEES			
Apis mellifera			Laidlaw and Page (1984)
worker nest-mates	≈0.25	g,E	
males to workers	0.50	g,E	
Augochlorella striata			Mueller *et al.* (1994)
females	0.76 < 0.79 < 0.81	b,D	
Bombus melanopygus			2 Owen and Plowright (1982)
worker nest-mates	0.75	g,E	
males to queen	0.81	g,E	
males to workers	0.60	g,E	
brood to queen	0.77	g,E	
brood to workers	0.62	g,E	
Exoneura bicolor			
cofounding females	0.48 < 0.60 < 0.72	b,A	1 Schwarz (1987)
overwintering females			
Beech	0.50 < 0.56 < 0.63	b,A	1 Blows and Schwarz (1991)
Cooks	0.47 < 0.58 < 0.69	b,A	1 Blows and Schwarz (1991)
Edgebrook	0.31 < 0.41 < 0.49	b,A	1 Blows and Schwarz (1991)
Grants	0.37 < 0.49 < 0.61	b,A	1 Schwarz (1987)
Grants	0.31 < 0.46 < 0.64	b,A	1 Blows and Schwarz (1991)
Moondara	0.39 < 0.51 < 0.62	b,A	1 Blows and Schwarz (1991)
PicnicM	0.05 < 0.27 < 0.48	b,A	1 Blows and Schwarz (1991)
Tysons	–0.09 < 0.22 < 0.53	b,A	1 Blows and Schwarz (1991)
Halictus ligatus			Richards *et al.* (1995)
workers	0.24 < 0.42 < 0.60	b,A	
new gynes	0.15 < 0.29 < 0.43	b,A	
gynes to queen	–0.27 < 0.11 < 0.49	b,A	
Lasioglossum hemichalceum			Kukuk and Sage (1994)
communal females	0.05 < 0.12 < 0.19	b,A	
Lasioglossum laevissimum			Packer and Owen (1994)
worker nest-mates	0.62 < 0.76 < 0.90	b,A	
gyne nest-mates	0.68 < 0.74 < 0.80	b,A	
worker to gyne	0.55 < 0.64 < 0.73	b,A	
Lasioglossum zephryum			
Kansas			3 Crozier *et al.* (1987)
worker nest-mates	.64 < 0.82 < 1.01	b,A	

Table 4.7 *Continued*

Species/classes	Estimate	Method	Notes and references
New York, Salmon Creek			Kukuk (1989)
first brood females	0.35 < 0.57 < 0.78	b,A	
New York, Robinson			Kukuk (1989)
worker nest-mates	0.23 < 0.43 < 0.63	b,A	
second brood females	0.54 < 0.72 < 0.90	b,A	
late summer females	0.33 < 0.53 < 0.73	b,A	
Melipona subnitida			Contel and Kerr (1976)
female nest-mates	0.75	g,A	
male to female	0.60	g,A	
Plebeia droryana			Machado *et al.* (1984)
worker nest-mates	0.75	g,A	
male to female	0.54	g,A	
WASPS			
Agelaia multiplicata			West-Eberhard (1990)
queen nest-mates	0 < 0.69 < 1	b,A	
worker nest-mates	0.13 < 0.27 < 0.40	b,A	
Cerceris antipodes			McCorquodale (1988)
nest-mate females	0.28–0.64	b,A	1, 12
Microstigmus comes			Ross and Matthews (1989a,b)
adult female nest-mates	0.57 < 0.63 < 0.70	b,A	1
Mischocyttarus basimacula			
female nest-mates	0.19 < 0.44 < 0.68	b,A	Strassmann *et al.* (1989)
	0.23 < 0.49 < 0.75	b,A	11 Queller *et al.*, (1992)
M. immarginatus			
female nest-mates	0.69 < 0.77 < 0.84	b,A	Strassmann *et al.* (1989)
	0.34 < 0.58 < 0.82	b,A	11, 13 Queller *et al.* (1992)
Parachartergus colobopterus			
female nest-mates	0.00 < 0.11 < 0.22	b,A	Queller *et al.* (1988)
worker nest-mates	0.07 < 0.31 < 0.54	b,A	Strassmann *et al.* (1991)
queen nest-mates	0.49 < 0.67 < 0.84	b,A	Strassmann *et al.* (1991)
Polistes annularis			
female nest-mates	0.15 < 0.30 < 0.46	b,A	Strassmann *et al.* (1989)
cofoundresses	0.47	b,A	Queller *et al.* (1990)
P. apachus-bellicosus			5 Lester and Selander (1981)
female nest-mates	0.43	g,A	
P. bellicosus			Strassmann *et al.* (1989)
female nest-mates	0.04 < 0.34 < 0.64	b,A	
P. canadensis			Strassmann *et al.* (1989)
female nest-mates	0.14 < 0.34 < 0.54	b,A	
P. carolinus			Strassmann *et al.* (1989)
female nest-mates	0.50 < 0.63 < 0.76	b,A	

Table 4.7 *Continued*

Species/classes	Estimate	Method	Notes and references
P. dominulus			
female nest-mates	0.42 < 0.65 < 0.86	b,A	Strassmann *et al.* (1989)
P. dorsalis			
female nest-mates	0.44 < 0.61 < 0.77	b,A	Strassmann *et al.* (1989)
P. exclamans			
female nest-mates	0.39	g,A	5 Lester and Selander (1981)
S. Texas			
female nest-mates	0.39 < 0.56 < 0.73	b,A	Strassmann *et al.* (1989)
E. Texas			
female nest-mates	0.48 < 0.69 < 0.90	b,A	Strassmann *et al.* (1989)
P. fuscatus			
female offspring-workers	0.47–0.65	g,A	Metcalf (1980)
P. gallicus			Strassmann *et al.* (1989)
female nest-mates	0.47 < 0.80 < 1.13	b,A	
P. instabilis			Strassmann *et al.* (1989)
female nest-mates	0.40 < 0.53 < 0.65	b,A	
P. instabilis			Tsuchida (1994)
female progeny	0.53 < 0.73 < 0.93	b,A	
P. metricus			
Illinois			4 Metcalf and Whitt (1977a,b)
female nest-mates	0.63	g,A	
Texas			Strassmann *et al.* (1989)
female nest-mates	0.39 < 0.57 < 0.75	b,A	
P. nimpha			Strassmann *et al.* (1989)
female nest-mates	0.17 < 0.54 < 0.91	b,A	
P. versicolor			Strassmann *et al.* (1989)
female nest-mates	0.19 < 0.37 < 0.55	b,A	
Polybia emaciata			
worker nest-mates	0.07 < 0.24 < 0.40	b,A	Strassmann *et al.* (1992)
queens	0.35 < 0.55 < 0.75	b,A	Queller *et al.* (1993b)
P. occidentalis			
female nest-mates	0.24 < 0.34 < 0.44	b,A	Queller *et al.* (1988)
workers	0.15 < 0.27 < 0.38	b,A	Queller *et al.* (1993a)
brood	0.02 < 0.27 < 0.55		
comb-mates	0.22 < 0.41 < 0.60		
non-comb-mates	0.08 < 0.33 < 0.58		
queens	0.42 < 0.57 < 0.73		
P. sericea			Queller *et al.* (1988)
female nest-mates	0.13 < 0.28 < 0.42	b,A	
Protopolybia exigua			Queller *et al.* (1993b)
workers in male-producing colonies	0.23 < 0.49 < 0.75	b,A	
queens	0.68 < 0.82 < 0.96	b,A	
Specius speciosus			Pfennig and Reeve (1993)
neighbouring females	0.01 < 0.13 < 0.25	b,D	1, 7

Table 4.7 *Continued*

Species/classes	Estimate	Method	Notes and references
Vespula maculifrons			
worker nest-mates	0.20 < 0.32 < 0.44	b,A	Ross (1986)
V. squamosa			
worker nest-mates	0.26 < 0.40 < 0.55	b,A	Ross (1986)
ANTS			
Aphaenogaster rudis			Crozier, (1977b)
worker nest-mates	0.75	g,A	
Camponotus ligniperda			
workers	–0.24 < 0.08 < 0.40	b,D	Gertsch *et al.* (1995)
Colobopsis nipponicus			
workers	0.64 < 0.75 < 0.85	b,A	Hasegawa (1994)
Conomyrma bicolor			Berkelhamer (1984)
worker nest-mates	0.33	b,A	
C. insana			Berkelhamer (1984)
worker nest-mates	0.65	b,A	
Formica aquilonia			
Ireland, Armagh			Pamilo *et al.* (1992)
worker nest-mates	0.7 < 0.12 < 0.18	b,A	
Finland, Bromarv			Pamilo *et al.* (1992)
worker nest-mates	–0.05 < 0.05 < 0.15	b,A	
Finland, Espoo			Pamilo (1982a)
queen nest-mates	–0.22 < 0.25 < 0.72	b,A	
worker nest-mates	–0.02 < 0.11 < 0.23	b,A	
Finland, Vantaa			Pamilo (1982a)
queen nest-mates	–0.28 <–0.08 < 0.13	b,A	
Scotland, Megernie			Pamilo *et al.* (1992)
worker nest-mates	–0.09 < 0.08 < 0.25	b,A	
F. argentea			Bennett (1986)
worker nest-mates	0.68 < 0.81 < 0.93	b,A	1
F. exsecta			Pamilo (1991d)
worker nest-mates	0.60	b,A	
F. fusca			Pamilo (1983)
worker nest-mates	0.29 < 0.57 < 0.85	b,A	
F. hewetti			Bennett (1986)
worker nestmates	0.45 < 0.59 < 0.74	b,A	1
F. lugubris			Pamilo *et al.* (1992)
Ireland, Tipperary			
nest-mate workers	0.28 < 0.48 < 0.68	b,A	
Finland, Pusula			
nest-mate workers	0.27 < 0.64 < 1.0	b,A	
Finland, Rovaniemi			
nest-mate workers	0.24 < 0.54 < 0.83	b,A	
Switzerland, Jura A			
nest-mate workers	0.07 < 0.25 < 0.43	b,A	
Switzerland, Jura B			
nest-mate workers	0.18 < 0.24 < 0.30	b,A	14

Table 4.7 *Continued*

Species/classes	Estimate	Method	Notes and references
F. podzolica			Bennett (1986)
nest-mate workers	–0.00 < 0.24 < 0.48	b,A	1
F. polyctena			Pamilo (1982a)
Siuntio			
nest-mate queens	0.08 < 0.47 < 0.87	b,A	8
nest-mate workers	0.21 < 0.29 < 0.37	b,A	6
F. pratensis			Pamilo *et al.* (1994)
workers	0.50 < 0.66 < 0.82	b,A	
F. pressilabris			Pamilo and Rosengren (1984)
nest-mate workers			
Espoo	0.03 < 0.29 < 0.55	b,A	
Tuusula	–0.09 < 0.07 < 0.23	b,A	
F. rufa			Pamilo *et al.* (1994)
workers	0.37 < 0.49 < 0.61	b,A	
F. sanguinea			Pamilo (1981)
Bemböle			
nest-mates			
queens	–0.13 < 0.10 < 0.33	b,A	
workers	0.07 < 0.19 < 0.32	b,A	
worker-queen	–0.02 < 0.12 < 0.27	b,A	
Skomakarskär			
nest-mates			
workers	0.44	b,A	
female brood	0.32	b,A	
female brood to workers	0.38	b,A	
F. transkaucasica			Pamilo (1982a)
nest-mate queens	–0.05 < 0.27 < 0.58	b,A	
nest-mate workers	0.16 < 0.33 < 0.50	b,A	
F. truncorum			Sundström (1989)
worker nest-mates			
monodomous			
Tvärminne	0.57	b,A	
Sibbo	0.51	b,A	
polydomous			
Inkoo	0.06	b,A	
Sibbo	0.14	b,A	
Hitis	0.03	b,A	
Harpagoxenus sublaevis			Bourke *et al.* (1988)
worker nest-mates	0.74	g,A	
Iridomyrmex humilis			Kaufmann *et al.* (1992)
nest-mate workers	–0.03 < 0.015 < 0.06	b,A	1
nest-mate queens	–0.15 < 0.03 < 0.21	b,A	1
I. pruinosum			Berkelhamer (1984)
worker nest-mates	0.79	b,A	
I. purpureus			9, Halliday (1983)
worker nest-mates (young colonies)	0.75	g,A	

Table 4.7 *Continued*

Species/classes	Estimate	Method	Notes and references
Lasius flavus			Boomsma *et al.* (1993)
female brood (MD)	0.22 < 0.40 < 0.58	b,A	
worker siblings	0.49 < 0.71 < 0.93	b,A	
L.neglectus			Boomsma *et al.* (1990)
worker nest-mates			
1987–8	0.03 < 0.14 < 0.24	b,A	1
1989	0.07 < 0.14 < 0.21	b,A	1
males to worker nest-mates	–0.03 < 0.01 < 0.05	b,A	1
L. niger			van der Have *et al.* (1988)
Kobbeduinen			
between female nest-mates	0.56	b,A	
males to workers	0.53	b,A	
Kooiduinen			
between female nest-mates	0.70	b,A	
males to workers	0.54	b,A	
Strandvlakte			
between female nest-mates	0.70	b,A	
males to workers	0.50	b,A	
Leptothorax acervorum			Douwes *et al.* (1987)
nest-mate queens	0.40	b,A	
L. acervorum			Stille *et al.* (1991)
locality A			
monogynous nests			
workers	0.55 < 0.65 < 0.74	b,A	
non-functional queens	0.19 < 0.88 < 0.99	b,A	
all polygynous nests			
functional queens	0.31 < 0.61 < 0.79	b,A	
non-functional queens	–0.01 < 0.60 < 0.88	b,A	
• polygynous nests with 2–4 functional queens			
functional queens	0.29 < 0.66 < 0.85	b,A	
non-functional queens	0.47 < 0.78 < 0.92	b,A	
workers	0.40 < 0.62 < 0.77	b,A	
• polygynous nests with ≥ 5 functional queens			
functional queens	0.18 < 0.59 < 0.83	b,A	
non-functional queens	–0.96 <–0.13 < 0.93	b,A	
workers	0.16 < 0.35 < 0.51	b,A	

Table 4.7 *Continued*

Species/classes	Estimate	Method	Notes and references
locality B			
monogynous nests			
workers	0.57 < 0.68 < 0.76	b,A	
non-functional queens	0.52 < 0.68 < 0.79	b,A	
all polygynous nests			
functional queens	0.01 < 0.33 < 0.58	b,A	
non-functional queens	–0.11 < 0.29 < 0.61	b,A	
workers	0.18 < 0.36 < 0.52	b,A	
• polygynous nests with 2–4 functional queens			
functional queens	0.01 < 0.33 < 0.58	b,A	
non-functional queens	–0.11 < 0.29 < 0.61	b,A	
workers	0.18 < 0.36 < 0.52	b,A	
• polygynous nests with ≥ 5 functional queens			
functional queens	–0.11 < 0.33 < 0.65	b,A	
non-functional queens	–0.64 < 0.26 < 0.86	b,A	
workers	0.06 < 0.36 < 0.61	b,A	
Leptothorax ambiguus			Herbers and Greco (1994)
workers	0.28 < 0.50 < 0.72	b,A	
Myrmecia pilosula			2 Craig and Crozier (1979)
worker nest-mates	0.02 < 0.17 < 0.32	b,A	
queen nest-mates	–0.02 < 0.29 < 0.61	b,A	
Myrmica gallienii			Seppä (1994b)
queens	–0.13 < 0.01 < 0.15	b,A	
workers	–0.16 < 0.02 < 0.21		
M. lobicornis			Seppä (1994a)
worker nest-mates	0.55 < 0.64 < 0.73	b,A	
M. punctiventris			Snyder and Herbers (1991)
worker nest-mates	0.58 < 0.77 < 0.96	b,A	
M. rubra,			Pearson (1982, 1983a)
site A			
1975 worker nest-mates	–0.20 < 0.11 < 0.41		
1977 worker nest-mates	–0.28 < 0.02 < 0.32		
1977 queen nest-mates	–0.62 <–0.15 < 0.32		
1978 worker nest-mates	–0.41 < 0.08 < 0.57		

Table 4.7 *Continued*

Species/classes	Estimate	Method	Notes and references
site B			
1977 worker nest-mates	0.17 < 0.54 < 0.92		
1978 queen nest-mates	0.30 < 0.67 < 1.0		
M. ruginodis			13 Seppä (1992, 1994a)
worker nest-mates			
Tvärminne	0.42 < 0.54 < 0.65	b,A	
Hanko	0.55 < 0.61 < 0.67	b,A	
M. sabuleti			Seppä (1994b)
workers	0.30 < 0.44 < 0.57	b,A	
M. scabrinodis			Seppä (1994b)
queens, Pusula	0.50 < 0.66 < 0.84	b,A	
workers, Pusula	0.03 < 0.21 < 0.39	b,A	
workers, Oravainen	0.26 < 0.44 < 0.62	b,A	
workers, Furzebrook	0.28 < 0.48 < 0.68	b,A	
Nothomyrmecia macrops			Ward and Taylor (1981)
worker nest-mates	0.25	b,A	
Rhytidoponera chalybaea			10 Ward (1983a)
type A colonies			
worker nest-mates	0.65 < 0.76 < 0.86	b,A	
males to workers	0.10 < 0.59 < 1.3	b,A	
type B colonies			
gamergate nest-mates	0.37 < 0.58 < 0.79	b,A	
worker nest-mates	0.17 < 0.34 < 0.51	b,A	
males to workers	0.31 < 0.38 < 0.44	b,A	
R. confusa			10 Ward (1983a)
type A colonies			
worker nest-mates	0.57 < 0.68 < 0.78	b,A	
males to workers	0.08 < 0.31 < 0.56	b,A	
type B colonies			
worker nest-mates	0.08 < 0.28 < 0.48	b,A	
males to workers	0.09 < 0.27 < 0.45	b,A	
Rhytidoponera sp. 12			
type B colonies			
1980 worker nest-mates	0.12 < 0.16 < 0.20	b,A	Crozier *et al.* (1984)
1981 worker nest-mates	0.10 < 0.16 < 0.23	b,A	Crozier and Pamilo (1986)
Solenopsis geminata			Ross *et al.* (1988)
female offspring	0.73 < 0.79 < 0.85	b,A	
S. invicta			Ross and Fletcher (1985b)
monogyne form			
virgin alate nest-mates	0.75	g,A	

Table 4.7 *Continued*

Species/classes	Estimate	Method	Notes and references
polygyne form			
worker nest-mates	–0.25 < 0.06 < 0.37	b,A	
queen nest-mates	–0.05 < 0.01 < 0.07	b,A	
S. richteri			Ross *et al.* (1988)
female offspring	0.65 < 0.74 < 0.82	b,A	
Tapinoma minutum			Herbers (1991)
worker nestmates	≈0.75	b,A	
TERMITES			
Reticulitermes flavipes			13 Reilly (1987)
unsexed nest-mate workers	0.35 < 0.57 < 0.79	b,A	

1. Limits are 2SE; these are approximate limits as we can assume *t*-distribution with *df* = number of colonies minus one.
2. Recalculated from the original data.
3. The occurrence of some nests with three or more genotypes indicates that relatedness is lower than 0.75.
4. Metcalf and Whitt (1977a,b) give more detailed information on the relatedness structures and inclusive fitnesses in nests with solitary foundresses or with cofoundresses. For example, the mean relatedness of female offspring is 0.46 to the α-foundress and 0.36 to the β-foundress.
5. Lester and Selander (1981) estimated maximum relatednesses based on the minimum number of matings and foundresses required to explain the observed offspring genotypes.
6. Not significantly different from zero (Pamilo, 1982a; randomization test).
7. Solitary, not eusocial.
8. Difference from zero cannot be tested by randomization test because of small number of heterozygotes (Pamilo, 1982a).
9. Older colonies become polygynous and relatedness is lower within them.
10. Type A *Rhytidoponera* colonies are queen-right and monogynous; type B colonies lack a queen but each has a number of mated workers (gamergates) able to lay both haploid and diploid eggs.
11. Relatedness of female nest-mates is given relative to the subpopulation of nests, not the entire population of nests, determined using the method of Queller and Goodnight (1989).
12. Communal, not eusocial.
13. Species with statistically significant inbreeding.
14. This ant population is morphologically placed in *Formica lugubris* but clustered with *F. aquilonia* on the basis of allozymes (Pamilo *et al.*, 1992), and hence may actually belong to the latter species.

matrilines, determined in a manner equivalent to the situation for different patrilines. To the extent that the workers can manipulate the production of sexuals, polygyny in worker production can have evolutionary consequences. Indeed, Ross (1988) found that, in laboratory colonies, coexisting queens of the fire ant *Solenopsis invicta* contribute more equally to the pool of workers than to the pool of sexual offspring. Some specialization in worker and alate production has been observed also in natural colonies of the ants *S. invicta* (Ross, 1993) and *Formica sanguinea* (Pamilo and Seppä, 1994).

If colonies persist for one generation and the queens are not mother and daughters, then the relatedness between the female offspring of the colony is, assuming single mating,

$$g_{WW} = \frac{3 + g_{QQ}(Q_E - 1)}{4Q_E} \tag{4.21}$$

and that of queen-produced males to the workers is

$$g_{MW} = \frac{1 + g_{QQ}(Q_E - 1)}{2Q_E} \tag{4.22}$$

where g_{QQ} is the relatedness between the queens.

Where a colony persists for many generations with the same number of queens (N_Q) which are recruited from its own daughter queens and which share the egg-laying evenly, it can be shown (Pamilo and Varvio-Aho, 1979) that

$$g_{WW} = \frac{3}{3N_Q + 1} \tag{4.23}$$

under the assumptions of random mating in the population and single mating per queen.

An important case is when the colony queens consist of a mother and her daughters (Pamilo and Varvio-Aho, 1979). In such cases, where the workers of different matrilines do not have the same relatedness to each other, the relative proportion of the offspring produced by the mother queen, P_Q, should be taken into account:

$$\begin{aligned} g_{WW} &= P_Q(3/8)(P_Q + 1) \\ &+ (1 - P_Q)\{P_Q(3/8) + (1 - P_Q)[3/(4Q_E) + (Q_E - 1)3/(16Q_E)]\} \\ &= (3/16)[(1 + P_Q)^2 + (3/Q_E)(1 - P_Q)^2] \end{aligned} \tag{4.24}$$

where Q_E is the effective polygyny among the daughter queens.

Comparison of eqn. (4.24) and (4.21) by setting $g_{QQ} = 3/4$ in the latter shows that relatedness within the colony declines when the foundress's daughters commence laying, and continues declining as her reproduction decreases (which is also intuitively expected). Ross (1993) shows how the number of queens and apportionment of reproduction between them affect the relatedness among the offspring.

4.4.2 *Polygyny and extensive replacement of reproductives occur in termites*

Until quite recently (e.g. Nutting, 1969; Brian, 1983:231–2), it was believed that colony founding by winged termite reproductives almost invariably involved a single pair. This picture has been reassessed, with polygyny known to occur in many species in a wide range of genera (Thorne, 1985). Sometimes this polygyny occurs at colony foundation, and is temporary, with the progressive elimination of individuals until only one pair is left; such transient polygyny seems to be the case in *Nasutitermes corniger* (Thorne, 1985). In other cases mature colonies may have large numbers of reproductives.

Several kinds of reproductive females and males occur in termites, as reviewed by Watson and Sewell (1981) and by Myles and Nutting (1988).

Alates or *imaginal reproductives* develop as winged individuals and normally disperse before mating and independent colony foundation by a pair. Colony-founding alates become primary reproductives and they lose their wings after the nuptial flight. A colony can also adopt new alates as secondary reproductives or adultoids, particularly after the death of the primary reproductives. These adultoids do not disperse and they inbreed with each other or with their parents inside the colony. By modification of the last nymphal moults, morphologically different alates, microimagoes, are produced as adultoids in some species of *Nasutitermes* (Roisin and Pasteels, 1985).

Neotenic reproductives lack wings and are either *nymphoid* (developing from nymphs) or *ergatoid* (developing from workers or pseudergates). They inbreed in the colony and become secondary reproductives, being either supplementary or replacement to the primary reproductives. Neotenics are commonly found in lower termites but are lacking in many higher termites. From their literature survey, Myles and Nutting (1988) found that the proportion of species in which neotenics are known is 17.2% in lower termites (104 out of 604 species) and 5.2% in higher termites (86 out of 1685 species).

Reproductive soldiers have been reported in some lower termites (Myles, 1988).

Dispersal and independent colony foundation are in all species associated with alate production and nuptial flights. Independent colony founding has been commonly regarded as haplometrotic (i.e. involving a single pair), and observations on several pairs cooperating in pleometrotic colony founding are few. Cases of pleometrosis have been reported for *Acanthotermes* (Roonwal and Rathore, 1975) and *Macrotermes* (Darlington, 1985). Darlington (1985) found that the number of reproductives in the pleometrotic groups of *Macrotermes michaelsoni* was drastically reduced after the colonies passed the founding stage, and it is not known whether or not pleometrosis leads to permanent polygyny in any termites.

In addition to independent colony foundation, new reproductives can be established in existing colonies by two means. A colony may recruit new reproductives (adultoids or neotenics) to supplement the old reproductives, or to replace them in orphaned colonies, or new reproductives can become functional in a distant part of a fragmented or otherwise widespread colony. The latter phenomenon is associated with colony foundation and short-distance dispersal by budding.

The replacement reproductives of lower termites are generally neotenics. They are produced in large numbers, and the number of functional neotenics is reduced by combats between them. In *Kalotermes* these fights continue until only one pair remains (Noirot, 1990), but in many other lower termites multiple neotenics coexist. In higher termites neotenics are rarer, and are completely absent in some groups. In the species where they do exist, they tend to be nymphoid and to form polygynous groups.

Polygyny for imaginal reproductives has been observed in many species of higher termites. Thorne (1985) lists 38 species and more species have been added to the list (Roisin and Pasteels, 1986). Most of these cases are anecdotal, based on single observations, and detailed studies are few. The observed proportion of polygynous nests is 100% in *Nasutitermes costalis* (14 nests, number of queens ranging from 5 to 97), 93% in *N. polygynous* (14 nests, number of queens up to 105) (Roisin and Pasteels, 1986), 65% in *N. princeps* (Roisin and Pasteels, 1985), at least 33% in *N. corniger* (76 nests with queens, number of queens up to 33) (Thorne, 1984) and 23% in *Macrotermes michaelsoni* (361 nests, up to seven queens) (Darlington, 1988). The coexisting queens are about the same age and size in *N. corniger*, *N. princeps* and *M. michaelsoni* but different age classes were found in *N. polygynous* and *N. costalis*. Whether the similarity of physogastry implies an equal egg-laying rate appears doubtful given the results of Roisin and Pasteels (1985), but more studies are needed to clarify the contributions of the queens.

Thorne (1984) reported that the polygynous nests of *Nasutitermes corniger* often had only one king, whereas Roisin and Pasteels (1985, 1986) found that the ratio of queens and kings is close to equality in the other polygynous *Nasutitermes* species.

The fecundity of neotenics is less than that of imaginal reproductives, and a number of neotenics may be required to maintain a colony which had required only one imaginal pair. *Macrotermes michaelsoni* has extremely large polygynous colonies, with over five million workers, and Thorne (1985) speculated that polygyny is necessary to maintain such a large colony size. Polygyny in *Nasutitermes* seems to be closely associated with polycaly (possession of multiple nests) and budding (Roisin and Pasteels, 1986; Thorne, 1985) and Thorne (1984) suggested that polygyny is advantageous in allowing rapid growth of young colonies and early production of sexuals. We will return to models for the evolution of polygyny in Section 4.4.3.

Polygyny for imaginal queens seems to be rare in lower termites, but Roonwal and Rathore (1975) report a case involving pleometrosis.

The consequence of the ability to replace reproductives is that relatedness levels in termite colonies may vary widely depending on the history of the colony. Because of the inbreeding induced by the mating of neotenics within the colony, relatedness between colony members may be higher than between an outbreeding alate and its offspring. Bartz (1979) suggested that termites may have in fact arisen by selection under conditions of inbreeding (but see also Pamilo, 1984b; Myles and Nutting, 1988).

4.4.3 *The evolution of polygyny*

The evolution of polygyny has attracted considerable attention (e.g. Baroni Urbani, 1968; Buschinger, 1968, 1970; Hölldobler and Wilson, 1977; West-Eberhard, 1978; Frumhoff and Ward, 1992; Keller, 1993; Bourke and Heinze, 1994). Phylogenetic analyses show that the level of polygyny has changed frequently in ants but has been very conservative in both bees and wasps (Ross and Carpenter, 1991; Frumhoff and Ward, 1992). The problem of shared reproduction has been generally recognized, but we still lack good alternative, testable hypotheses. Polygyny appears problematic in terms of adaptation, because it seems that the reproductive output per single queen is often a diminishing function of the number of coexisting queens (Michener, 1974:245).

Nonacs (1988) and Pamilo (1991b) have classified the hypotheses into broad categories, and the following list is modified from the latter treatment.

1. *Mutualism*. Cooperation of queens increases their personal reproductive output.
    - 1(a). The fecundity of queens is higher in polygynous than in monogynous colonies
    - 1(b). The survival of polygynous colonies is higher than that of monogynous colonies, leading to an overall higher fitness of queens participating in polygynous associations even if their fecundity is lower than that of successful monogynous queens.

2. *Kin selection*. Cooperation does not necessarily increase the personal fitness of queens (the number of their own offspring) but increases the inclusive fitnesses of either queens or workers.
    - 2(a). Asymmetric relationships. An existing colony is invaded by new queens which are related to the old one. We assume that this is the best choice for these new queens.
        - (i) The old queen accepts the joining queens, provided that her inclusive fitness increases.
        - (ii) The workers accept the joining queens, provided that their inclusive fitness increases, although that of the old queen may decrease.
    - 2(b). Symmetric relationships. A colony has many foundresses, or an existing colony is taken over by a group of joiner queens. Inclusive fitness may be increased by allowing this association to continue even though the personal fitnesses may be lower than in monogynous colonies.

(i) Persistence of the association results from selection at the queen level, provided that this enhances their average inclusive fitness.

(ii) Persistence of the association results from selection at the level of the new daughter workers, provided that this enhances their average inclusive fitness.

3. *Parasitism.* New queens gain by joining an existing nest, whereas the old colony members suffer a loss in fitness.

The first hypothesis covers the hypotheses of mutualism and colony-level selection of Nonacs (1988), because both of them seem to include increased success but at different stages of colony cycle. Hypotheses 1 and 2 differ from each other in that under kin selection the personal fitnesses of the old colony members decrease, but polygyny is favoured by inclusive fitness effects.

The above hypotheses were classified according to the question: To whom is polygyny advantageous? Alternatively, we could try to classify hypotheses on the basis of ecological conditions favouring polygyny (Herbers, 1993). We will next combine these two approaches and briefly examine the situations in which polygyny could be selected for.

Mutualism

Nest founding by a cooperative group of females is well known in ants (Hölldobler and Wilson, 1977), wasps (West-Eberhard, 1978) and termites (Thorne, 1984, 1985). Such pleometrosis leads to primary polygyny. The cofoundresses in polistine wasps are thought usually to be close kin (West, 1967; Noonan, 1981; see also Table 4.8 below), but in ants they are often unrelated (Bartz and Hölldobler, 1982; Hagen *et al.*, 1988; Nonacs, 1988; Rissing *et al.*, 1989).

Pleometrosis by unrelated females has been considered adaptive chiefly because of a high initial growth rate and improved survival compared to single-foundress colonies. The queens in pleometrotic associations, however, face the threat of later being killed. Hölldobler and Wilson (1977), when reviewing polygyny in ants, found that no cases of pleometrosis are known to lead to permanent polygyny in ants. Possible primary polygyny has since been reported in the ants *Atta texana* (Mintzer, 1987), *Pheidole morrisi* (S. Cover in Hölldobler and Wilson, 1990:218), and *Acromyrmex versicolor* (Rissing *et al.*, 1989). In termites the queen associations seem to last longer (Thorne, 1984, 1985). An interesting point is that the queens are generally mutilated by workers, seldom by their fellow queens.

If a pleometrotic association has initially n queens, the probability of surviving to be the final queen is $1/n$ for any one. Pleometrosis should be favoured if $1/n$ is larger than the relative success of a haplometrotic colony. Thorne (1984) has considered this situation in an ecological model based on termite life-patterns. The model examines the relative success of queens in colonies with different numbers of queens, and its parameters are the probabilities of surviving from one age to the next, and the fecundity at each age in terms of the numbers of reproductives produced. Under Thorne's model, colonies with larger numbers of queens do not have larger rates of alate production, but this is a conservative restriction. The finding is that an increase in either the survival rate or fecundity may select for polygyny over monogyny. Roisin (1987) pointed out that Thorne's model reduces to a requirement that a pleometrotic colony with n queens be at least n times as productive as a haplometrotic one if pleometrosis is to be favoured.

If the queens are related, the conditions determined by Thorne are relaxed to some degree. The worker force of a polygynous colony consists of different matrilines and we can expect intra-colony competition between them. If one of the matrilines becomes very weak, the other workers could kill its queen without much resistance. It is not necessary for the workers actually to measure the relative strengths of the matrilines within the colony and to identify the corresponding queen, because they reach the same goal simply by killing the least productive queen of the colony (Forsyth, 1980). In monogyne *Solenopsis invicta* the workers, in fact, do tend to kill the queens on the basis of their ovarian development as reflected by the degree of physogastry (Fletcher and Blum, 1983).

What is actually found for foundress associations? The ease with which unrelated queens could be experimentally induced to associate in colony founding made it seem likely that natural associations are not made up of relatives (Strassmann, 1989). In addition, as noted above, behavioural observations appeared to show queens associating randomly. Using allozyme markers, Rissing *et al.* (1989) showed that queens of *Acromyrmex versicolor* not only associate apparently at random with respect to relatedness but that a division of labour develops so that one queen is almost certain to be lost to predation because she takes up a foraging role. In *A. versicolor*, as in most such cases (Hölldobler and Wilson, 1990:217–20), all queens but one are eliminated once the first workers emerge, generally by the workers themselves selecting the queen with the highest fertility (and hence most likely to be the mother of most of them).

There is a counter-example to this generally kinless picture: Nonacs (1990) found that kin pairs and solitary queens of *Lasius pallitarsis* produced workers whereas non-kin pairs of queens did not. Such cases are liable to occur only where nest-mates have a fairly high chance of finding each other among the large numbers of queens settling from a mating flight, and Nonacs (1990) notes that 'Many alates often fly less than 20 m from their natal nest before shedding their wings . . .'.

Reeve and Nonacs (1992) suggest that the foundress associations must have some type of social contract which guarantees mutual benefits to both the dominant and subordinate females. They found in experimentally manipulated nests of *Polistes fuscatus* that the subordinate foundresses increased aggression when their reproductive-destined eggs were removed but not when their worker-destined eggs were removed from the nest cells. Reeve and Nonacs interpret such behaviour as a retaliation against the dominant foundress thought to be cheating and breaking the social contract (see also Strassmann, 1993).

As expected, founding colonies which are polygynous do indeed have higher survival rates than monogynous colonies of the same species in ants (Walloff, 1957; Wilson, 1974b; Bartz and Hölldobler, 1982; Rissing and Pollock, 1987), and wasps (Gamboa, 1978; Gibo, 1978). In various ant species studied, only the largest colonies will survive, because once the workers emerge there is extensive raiding between colonies until only one per locality has accumulated all the brood and all the workers (as in *Solenopsis invicta* (Tschinkel, 1992)). The queens of the unsuccessful colonies are excluded. In other cases, territorial battles will achieve much the same result. Strassmann (1989) and Hölldobler and Wilson (1990:217–20) review the biology of this situation. Under such circumstances, a solitary queen is doomed in competition with queens whose egg-laying has been augmented by those of others, and Nonacs (1989) has argued that it is just not cost-effective for queens to seek out nest-mates with which to cooperate in nest founding.

Kin selection

We have a special case of mutualism when the queens in a pleometrotic association are related to each other. This is, of course, just kin selection. The threshold for cooperation is likely to decrease with increasing relatedness, because there are positive inclusive fitness effects.

Relatedness patterns under polygyny are often asymmetric in the sense that new queens try to join already existing nests (secondary polygyny). The asymmetry can exist within a generation (foundress and joiners) or it can include successive generations (e.g. mother and daughters). The question is: under which conditions should the old colony accept these additional queens. The likelihood of a new queen successfully founding her own colony is normally very small. Therefore we can assume that the new queens benefit from joining existing colonies, whereas the old queen is likely to suffer a reduction in the number of offspring she can produce. This conflict could lead the old queen to resist new queens and to prevent even her own daughters staying in the nest. She could do this either by her own behaviour or by manipulating the workers to oppose new queens. As noted by Rosengren and Pamilo (1983), conflict is possible when the options of the old queen and workers disagree.

Assuming both monogyny and monandry, the relatednesses of the offspring to the queen are $g_{FQ} = 0.5$ for daughters and $g_{MQ} = 1$ for sons. If one of the queen's daughters starts laying eggs, these offspring will be related to the old queen by 0.25 and 0.5, respectively, i.e. the relatednesses are reduced by one half. The relatednesses of the old queen's offspring to the workers are $g_{FW} = 0.75$ and $g_{MW} = 0.5$, whereas the relatednesses of the offspring produced by a sister would be 0.375 and 0.75, respectively. Both the old queen and her worker offspring suffer a 50% loss in the relatedness of the colony's new daughters produced by the daughter queen. The old queen also loses 50% in the relatedness of sons, whereas the workers gain 50%. It is, therefore, easier for the workers than for the old queen to accept such a daughter queen as a new reproductive in the colony (Rosengren and Pamilo, 1983). This situation is very like that of worker reproduction (see Section 4.2): it becomes more difficult for workers to accept the new queens if their mother queen has mated more than once.

Based on the above suggestions, Nonacs (1988) modelled the evolution of polygyny. He examined the conditions under which the old queen or workers accept new queens, provided that the population has a 1:1 sex allocation ratio and that the new queens are always daughters of the old monogynous queen. Pamilo (1991b) extended this model by relaxing these assumptions.

Following the model presented in Chapter 2 (eqn. 2.7)), the inclusive fitness of the queen in a monogynous colony where she produces all the offspring is proportional to

$$V_Q = \frac{f}{F} + \frac{m}{M} \tag{4.25a}$$

where F and M refer to the population sex investment ratios and f and m to the sex investment ratios among the queen's own brood. If the colony recruits n new queens which are daughters of the queen, and which will share the egg-laying with the old queen ($1/(n + 1)$ per each queen), the fitness of the queen will become

$$V_Q = K\left(\frac{1}{n+1}\right)\left[\frac{f}{F} + \frac{m}{M} + n\left(\frac{0.5f}{F} + \frac{0.5m}{M}\right)\right]$$

$$= K\left(\frac{0.5n+1}{n+1}\right)\left(\frac{f}{F} + \frac{m}{M}\right) \tag{4.25b}$$

where K is the relative productivity of the polygynous colony, compared to the monogynous colony. The old queen favors polygyny when the fitness V_Q given by eqn. (4.25b) exceeds that given by eqn. (4.25a), i.e. when

$$K > \frac{2n+2}{n+2}. \tag{4.26}$$

The inclusive fitness of the workers in the initial monogynous and monandrous colony is

$$V_W = \frac{1.5f}{F} + \frac{0.5m}{M} \tag{4.27a}$$

and after the colony has adopted n of their sisters as new queens sharing the reproduction with the old queen

$$V_W = K\left(\frac{1}{n+1}\right)\left[\frac{1.5f}{F} + \frac{0.5m}{M} + n\left(\frac{0.75f}{F} + \frac{0.75m}{M}\right)\right]. \tag{4.27b}$$

The conditions under which the workers favour polygyny are obtained by setting eqn. (4.27b) greater than eqn. (4.27a). These conditions now depend on the sex ratios. If the population sex investment ratio is controlled by the queens, i.e. $F = M = 0.5$, we have

$$K > \frac{1 + 2f + (1 + 2f)n}{1 + 2f + 1.5n} \tag{4.28a}$$

and when the population sex ratio is controlled by workers, i.e. $F = 3M = 0.75$, we obtain

$$K > \frac{2 + 2n}{2 + (3 - 2f)n} \tag{4.28b}$$

as a condition for polygyny to evolve.

Comparing the conditions given by eqns. (4.26) and (4.28), we see that the workers accept the daughter queens more readily than does the old queen. Depending on the sex ratios of both the population and of the colony in question, workers can favour new queens even if the colony productivity decreases. This result follows from the constraint that the colony keeps its old queen — it might be better for the workers to replace the old queen with a new one. For that reason we should also check the conditions under which the workers do not prefer the old queen to be superseded. If the relative productivity of a colony with the old queen and n of her daughters, compared to a colony which killed the old queen, fills the inequality

$$K > (n + 1)\frac{(0.75f/F + 0.75m/M)}{[1.5f/F + 0.5m/M + n\,(0.75f/F + 0.75m/M)]}$$

then the workers should keep both the old queen and her daughters. Specifying again the population sex ratios, this condition becomes, when $F = M = 0.5$,

$$K > \frac{1.5 + 1.5n}{1 + 2f + 1.5n} \tag{4.29a}$$

and when $F = 3M = 0.75$,

$$K > \frac{3 - 2f + n\,(3 - 2f)}{2 + n\,(3 - 2f)}\,. \tag{4.29b}$$

If the additional queens do not affect colony productivity, i.e. $K = 1$, conditions (4.29) and (4.28) cannot be satisfied simultaneously. The thresholds for $n = 1$ are presented in Fig. 4.6, and it is seen that for both population sex ratios studied (1:1 and 3:1) there is a point where the threshold for workers keeping both the old and new queens is $K = 1$. The K values between the thresholds for the queen and for the workers represent the area of queen–worker conflict over polygyny.

A few remarks can be made about the above results. The willingness of workers to accept sisters as new queens increases when the colony produces a male-biased sex ratio (Fig. 4.6). Colony-level variation of sex ratios is discussed in Chapter 6, where we show that there tend to be systematic biases in the sex ratios produced by individual colonies. Workers in nests with relatively male-biased colony sex ratios are more likely to favour adopting new queens. The likelihood that such workers will be selected to accept new queens is increased if the new queens produce offspring whose relatedness to the workers is higher than that used when deriving the above conditions, or if the old mother later produces offspring less related to the workers than assumed above (Rosengren and Pamilo, 1983). The first condition is met if new queens produce a larger proportion of males and the old queen a larger proportion of daughters (a situation analogous to worker reproduction), or when there is intranidal mating and the new queens are inbred. The second condition is met if there is polyandry accompanied by sperm clumping, so that the sperm of different males is used at different times, i.e. if the new queens are full sisters of the contemporary workers but later daughters of the queen consist largely of their half-sisters. The honey-bee data indicate that sperm clumping is negligible (Crozier and Brückner, 1981; Page, 1986). The effect of sperm mixing is to

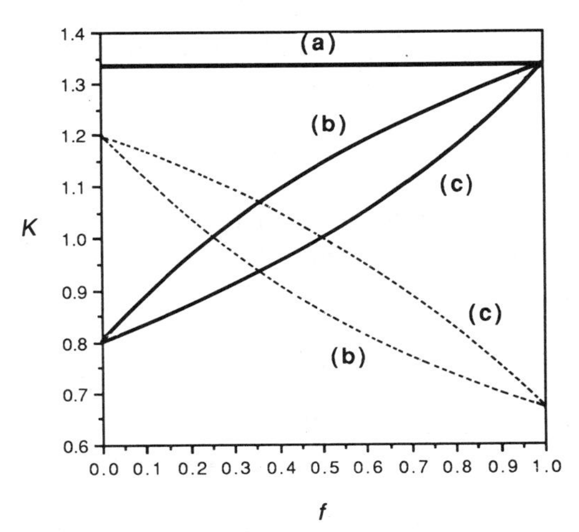

Fig. 4.6 Curves showing the threshold values of K above which selection favours a colony keeping two queens and below which retention of only one queen is favoured. K is the relative productivity of a polygynous colony relative to that of a monogynous colony. The curves are shown as functions of the colony sex allocation, f. The potential new queen is a daughter of the old queen, who is assumed to have mated only once. The solid curves give the thresholds for accepting a new such queen, and the broken lines give the thresholds for retention of the old queen after a new queen has been accepted. The curves further vary depending on whether (a) the old queen makes the decision, (b) the workers make the decision and the population sex allocation is $F = M = 0.5$, or (c) the workers make the decision and the population sex ratio is $F = 3M = 0.75$. The decision of the old queen on whether or not to accept the new queen does not depend on the population sex ratio.

bring the threshold value K of the workers closer to that of the queen (eqn. (4.26)). Male production by workers has a similar effect.

The above model assumes that the daughters would produce no offspring if they failed to join the maternal colony. This, of course, cannot apply to all daughters, but the model does apply to those which try to stay in the natal nest, provided that a majority do disperse.

After colonies have become polygynous, the relatednesses will change and the expected sex ratio under worker control is no longer 3:1. Consequently, the conditions given above for K will be altered. If the colony has n coexisting monandrous queens with relatedness g_{QQ} among them, the colony should, on average, favour an addition of one new daughter queen, produced by any of the n queens, if

$$K > \frac{2n+2}{2n+1} \qquad (4.30a)$$

under queen control, and

$$K > \frac{(1+2f)(n+1)+(n+1)(n-1)g_{QQ}}{(1+2f)n+(3/2)+(n+0.5)(n-1)g_{QQ}} \qquad (4.30b)$$

under worker control, where f denotes the colony investment in females. Male bias of the colony sex ratio would still have some effect on the acceptance threshold of workers. The thresholds of eqn. (4.30) are shown in Fig. 4.7 for several values of g_{QQ}. The threshold is always lower for workers than for queens, as already pointed out by Nonacs (1988) in a slightly different analysis. Polyandry is expected to alter the situation. It is also apparent that the threshold for adding new queens becomes lower when the queens are related and when there are many of them. On the other hand, if n is already large, adding one new queen is not likely to affect the colony's productivity very much.

Whereas the workers tend to accept daughters of the original mother queen, they tend to reject additional queens if they are sisters of the old queen (Nonacs, 1988). In that case the threshold for the old queen to accept these new queens is the same as given in eqn. (4.26), whereas that for her workers is

$$K > \frac{16\,(n+1)}{5n+17} > \frac{2n+2}{2n+1}\,. \qquad (4.31)$$

There is a tendency among the workers to eliminate these other queens.

There are practically no data to evaluate whether polygyny is favoured by queens or workers, or by both groups. There are some observations on ants indicating that the workers try to prevent daughter queens from flying off, while they do not prevent the males (Rosengren and Pamilo, 1983). This points to a potential role of the workers. If polygyny is initially controlled by workers, we would expect that it is associated with colonies accepting daughter queens, queens being monandrous, the population sex ratio being female biased, and the colony sex ratio being more male biased than the population average.

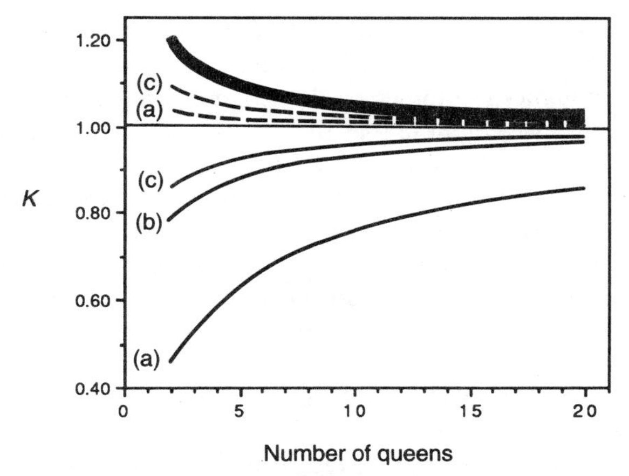

Fig. 4.7 Thresholds of K above which selection favours an already polygynous colony accepting one further new queen which is a daughter of one of the existing queens. K is the ratio of the productivity of a colony with $n + 1$ queens to that with n queens. n is the number of existing queens, whose relatedness to each other is (a) 0.75, (b) 0.25, or (c) 0. The thin solid curves represent the cases where the decision is under worker control and the colony sex allocation is $f = 0$; broken lines represent worker decision case where $f = 0.5$, and the thick solid line represents the case where the queens make the decision. The population sex allocation is assumed to be $F = M = 0.5$.

The importance of workers in accepting or rejecting new queens is clearly seen in genetic studies of the fire ant *Solenopsis invicta.* Ross (1992) found that a specific homozygous genotype of the allozyme locus phosphoglucomutase is absent among the queens of polygynous colonies although the genotype is expected to occur under Hardy–Weinberg conditions given the prevailing allele frequencies and because it does occur in alates at these expected frequencies. A reasonable explanation is that the workers reject these homozygous queens on the basis of a physiological change associated with the enzyme phenotype. A further dimension is provided by the fact that queens with this genotype have no disadvantage in monogynous colonies, indicating that perhaps this marker has the effect of a reproductive isolating mechanism (Keller and Ross, 1993a).

Polygyny has often been considered in terms of specific ecological situations. From the point of view of the new queens, the problem can be presented in a form of different dispersal strategies (Rosengren and Pamilo, 1983). The queens of polygynous ants seem often to be poor dispersers. From the point of view of the old colony, the life histories of individual queens and colonies become important. Joining queens can start producing sexual brood earlier than independent foundresses, and the probability of the old ones surviving to the next season can be small. These factors are expected to favour recruiting new, related queens because that can increase the inclusive fitnesses of both the old queen and the workers (Pamilo and Rosengren, 1984). Nonacs (1988) concluded, on the basis of a quantitative model, that polygyny is favoured through its effects on inclusive fitnesses when the queens are short lived but colonies long lived. In monogynous colonies the life span of the colony tends to equal that of the queen, but we can note that the age of a colony probably correlates with the investment made in the nest structures. It would therefore be adaptive for the colony to adopt related queens, if the old queen is expected to die soon and if building new nests is very expensive, significantly reducing the success of the queens in founding colonies independently. The life history model of Thorne (1984) shows that polygyny in termites can be favoured even if the queens are not related. The same principle is strengthened if the queens are close kin.

In their review of polygyny in ants, Hölldobler and Wilson (1977) concluded that there are two broadly defined groups which commonly show polygyny. One group consists of species specialized in patchy habitats. In such species, the dispersal risks are high and the established territory of a colony can be considered as a very valuable investment. Resources may be patchily distributed not only because of pre-existing heterogeneity in nature, but also because a colony itself may create artificial patches by generating food sources (e.g. tending aphids), by removing competitors, by developing efficient foraging patterns (building trails and a 'colony tradition' through recruitment), and sometimes by cooperative exchange between different nests (Rosengren and Pamilo, 1983). The other group defined by Hölldobler and Wilson (1977) includes species with easily fragmenting colonies. Polygyny could guarantee that each fragment is likely to have at least one queen.

Studies of genetic relatedness support the kin selection hypothesis of polygyny in that the coexisting (as against cofoundress) queens are generally kin, although not necessarily very close kin (Table 4.8). MtDNA variation shows that colonies of the ant *Leptothorax acervorum* adopt alien queens (Stille and Stille, 1992), although the average relatedness among coexisting queens is significantly positive (Stille *et al.*, 1991) and the microgeographic distribution clearly shows clustering of mitochondrial haplotypes (Stille and Stille, 1993). The high level of variation in the mtDNA molecule (13 haplotypes in a small area, using four four-base restriction enzymes) makes it a very useful genetic marker for tracing matrilines in this species.

The most promising genetic approach to evaluating the adaptive significance of polygyny, however, is the direct estimation of the inclusive fitnesses concerned. Previous efforts at estimating inclusive fitness have used the relatednesses and reproductive successes of individuals in natural populations. It is clearly desirable to manipulate the various individuals concerned to determine the consequences of the various reproductive options. Initial efforts in this area include the manipulation of founding queen ant groups (e.g. Rissing *et al.* 1989) and of food supplies to overwintering *Polistes* females (Queller and Strassmann, 1988).

Table 4.8. Relatednesses (g_{YX}) between coexisting queens or other mated egglayers in eusocial Hymenoptera. The estimates are given with 95% confidence intervals if available. The data are either allozyme frequencies (A) or field observations (N), analysed through regression (b) or pedigree (g) methods, or simply used to infer relatedness (i).

Taxa	Estimate	Method	Notes
WASPS			
Parachartergus colobopterus	0.49 < 0.67 < 0.84	b,A	Strassmann et al. (1991)
Polistes annularis	0.47	b,A	Queller *et al.* (1990)
Polybia emaciata	0.35 < 0.55 < 0.75	b,A	Queller *et al.* (1993b)
P. occidentalis	0.42 < 0.57 < 0.73	b,A	Queller *et al.* (1993a)
Protopolybia exigua	0.68 < 0.82 < 0.96	b,A	Queller *et al.* (1993b)
ANTS			
Acromyrmex versicolor	–0.18 <–0.12 <–0.07	b,A	Hagen *et al.* (1988)
Formica aquilonia			Pamilo (1982a)
Espoo population	–0.22 < 0.25 < 0.72	b,A	
Vantaa population	–0.28 <–0.08 < 0.13	b,A	
F. exsecta	–0.31 < 0.09 < 0.49	b,A	Pamilo and Rosengren (1984)
F. polyctena			Pamilo (1982a)
Siuntio population	0.08 < 0.47 < 0.87	b,A	
F. pressilabris	0.50 < 0.83 < 1.15	b,A	Pamilo and Rosengren (1984)
F. sanguinea	–0.13 < 0.10 < 0.33	b,A	Pamilo (1981)
F. transkaucasica	–0.05 < 0.27 < 0.58	b,A	Pamilo (1982a)
Iridomyrmex humilis	–0.15 < 0.03 < 0.21	b,A	Kaufmann *et al.* (1992)
I. purpureus			Hölldobler and Carlin (1985)
cofoundresses	≈0	i,N	
adoptive queens to residents	> 0	i,N	
Lasius pallitarsis			Nonacs (1989)
cofoundresses	≈0	i,N	
Leptothorax acervorum	0.40	b,A	Douwes *et al.* (1987)
functional queens			Stille *et al.*(1991)
locality A	0.31 < 0.61 < 0.79	b,A	
locality B	0.01 < 0.33 < 0.58	b,A	
Myrmecia pilosula	–0.02 < 0.29 < 0.61		1 Craig and Crozier (1979)
Myrmecocystus mimicus	≈0	i,N	Bartz and Hölldobler (1982)
Myrmica gallienii	–0.13 < 0.01 < 0.15	b,A	Seppä (1994b)

Table 4.8 *Continued*

Taxa	Estimate	Method	Notes
M. rubra,			
site A, 1977	−0.62 <−0.15 < 0.32	b,A	Pearson (1982, 1983a)
site B, 1978	0.30 < 0.67 < 1.0	b,A	Pearson (1982, 1983a)
Tvärminne	0.47 < 0.67 < 0.87	b,A	Seppä (1994b)
Pusula	−0.29 < 0.10 < 0.49	b,A	Seppä (1994b)
M. ruginodis	0.26 < 0.41 < 0.55	b,A	Seppä (1994a)
M. scabrinodis	0.50 < 0.66 < 0.84	b,A	Seppä (1994b)
Rhytidoponera chalybaea	0.37 < 0.58 < 0.79	b,A	2 Ward (1983a)
Solenopsis invicta	−0.05 < 0.01 < 0.07	b,A	Ross and Fletcher (1985b)
Veromessor pergandei			
cofoundresses	0.033	b,A	Hagen *et al.* (1988)

1. Recalculated from the original data.
2. Gamergates.

Parasitism

We argued above that the new queens are likely to gain by joining existing nests, and we then worked out the thresholds of when the colony should accept extra queens. There is an area of conflict when the real value of K is smaller than the threshold where the colony would accept the joiners. The new queens can in such a case be considered as parasites of the colony (Elmes, 1973). Although the threshold for accepting additional queens becomes smaller with an increasing level of polygyny, the number of queens in some ant colonies is so high that it is hard to believe that the extra queens actually increase the colony's productivity. For example, nests of *Formica uralensis* may contain 500 mated queens (Rosengren and Pamilo, 1983).

If new queens do indeed decrease the inclusive fitness of the other individuals of the colony, why would the colony let them come in? This is a difficult question, but we should remember that interspecific social parasitism is common in ants, bees and wasps (Wilson, 1971:349–88). (Curiously, socially parasitic termites appear to be unknown.) An interesting point is that the parasites are often phylogenetically close relatives of the host species; this is known as Emery's Rule (Emery, 1909). It has been suggested that social parasites could arise by sympatric speciation (see Buschinger, 1986, 1990; Bourke and Franks, 1991), although allopatric modes cannot be excluded.

Whatever the origin of parasitic species, their existence shows that foreign queens manage to insinuate themselves into established colonies. It is certainly possible that this behaviour originated as intraspecific parasitism associated with polygyny (or alternatively, as nest usurpation, because the parasite queens often kill the host queens). Reflecting this last interpretation, Bolton (1986) gives to secondary polygyny (the addition of new queens to existing colonies) the term autoparasitism.

The parasitism hypothesis of polygyny does not rule out relatedness. In fact, if the new queens are related to the other individuals of the colony, it may be easier for them to intrude. 'Parasitism' may also arise as an error, because the workers may not be very good at counting. If the number of queens is less than optimal, we do not know whether the workers apply the principle 'Accept six more queens', or 'Accept new queens'.

4.5 The organization of colonies and populations

4.5.1 *The occurrences of population viscosity and inbreeding vary widely in social insects*

In Section 2.8 we showed how population structure can affect sex allocation. Inbreeding and dispersal of the two sexes are important determinants of the optimal sex ratio (Hamilton, 1967; Cannings and Cruz-Orive, 1975; Bulmer, 1986). These features are relevant not only to sex ratios but also to the operation of kin selection in general, because they affect the genetic differentiation between groups or subpopulations where social interactions take place (e.g. Wade, 1985).

It is also relevant to separate dispersal taking place after mating from that taking place before mating. Mating before dispersal is much more likely to lead to breeding between relatives than mating after dispersal.

Viscosity, a somewhat nebulous term first introduced by Hamilton (1964b), may be operationally defined as the tendency of related individuals to occur together. It can also be considered as a strong isolation by distance effect (Wright, 1943). We thus have two different patterns of genetic differentiation (Pamilo, 1983).

1. Subdivision refers to population structure, where the population is physically subdivided into separate subpopulations. Limited gene flow between such subpopulations leads to genetic differentiation between them.

2. Viscosity, or isolation by distance, refers to continuous populations in which limited dispersal by individuals leads to gradual genetic differentiation (gene frequency gradients) within the population but without physical separation of identifiable subunits.

We should also note that limited dispersal of the two sexes need not always lead to close inbreeding: in highly polygynous species intranidal mating (mating between individuals born in the same nest), can involve distant relatives.

When trying to evaluate the occurrence of inbreeding and viscosity, it is necessary to take into account the geographical scale being examined. For example, in the ant *Formica transkaucasica*, the inbreeding coefficient estimated as a departure from binomial genotype frequencies within a 25×25 m^2 study plot was zero (55 nests), whereas the estimate within the entire 0.2 km^2 study population was 0.12 (Pamilo, 1982a, 1983). This comparison suggests that the total study area exceeds the range of dispersal of the reproductives. If we take a small enough area, the population appears panmictic (random mating), but the population as a whole is not genetically homogeneous. It is possible to describe viscosity by measuring genetic relatedness between neighbouring colonies or autocorrelation of allele frequencies as a function of distance. Such studies have shown viscosity in the ants *Formica sanguinea* (Pamilo, 1981) and *Rhytidoponera* sp. 12 (Crozier *et al.*, 1984; Crozier and Pamilo, 1986), although not in all populations of the same species or in all years in the same population.

The impression from a voluminous literature (see Wilson, 1971; Crozier, 1980; Hölldobler and Bartz, 1985) is that inbreeding (in the sense of a departure from random mating within a population) is extremely rare among social Hymenoptera, as also seen from direct estimates (Table 4.9). Among wasps and bees, the opportunities for random mating within the population are obvious due to the powerful flying abilities of both sexes. The great distances flown by both sexes to the mating area in *Apis mellifera* (Seeley, 1985:67–8) are a case in point. Among ants, Kannowski (1963) noted that female ants of the subfamily Formicinae tend either to join *aerial swarms* of males, or to attract males to them in *ground swarms* (presumably by pheromones). Brian (1983:320) and Hölldobler and Bartz (1985) bravely and largely successfully attempted to bring order to diversity by generalizing Kannowski's (1963) observations to all ants, and Hölldobler and Bartz (1985) renamed Kannowski's categories. In the *female calling syndrome* (FCS), females on the ground attract males pheromonally. In the *male aggregation syndrome* (MAS), males aggregate, sometimes marking their aggregation sites with attractant pheromones, and females fly to these sites. Hölldobler and Bartz (1985) further suggested a link between single mating (by females) with FCS and multiple mating with MAS, but there are certainly many single-mating MAS species, such as *Pheidole sitarches*, *Solenopsis invicta* and *Iridomyrmex purpureus* (see Page, 1986, and Table 4.1).

It is clear that inbreeding is likely not to occur under the MAS, but one could imagine it under the FCS because the males might come from close by, perhaps even from the same nest. Such direct genetic evidence as there is (Table 4.8) shows that detectable inbreeding is rare even in FCS species. Critical interpretations of the existing data are difficult to make because the observed inbreeding coefficients depend on the geographical scale of the study, as explained above.

Table 4.9. Social insect species for which there are genetic data about the level of inbreeding. Unless indicated by *, all species listed have inbreeding levels not significantly different from zero. If possible the presence of the female calling (F) or male aggregation (M) syndromes of Hölldobler and Bartz (1985) are indicated. In some species, significant microgeographic variation was found (S). The data are from allozyme (A) or external morphological (E) markers.

Species	Notes	References
BEES		
Bombus melanopygus	E	Owen and Plowright (1982)
Lasioglossum zephyrum	A, S	Crozier et al. (1987)
Melipona subnitida	A	Contel and Kerr (1976)
WASPS		
Mischocyttarus basimacula	A	Queller et al. (1992)
*M. immarginatus**	A	Queller *et al.* (1992)
Parachartergus colobopterus	A	Queller *et al.* (1988), Strassmann *et al.* (1991)
*Polistes bellicosus**	A	Davis *et al.* (1990)
P. carolinus	A	Davis *et al.* (1990)
P. exclamans	A, S	Davis *et al.* (1990)
P. metricus	A	Metcalf and Whitt (1977a)
*P. metricus**	A	Davis *et al.* (1990)
P. fuscatus	A	Metcalf (1980)
Polybia occidentalis	A	Queller *et al.* (1988)
P. sericea	A	Queller *et al.* (1988)
Ropalidia marginata	A	Muralidharan *et al.* (1986)
ANTS		
Aphaenogaster rudis	M?, A	Crozier (1977b)
Conomyrma bicolor	A	Berkelhamer (1984)
C. insana	A	Berkelhamer (1984)
Formica aquilonia	M?,A	Pamilo (1982a)
*F. exsecta**	M?,A	Pamilo and Rosengren (1984)
F. polyctena	M?,A	Pamilo (1982a)
F. pressilabris	M?,A	Pamilo and Rosengren (1984)
F. sanguinea	M?,A, S?	Pamilo (1981, 1983)
F. transkaucasica	M?,A	Pamilo (1982a)
Iridomyrmex pruinosum	A	Berkelhamer (1984)
I. purpureus	M, A	Halliday (1983)
Lasius niger	M?,A	van der Have et al. (1987)
Leptothorax acervorum	F?,A	Stille *et al.* (1991)
Myrmecia pilosula	A	Craig and Crozier (1979)
Myrmica rubra	A	Pearson (1983a)
*Myrmica ruginodis**	A	Seppä (1992)
Nothomyrmecia macrops	F, A	Ward and Taylor (1981)
Rhytidoponera chalybaea	F, A	Ward (1983a)
R. confusa	F, A	Ward (1983a)
R. sp. 12 (nr. *mayri*)	F, A, S	Crozier *et al.* (1984)
Solensopsis invicta	M, A	Ross and Fletcher (1985a)

Table 4.9 *Continued*

Species	Notes	References
TERMITES		
Incisitermes schwarzi (primary reproductives)	A	Luykx (1985)
*Reticulitermes flavipes**	A	Reilly (1987)

An important exception to the general rule of a lack of inbreeding in social Hymenoptera is provided by socially parasitic ants, in which mating occurs within the nest prior to the females dispersing (Wilson, 1971:374–6). The ease with which species can shift from outbreeding to apparent inbreeding is shown by the slave-maker genus *Epimyrma*, in comparisons between those species producing many workers (*E. ravouxi*, *E. algeriana*, and *E. stumperi*) and those that are 'degenerate' in that they produce very few or no workers (*E. kraussei*, *E. corsica*, and a new *Epimyrma* species). The members of the first group show an outbreeding system whereas the reproductives of the degenerate slave-makers mate in the nest before dispersal (Winter and Buschinger, 1983; Jessen, 1987; Buschinger, 1989). The risk of concluding that inbreeding is *necessarily* strong in parasitic species is shown by the fact that, although Winter and Buschinger (1983) state that *E. kraussei* colonies have only a single queen, those of another social parasite, *Anergates atratulus*, sometimes have several which may therefore have been derived from different parent nests (see Crozier, 1977a). Buschinger correctly notes that the adoption of the inbreeding life-pattern requires a sex determination system which is not based on heterozygosity.

Termite species are also characterized by wide-ranging mating flights well-adapted for both dispersal and outbreeding (Nutting, 1969). The primary reproductives (those founding the colony after a mating flight) are therefore expected to be unrelated and to yield outbred progeny. But in many species there are further generations of reproductives that mate with their nest-mates. Despite the apparent long-continued presence in a minority of species of multiple primary reproductives (Thorne, 1982), inbreeding does seem likely within the colonies of many termite species. In fact, Reilly (1987) found significant inbreeding in *Reticulitermes flavipes*. Rosengaus and Traniello (1993) suggest that immunity to local pathogens may select for inbreeding and restricted dispersal in termites. Outbreeding in the termite *Zootermopsis angusticollis* increased mortality in the mating pair, probably because of exposure to new pathogen strains transmitted in mating.

A system of cyclic inbreeding similar to that in termites is found in the ant *Technomyrmex albipes* (Yamauchi *et al.*, 1991). A colony is founded by a single winged female who is soon replaced by her wingless daughters. These daughters inbreed with the wingless males produced in the same colony. Such cycles of inbreeding continue during the colony growth and after several generations of intra-colonial matings the colony produces winged sexuals that disperse and establish new colonies.

Evidence on the tendency of social insects to avoid incest is scanty, and indicates considerable variation between species. Wesson (1939) concluded, on the basis of very few observations, that queens and males of the slave-maker *Harpagoxenus americanus* prefer to mate with unrelated females, but Mintzer (1982a) found no tendency of *Pogonomyrmex californicus* sexuals to avoid mating with nest-mates. The behaviour reported by Wesson for a slave-maker contrasts strongly with the habitual inbreeding reported for many social parasites. Well-supported behavioural evidence for inbreeding avoidance in social insects is found in the bee *Lasioglossum zephyrum* (Smith, 1983), in two *Bombus* bumble-bees (Foster, 1992), in *Polistes* wasps (Ryan and Gamboa, 1986) and in the Argentine ant *Iridomyrmex humilis* (Keller and Passera, 1993).

4.5.2 *Nest-mate recognition: cornerstone of colony integration*

In spite of conflicts within colonies, a colony is the basic functional unit in eusocial insects in the sense that the cooperative effort of colony members affects positively reproductive output and colony maintenance. Individuals work for the colony and they also defend it at different levels. For that purpose it is adaptive to be able to distinguish nest-mates from non-nest-mates. A still closer distinction of kin is required for intracolony interactions.

Kin recognition can potentially affect sex allocation patterns, such as by enabling workers to determine the genetic structure of the colony they are in. For example, *Lasius niger* colonies differ in sex allocation pattern depending on whether or not the queen has mated more than once, and it is plausible to relate this to the detection by workers of odour-cue diversity (van der Have *et al.*, 1988).

There has been for a long time good evidence for colony recognition in many species, and there is a long history of studies on colony odours. Although the environment can contribute to that odour, it has been shown recently that innate, genetic determinants are highly important (reviewed by Waldman *et al.*, 1988). Genetic variation in kin recognition cues has been demonstrated by crossing experiments in bees (Greenberg, 1979), ants (Mintzer, 1982b; Provost, 1991) and termites (Adams, 1991).

The conceptual framework for considering the innate components of colony odour was given its present form by the terminology of Lacy and Sherman (1983) of *labels*, *templates* and *referents*. A label is any characteristic, such as particular pheromone or morphological characteristic, which can be used by one individual in classifying another with regard to degree of kinship. A template is an internal representation of the labels expected in kin within the mind of the discriminating individual. A referent is the source of the labels used in template construction.

Depending on the referents and the way in which templates are constructed, there are several possible modes of kin recognition (Crozier, 1987b, 1988). With regard to template formation, the following possibilities may be distinguished.

I. Gestalt model (Crozier and Dix, 1979). Genotypes under the gestalt model are not individually distinguished but rather the individual labels and their proportions within the social group form a blend that is used in recognition. This model requires that labels be transferred between all group members. A colony made up of two genotypes, AA and AB, in equal numbers, would form a template of 75% A and 25% B, and be quite distinct from one made up of BB and AB individuals. Naturally, it is possible that some individuals (e.g. queens) may contribute more than others to the group odour.

II. Individualistic models (Crozier and Dix, 1979). The alternative to the transfer of labels, the retention by individuals of labels without mixing, yields a family of models describable as individualistic. Crozier and Dix (1979) did not incorporate the possibility of learning in their individualistic model, whereas it is now known that social insects generally learn their templates rather than having them innate. In order of decreasing stringency, the possible individualistic models are as follows.

II(A). *Genotypic identity*. Individuals accept others as kin only if they match exactly a genotype present in the template. If the template has a single genotype, AB, then AA, BB, AC, and CC individuals would be rejected.

II(B). *Foreign-label rejection.* Individuals accept others as kin only if they do not possess any labels not included in the template. Thus a template formed from an AB genotype would accept AA and BB individuals but reject AC, BC and CC individuals.

II(C). *Common-label acceptance.* Individuals accept others as kin if they possess a label included in the template. For example, an AB template would lead to acceptance of AA, BB, and AC individuals, but CC individuals would be rejected. This model was termed *habituated-label acceptance* by Getz (1982), who derived it; the term used here and by Crozier (1986) seems more consistent with those used for the other individualistic models.

With regard to the referents involved in the formation of an individual's template, these may be either all adults in the colony, the larvae, or particular adults (such as queens, and also kings in termites), or solely the individual concerned. It is likely that, if there are several systems of labels (such as those encoded by different loci), then they may differ in their referent and template formation characteristics. For example, both queen- and worker-derived labels occur in the kin-recognition system of *Camponotus* ants (Carlin and Hölldobler, 1983, 1986, 1987; Carlin *et al.*, 1987), although queen-derived labels appear to be of negligible importance in some other ants (Mintzer, 1982b; Crosland, 1990).

Empirical work on social insect kin-recognition may therefore be summed up as involving the determination of the mode and ontogeny of template formation, and the referents involved in this formation. Considerable diversity has been found so far, including examples of both gestalt (e.g. Stuart, 1988) and individualistic (Mintzer and Vinson, 1985) cases, and a variety of referents. Where the referents are outside the worker caste, then the template-formation mechanism must be either gestalt or the individualistic with common-label acceptance. The cues are commonly learned within a short period immediately after eclosion, at least in ants (Ichinose, 1991; Stuart, 1992).

Less attention has been paid to the question of the relative importance of innate as against environmentally derived labels (Figure 4.8) in natural populations, although it has been known for a long time that both are important. An exception is the study of Stuart (1987) finding that kin-recognition in *Leptothorax curvispinosus* is 'largely maintained by transient environmentally-based nestmate recognition cues'.

It is worth stressing that when the template is learned, as seems likely to be the usual case, individuals are mainly learning to distinguish nest-mates from non-nest-mates. The extent to which they distinguish between kin of different degrees depends then on colony make-up. Experimentally, it has generally been easy to induce individuals to accept unrelated individuals as 'kin', as in the classic adoption studies on the bee *Lasioglossum zephyrum* (Greenberg, 1979; Buckle and Greenberg, 1981; Michener and Smith, 1987).

The genetic architecture of kin recognition in social insects has been little studied empirically, and neither have the substances and their sources actually making up the labels, although there is now suggestive evidence that these may be cuticular hydrocarbons originating in postpharyngeal glands in *Camponotus*

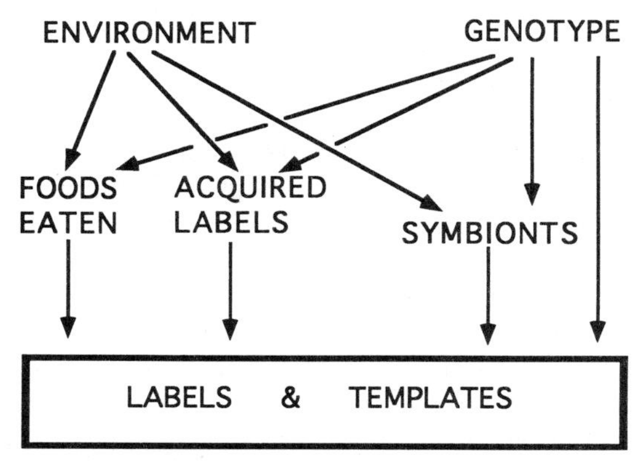

Fig. 4.8 Possible sources of labels for the kin recognition system of a social insect. (From Crozier, 1988; see also Carlin and Hölldobler, 1986.)

ants (Bonavita-Cougourdan *et al.*, 1987) or the Dufour's gland of *Evylaeus* bees (Hefetz *et al.*, 1986).

Theoretical work on kin recognition has passed from elucidation of the various possible models to the question of whether they can indeed be maintained by selection. In the case where agonistic encounters lead to costs, Crozier (1986) showed by a population-genetic analysis that selection against the rarer labels will lead to their loss and the consequent elimination of kin recognition abilities. The apparent paradox of the widespread occurrence of kin recognition seems to have been solved by Grosberg and Quinn (1988, 1989), who carried out a games-theoretic analysis showing that rare labels will be selected for by the occurrence of occasional indiscriminate mutants. These indiscriminate mutants do not bear the cost of maintaining the means of discrimination or aggression, and will increase in frequency if the population is made up only of potentially discriminating individuals of the same label phenotype as the mutants. In this situation, discriminating individuals with rare labels will be of higher fitness than average due to their ability to sequester resources from the indiscriminate mutants. Grafen (1990) has plausibly conjectured that the same result occurs under a model in which matching individuals behave altruistically to each other, rather than non-matching individuals behaving agonistically, as in the Grosberg–Quinn analyses.

4.5.3 *A colony can occupy many nests*

The nests of social insects are not always equivalent to single colonies, but rather may be linked together so that we can speak of a colony occupying many nests. Such colonies are termed *polydomous*, or *polycalic*. The nests in a polydomous colony will exchange workers, as well as food, brood, and queens, but may still have some identity within the colony. Relations between nests of a polydomous colony can be mutually competitive, indifferent or cooperative. The term polycalic has been recommended to describe the last type of internest relations (Rosengren and Pamilo, 1983). On occasion warfare may break out between nests that previously exchanged individuals apparently amicably, as found, for example, in mound-building *Formica* ants (Mabelis, 1979; Rosengren and Pamilo, 1983).

Although polydomous colonies are frequently polygynous as well, the correlation is far from perfect. Where monogynous colonies are polydomous, the general pattern is for the nest with the queen to be larger and the other nests to be smaller, so-called satellite nests. Such satellite nests are known from ants of several subfamilies (e.g. in *Paratrechina flavipes*, Ichinose, 1986, in *Aphaenogaster cockerelli*, Hölldobler and Carlin, 1989, in *Myrmica punctiventris*, Snyder and Herbers, 1991, in *Colobopsis nipponicus*, Hasegawa, 1992), as well as from some polistine wasps (e.g. *Ropalidia revolutionalis*, Itô, 1987a). (But satellite nests of polistine wasps appear usually to acquire their own mated egg-layer, as reported for *Polistes exclamans* (Strassmann, 1981), *P. fuscatus* (Page *et al.*, 1989b) and *Ropalidia fasciata* (Itô, 1986)).

The pattern of a division between the queen-right and satellite nests of a colony has an impact on sex allocation. In *Hypoclinea quadripunctata*, for example, according to Torossian (1967a,b, 1968a,b) queen-laid eggs never produce males, the presence of a queen inhibits worker oviposition, and the males are produced by workers in the satellite nests.

The progression from monodomous to polydomous social systems can lead to *unicoloniality*, the complete absence of colony distinction. Each population of such species consists of one huge, widely dispersed colony. These species pose an important evolutionary problem, in that the lack of selection at the colony level within populations is expected, at first sight at least, to lead to escalating reproduction of reproductives and the loss of social efficiency (Sturtevant, 1938; Crozier, 1977a, 1979). Crozier (1979) and Hölldobler and Wilson (1977) independently suggested that unicolonial species may occupy transient habitats, with selection at the population level being strong enough to counter the lack of colony-level selection. The time-scale for the population extinction supposed to drive such a selection scheme differs markedly from species to species.

Students of unicolonial populations should endeavour not only to confirm them as such (see below) but also to involve both interpopulation genetic differentiation (the equivalent to the inter-colony relatedness of multicolonial species) as well as the tendency of relatives to assort together within the population. It may be, for example, that workers preferentially aid their own mother to reproduce, avoiding other queens. Such studies would involve not only genetics but also careful observation.

Although the concept of unicoloniality is clear enough, it is uncertain as to how many of the apparent cases are actually supercolonial instead. *Supercoloniality* refers to the presence of very large multinested and polygynous colonies. Such supercolonies are hostile to other such colonies, and may be geographically separate from them. For example, mature colonies of the Australian meat ant *Iridomyrmex purpureus* are polygynous and polydomous and can occupy territories of 10 ha, while engaging in strong territorial warfare at the boundaries (Greenslade and Halliday, 1983).

In species with single-nest colonies, increasing intensity of aggression in encounters between individuals with the distance between their nests has been recorded for the ants *Rhytidoponera metallica* (Haskins and Haskins, 1979) and *Myrmica rubra* (Brian, 1983:250), and for the bee *Lasioglossum zephyrum* (Kukuk and Decelles, 1986), but tests failed to demonstrate this relationship in the supercolonial Okinawan ant *Polyrhachis dives* (Yamauchi *et al.*, 1987). The relationship between agonism and geographic distance is plausibly suggested by Kukuk and Decelles (1986) to result from increasing genetic differentiation, and Crozier *et al.* (1987) confirmed using allozymes that *L. zephyrum* does indeed show increasing genetic differentiation on a microgeographic scale.

Wilson (1971:457) notes that unicolonial, or supercolonial, species frequently occur as a sibling species with a monogynous multicolonial one. This occurrence weakens the view that unicolonial species evolved from polygynous ones. Likely to be most informative is the evolution of supercolonial *Solenopsis invicta* populations ('polygyne') from the usual single-nest monogynous type ('monogyne'). The supercolonial populations occur at scattered locations in the southeastern United States, and allozyme data indicate that they are more closely related to nearby monogyne populations than to other polygyne ones (Ross *et al.*, 1987). Workers from different nests of the same polygyne population show no hostility to each other, and the nests exchange workers (Bhatkar and Vinson, 1987), but tests involving workers from *different* polygyne populations have yet to be reported.

What forces have led to the evolution of polygyne fire ants? One suggestion is that this event has been fortuitous, resulting from the loss of kin-recognition label variability due to the genetic bottleneck involved in the establishment of these immigrant populations. However, it is more likely that the selective conditions in their new habitat simply favour the accretion of further queens according to the requirements outlined above (Section 4.4).

Although the balance of evidence favours the *de novo* origin of polygyne fire ants in the United States, the discovery that this form also occurs in South America (Jouvenaz *et al.*, 1989; Ross and Trager, 1990) impels caution. What is required is a genetic comparison of populations in both continents, perhaps using molecular markers to see whether the South American polygyne populations are more closely related to neighbouring monogyne populations than to the North American populations of either form.

There is increasing evidence that dispersal between highly polygynous and polydomous colonies can be restricted. Genetic differentiation is greater among populations with polydomous colonies than among those with monogynous and monodomous colonies of the same species (*Formica truncorum*, Sundström, 1993) or of related *Myrmica* species (Seppä and Pamilo, 1995). The alate queens in monogynous and monodomous colonies of the ants *Formica truncorum* and *Solenopsis invicta* are heavier than those produced in polygynous colonies (Porter *et al.*, 1988; Sundström, 1993). At least in *S. invicta* this difference is due to the type of colony rather than to genetic differences. The alate queens produced in polygynous *S. invicta* colonies are largely incapable of independent colony foundation and have to join existing nests. The queens originating from monogynous colonies are not accepted in polygynous colonies because they are much heavier than the other queens and initiate oogenesis rapidly. The same is true for one *Pgm* genotype, Pgm^{3a}/Pgm^{3a} irrespective of the colony of origin. Queens with this genotype are eliminated by workers in polygynous colonies. Gene flow between the two social forms is mediated by males, and the reproductive options available for new queens are culturally transmitted depending on the type of natal colony (Keller and Ross, 1993b; but see also Berrigan *et al.*, 1994 and the response by Keller and Ross in the same issue of *Science* concerning the possibility that heterozygote advantage in egg-laying rate might partly explain the maintenance of the allele Pgm^{3a} in polygynous societies).

4.6 Summary

The simplest colony structure is that of a family comprising a once-mated queen and her worker progeny, but there are many other types of colony structures. Species, and even in some cases populations within species, vary in whether a queen mates once (monandry) or more than once (polyandry), whether colonies have but one queen (monogyny) or more than one queen (polygyny). Similarly, colonies may occupy only one nest (monodomy) or more than one nest (polydomy).

In many social Hymenoptera, some or all of the males are produced by workers. In particular, if all males arise from worker-laid eggs, then the reproductive values of males and females are equal. The likelihood of worker reproduction evolving is affected by the mating biology of the queens. If the queens mate more than twice, then workers are selected to prevent each other from reproducing. If the queens mate fewer than twice, workers are selected to be tolerant of each other's reproductive activities. Worker reproduction should evolve even if a considerable cost in colony efficiency arises from it. Polygyny also tends to select for worker intolerance of each other's reproduction.

The causes of polyandry are not completely clear but we distinguish three hypotheses as worthy of further study. One is that the maximization of genetic variation within colonies is selected for because it is needed to allow the full range of castes. The second is that genetic variation for resistance to pathogens protects a colony against disease. The third is that the load of diploid males due to the sex determination system causes variation in average growth rates between monandrous and polyandrous strains. Depending on the ecological conditions, the disease resistance and load hypotheses predict monandry or polyandry. Within these broad limits, all three hypotheses generate testable predictions.

Polygyny leads to considerable potential complexity in the relatedness structure of colonies, with a general decline in nest-mate relatedness as generations pass. The factors governing the occurrence of polygyny, and of the number of queens per colony, have been hypothesized to include kin selection, mutualism, and parasitism. The efficiency thresholds governing whether or not further queens should be added to the colony vary for the queens and workers already resident, and depend critically on the relatedness of the various nest-mates, and the population and colony sex ratios. The association of young queens to found a new colony is forced by the need to produce as large a worker force as possible in order to survive the stage of intense competition between young colonies; such associations have a strong tendency to be formed of unrelated queens.

Polygyny can be combined with polydomy. As the extent of polydomy increases, it approaches the condition termed unicoloniality, in which there are no colony boundaries. Although the concept of unicoloniality is clear enough, its applicability to natural populations is open to doubt. A case in point is the polygynous form in introduced populations of the fire ant *Solenopsis invicta* in the southern United States. While the monogynous and polygynous forms may yet prove to be different species, theory suggests that polygyny can arise spontaneously within a monogynous population and there is genetic evidence to support this interpretation in *S. invicta*. The distinction between a unicolonial population and a single colony is also made more difficult by the tendency of highly polygynous and polydomous species to show considerable population viscosity and inbreeding.

5

Intra-colony conflicts over sex-allocation

The relatedness asymmetries to which social Hymenoptera are prone lead inevitably to a conflict of interest over the basic function of the colony: the reproduction of the genetic material. The queens and workers will be selected to favour different patterns of investments between the new queens and males, and colony maintenance. Factors affecting the relatedness structure of colonies and of populations will inevitably affect the results of this conflict. Such factors include whether or not the workers produce any of the males (thus affecting relatedness structure) and whether queenless colonies contribute significantly to male-production (thus devaluing males for the queen-right colonies). The sex ratio among reproductives is not the only arena of conflict between queens and workers: the question of the division of resources actually involves a three-way split, in that resources have to be allocated not only to queens and males but also to colony maintenance (further workers). In those species whose colonies reproduce by splitting ('fission'), there will also be the potential for disagreement over the timing of this event, and the relative investment in males and the daughter colonies. The central problem of social insect sex allocation is thus 'Who wins?' in the evolutionary struggle between queens and workers.

5.1 The optimal sex ratio in monogynous colonies

As shown in Section 2.2, the optimal population-wide sex allocation is predicted by a simple equation (2.10) that depends on the genetic relatednesses and reproductive values. We also showed in Section 2.2 that there is a conflict between the queens and workers of monogynous colonies in how to allocate the available resources between daughters and sons. We will in this chapter further extend this basic model to cover situations where the relatednesses and reproductive values depart from those of the simplest model and where the fitness returns are not necessarily linear functions of investments. As we will frequently refer to the equation giving the population equilibrial sex allocation, we repeat it here. If the population investment in males is denoted by M, the population investment in females by F, the sex-specific reproductive values of females and males by ν_F and ν_M, and the coefficients of genetic relatedness to the individual controlling sex allocation (I) by g_{FI} and g_{MI}, we have at equilibrium

$$\frac{g_{FI}\nu_F}{F} = \frac{g_{MI}\nu_M}{M} \tag{5.1}$$

where $1/F$ and $1/M$ are proportional to the average mating successes of females and males, respectively.

The reproductive value has been defined as the contribution of an individual to the asymptotic gene pool, i.e. the reproductive value tells us how many copies of a gene there will eventually be in the population in the remote future (Oster *et al.*, 1977). These reproductive values were shown in Section 4.2 to depend on the extent of worker reproduction in the whole population and are then the same for all females and all males. The values of genetic relatedness in eqn. (5.1) depend on the type of the colony and on the individual I who is controlling sex allocation in the colony. The values of relatedness are, of course, different when looked at from the view point of the queen or of the workers, and so are the optimal sex investment frequencies. The fundamental question in social insect biology is whether the workers can increase their inclusive fitness by biasing sex allocation towards their own optima. We will therefore analyse in this chapter optimal sex allocation separately for queens and for workers in order to see how large is the conflict. At the end of the chapter we try to see whether the available data from natural populations indicate which party has won the conflict. We start here by analysing optimal sex allocation and queen–worker conflicts in monogynous colonies, assuming that workers may not only try to bias the sex ratio but that they can also take part in male production.

Although workers of some ant genera, such as *Solenopsis*, lack functional ovaries (Wilson, 1971:320; Fletcher and Ross, 1985), and those of higher termites appear to lack gonads (Wilson, 1971:195), the workers of most social insects retain the capacity for reproduction. This capacity is of special significance in Hymenoptera, because workers may reproduce even if unmated, yielding males.

Worker production of males has been found in a variety of species (Table 4.3). These species vary between those in which workers produce male-destined eggs even in the presence of the queen, and those in which the queen's presence leads to inhibition of worker oviposition or to its redirection towards the production of trophic eggs, eggs made only to be eaten (e.g. Brian, 1983:150–2; Crespi, 1992a; Kukuk, 1992).

Worker-laid and male-producing eggs enter twice into discussions of sex ratio.

Firstly, within colonies there is a potential conflict of interest. If a worker can substitute one of her own eggs for an egg laid by another individual, even her mother, she gains in inclusive fitness because her own son is more closely related to her than her brother would be. These conflicts of interest are mediated by the relatednesses between the various offspring of the colony and the workers, which in turn affects the preferred investment ratios of the various parties.

Secondly, if worker egg-laying is released by the death of the queen, then there is a fraction of the population of colonies which are dying out and produce only male reproductives. Selection then favours a more female-biased sex ratio among the queen-right colonies than would be the case if the queenless colonies failed to produce reproductives, a phenomenon known as 'sex ratio compensation' (Taylor, 1981).

5.1.1 *Worker male-production affects the preferred investment ratios*

When the workers produce some, or all, of the males, the coefficients of genetic relatedness and the reproductive values will be affected. When the reproductive values and genetic relatednesses have been determined, we can calculate the equilibrium sex ratios using eqn. (5.1). The relatedness of a daughter to her mother is 0.5 in a monogynous colony. Of the males, a proportion $1 - \psi$ are sons with relatedness 1 to the queen, and a proportion ψ are grandsons of the queen with relatedness to her of 0.5. From these values and the sex-specific reproductive values of eqn. (4.2) we obtain the sex allocation ratio optimal for the queen:

$$\frac{(1/2)[2 - \psi/(3 - \psi)]}{F} = \frac{(1 - \psi/2)[1 + \psi/(3 - \psi)]}{M}$$

which, when solved for M, yields

$$M^* = 0.5. \tag{5.2a}$$

The optimal sex ratio for the queens is thus 1:1 irrespective of the value of ψ. This result was first found by Benford (1978) using a slightly different approach. The result differs from that of Trivers and Hare (1976), because their arguments were based on the matrix **P** instead of $\mathbf{P}_\infty$ (Pamilo, 1991a; see Section 2.4).

From the standpoint of the workers, the optimal sex ratios are different. The genetic relatedness of an average female offspring to a worker is 3/4 (full sisters) and that of a male offspring is $(1 - \psi)/2 + (3/4)\psi = 0.5 + \psi/4$. This latter relatedness is obtained when we note that a proportion $(1-\psi)$ of the new males are brothers of the workers (relatedness 1/2) and a proportion ψ are nephews (relatedness 3/4). From these values we now obtain the equilibrium equality:

$$\frac{(3/4)[2 - \psi/(3 - \psi)]}{F} = \frac{(1/2 + \psi/4)[1 + \psi/(3 - \psi)]}{M}$$

from which can be derived

$$M^* = 0.25 + \frac{(0.75)\psi}{4 - \psi}. \tag{5.2b}$$

This optimal proportion of males ranges from 0.25, with $\psi = 0$, to 0.5 for $\psi = 1$, a result first obtained by Benford (1978).

5.1.2 *Sex ratio compensation*

In many species of eusocial Hymenoptera, the workers can and may start laying haploid eggs after the colony loses its queen. These orphaned colonies therefore contribute to the male-production of the population. The situation resembles that of worker-reproduction analysed above, but with the difference that the orphaned colonies can only produce males and it is only the queen-right colonies that can adjust their sex allocation ratio. This led Metcalf (1980), Owen *et al.* (1980) and Forsyth (1981) to expect that the queen-right colonies shift their sex allocation in favour of females as a response to the males produced by the orphaned colonies. This phenomenon was called sex ratio compensation by Taylor (1981), who first analysed this case. It is logical to define a parameter corresponding to ψ above, the proportion of males produced by the workers. We use a further symbol, β_m, the fraction of the population's *males* arising from orphaned nests, and assume that all reproduction in the queen-right nests is by their queens. Note that this definition of β_m differs from that used originally by Taylor (1981) and follows the notation of Pamilo (1991a). It is also convenient to define another parameter, namely the proportion of all *sexual* production due to the orphaned nests, without any attention to sex. We denote this by β_s.

Under this notation, the elements of the transition matrix **P** become $p_{FM} = 0.5 - \beta_m/4$ and $p_{MM} = \beta_m/2$ (Pamilo, 1991a). The elements p_{FF} and p_{MF} remain unaltered and we easily obtain from eqn. (2.14) that $v_F/v_M = 2 - \beta_m$. This result is of the same general form as that $(2 - \psi)$ obtained for worker reproduction in queen-right colonies. The sex-specific reproductive values are, recalling that the $v_F + v_M = 3$,

$$v_F = 2 - \frac{\beta_m}{3 - \beta_m},$$

$$v_M = 1 + \frac{\beta_m}{3 - \beta_m}.$$

With these results, we can now examine the optimal sex allocation in the queen-right colonies, assuming that they are monogynous and that the queens are monandrous. The appropriate relatednesses under queen control are $g_F = 0.5$ and $g_M = 1$, and under worker control they are $g_F = 0.75$ and $g_M = 0.5$. Only in the queen-right colonies is the colony sex allocation, m, free to vary. The optimal sex investment in the queen-right colonies, m^*, is found by solving the equilibrium equation:

$$\frac{g_F v_F}{(1 - \beta_s)f} = \frac{g_M v_M}{\beta_s + (1 - \beta_s)m}.$$

Solving this, and using the connection between the two parameters for male production, $\beta_m = \beta_s/[\beta_s + (1-\beta_s)m]$, we obtain

$$m^* = 0.5 - \frac{\beta_s}{4(1-\beta_s)}$$

$$= 0.5 - \frac{\beta_m}{2(4-3\beta_m)} \qquad \text{when } \beta_s \leq 2/3$$

$$m^* = 0 \qquad \text{when } \beta_s > 2/3 \qquad (5.3a)$$

under queen control, and

$$m^* = 0.25 - \frac{3\beta_s}{8(1-\beta_s)}$$

$$= 0.25 - \frac{3\beta_m}{(8-5\beta_m)} \qquad \text{when } \beta_s \leq 2/5$$

$$m^* = 0 \qquad \text{when } \beta_s > 2/5 \qquad (5.3b)$$

under worker control.

This result was first developed for the queen control case by Taylor (1981) using a slightly different approach, and extended to the worker control case by Nonacs (1986a). As Taylor noted, there is sex ratio compensation in the queen-right colonies, which produce more female-biased sex ratios than otherwise expected in these colonies. However, the mean sex ratio in the whole population turns out to be more male biased than expected without orphaned colonies. The overall sex investment in the whole population will be $\beta_s + (1-\beta_s)m^*$, and this gives

$$M = 0.5 + \frac{\beta_m}{2(4-\beta_m)} \qquad (5.4a)$$

under queen control, and

$$M = 0.25 + \frac{3\beta_m}{4(8-3\beta_m)} \qquad (5.4b)$$

under worker control (Pamilo, 1991a; Fig. 5.1). We can now compare these values with those given by eqns. (5.2a) and (5.2b) by assuming that the proportion of worker-produced males is the same, i.e. by setting $\beta_m = \psi$. Clearly, male-production in orphaned colonies leads to somewhat more male-biased population sex ratios than worker reproduction in queen-right colonies.

Estimates of β_s in natural populations are harder to make than might be expected and are rare. For example, Bourke *et al.* (1988) estimated β_s at 0.098 in *Harpagoxenus sublaevis* as the proportion of total production made up by males in colonies in which a queen was not found and which produced only males in the next generation. But this procedure overestimates β_s, because queens take two years to develop and hence colonies still continue to produce the foundress's progeny over that period. Classing as queenless colonies only

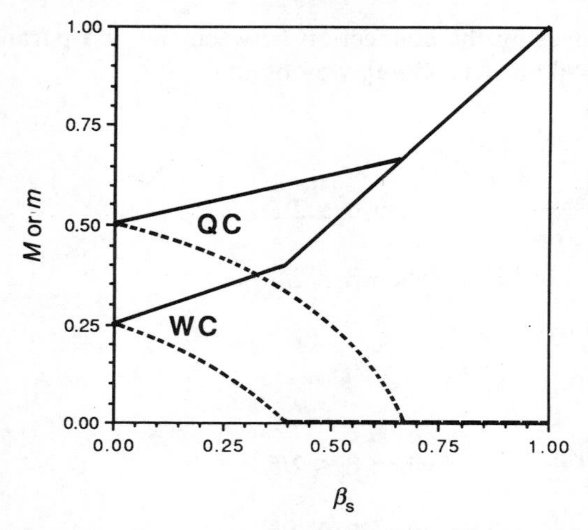

Fig. 5.1 Population sex allocation (M) and that of queen-right colonies (m) as a function of the proportion (β_s) of the production of sexuals due to orphaned colonies. Solid lines show the sex ratio in the whole population (M). Dashed lines show the sex ratio for queen-right colonies (m). Values are given for both queen control (QC) and worker control (WC). All males in queen-right colonies arise from queen-laid eggs (i.e. $\psi = 0$). Based on eqns. (5.3) and (5.4).

those for which a queen was not captured and which had all-male broods both at time of capture and a year later produces an estimate of $\beta_s = 0.0009$, probably an underestimate and certainly inaccurate. A higher estimate is obtained for *Epimyrma ravouxi* from the data of Winter and Buschinger (1983), of $\beta_s = 0.11$. From eqns. (5.4), these values are too low to make much difference to the sex allocation of queen-right colonies. However, for *Apterostigma dentigerum*, Forsythe's (1981) data yield a value of $\beta_s = 0.22$, predicting a value for m^* in queenright colonies of 0.43 under queen control and 0.15 under worker control. The observed m^* of 0.38 falls between the predicted queen- and worker-control values, and is significantly different from either (t-test, Sokal and Rohlf, 1981:450). *Apterostigma* colonies are very small (about 35 individuals in queen-right colonies), so that the queen makes up a high proportion of the biomass, meaning that she would have more capacity for control by personal intervention than in a colony with thousands of individuals.

The situation in *Rhytidoponera* differs from that in the species just discussed. A few *Rhytidoponera* species (e.g. *R. purpurea*) have normal queen-right colonies, many species do not have morphologically distinguished queens at all, and in some species a proportion of the colonies lack queens whose role is taken over by mated workers (gamergates), as in the *R. impressa* group. The gamergates produce new workers and males, but are not perfectly analogous to the orphaned colonies discussed in this section because worker-production can lead to an unknown extent to further queenless colonies by budding or fission (Ward, 1981). The proportion of queenless colonies varies markedly from population to population, and between species (Ward, 1983b), reflecting either differing mortality schedules for the queenless colonies or variation in success in colony founding. Nevertheless, because queenless colonies produce males almost exclusively, they can probably be treated the same as the orphaned colonies in this section with respect to their effects on queen-right colonies. Queenless colonies of R. confusa produce from 0.09 to 0.88 of the sexual biomass, depending on population, and this is reflected in changes in the allocation pattern of queen-right colonies (Ward, 1983b; Fig. 5.2).

5.1.3 *Joint effects of worker male-production and sex ratio compensation*

Workers can, of course, produce males in both orphaned and queen-right colonies, and we could have taken both of these into account simultaneously in the first place. However, it seemed preferable to present the two cases separately in order to explain the underlying principles and the techniques used

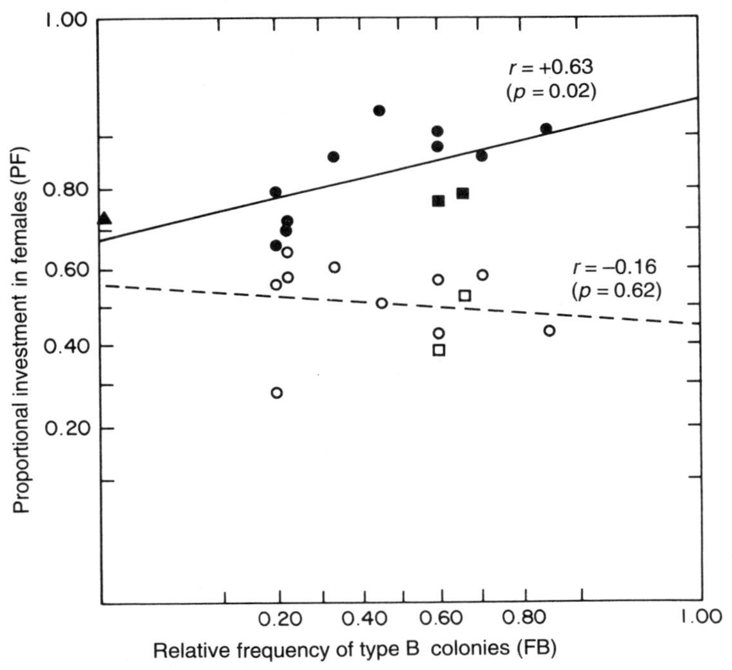

Fig. 5.2 Proportional investment in females of queen-right (closed symbols) or queenless (open symbols) colonies as a function of the proportion of queenless colonies in the same population for species in the *Rhytidoponera impressa* group: circles, *R. confusa*; squares, *R. chalybaea*; triangles, *R. purpurea.* Reproduction in the queenless colonies is carried out by gamergates (mated, egg-laying workers). (From Ward (1983b), reprinted by permission.)

in the analysis. If we now combine the two sources of worker-produced males and let $\beta = \beta_S$ and m be the proportional investment in males in the queen-right colonies, we find (Pamilo, 1991a; see Appendix at the end of this chapter) that the reproductive values are

$$v_F = 2 - \frac{\beta + (1-\beta)m\psi}{2\beta + (1-\beta)m(3-\psi)} \tag{5.5a}$$

and

$$v_M = 1 + \frac{\beta + (1-\beta)m\psi}{2\beta + (1-\beta)m(3-\psi)} \tag{5.5b}$$

The optimal sex ratio within the queen-right colonies is then obtained by solving

$$\frac{g_{FI}v_F}{(1-\beta)f} = \frac{g_{MI}v_M}{\beta + m(1-\beta)}$$

where the relatednesses are, to the queen

$$g_{FQ} = 0.5 \text{ and } g_{MQ} = 1 - \frac{\psi}{2},$$

and to the workers

$$g_{FW} = 0.75 \text{ and } g_{MW} = \frac{1-\psi}{2} + \frac{3\psi}{4} = 0.5 + \frac{\psi}{4}.$$

Using these values, the optimal sex ratios in the queen-right colonies (m^*) become, under queen control

$$\begin{aligned} m^* &= 0.5 - \frac{\beta}{2(1-\beta)(2-\psi)} && \textit{when } \beta \le \frac{2-\psi}{3-\psi} \\ m^* &= 0 && \textit{when } \beta > \frac{2-\psi}{3-\psi} \end{aligned} \tag{5.6a}$$

and, under worker control

$$\begin{aligned} m^* &= 0.25 + \left(\frac{3}{4}\right)\left[\frac{(1-\beta)\psi - 2\beta}{(4-\psi)(1-\beta)}\right] && \textit{when } \beta \le \frac{2+\psi}{5+\psi} \\ m^* &= 0 && \textit{when } \beta > \frac{2+\psi}{5+\psi} \end{aligned} \tag{5.6b}$$

From these sex ratios, we can obtain those derived at the start of this section by setting $\beta = 0$ or $\psi = 0$. Some other cases are presented in Fig. 5.3.

5.1.4 *Some additional problems associated with worker reproduction*

There are (at least!) two additional problems that should be added to the models of worker reproduction. First, the workers may in some cases produce not only males but also new females. Second, the models in the previous sections of this chapter apply to a population with discrete (non-overlapping) generations.

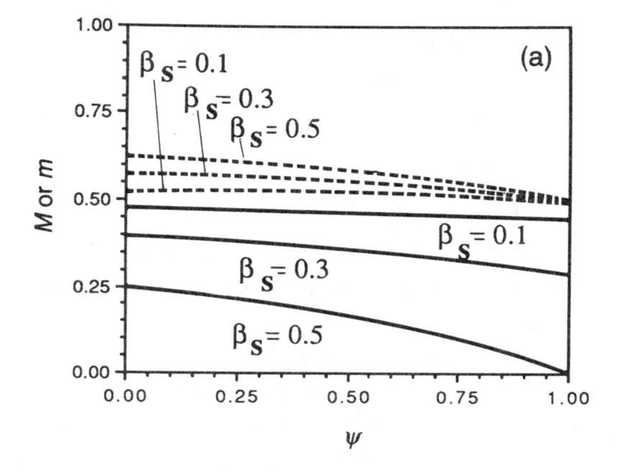

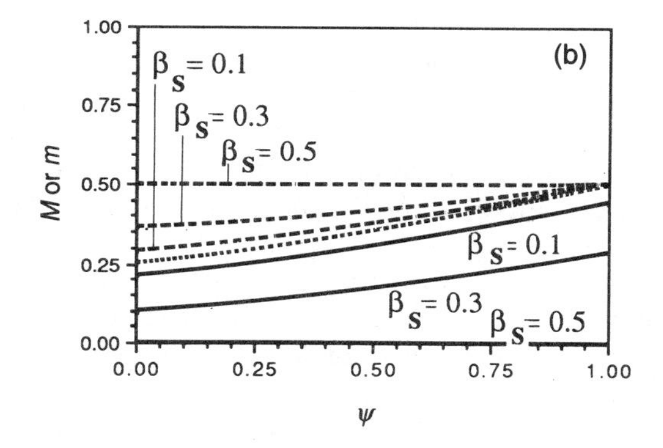

Fig. 5.3 Sex allocation as a function of ψ (the proportion of the males from queen-right colonies derived from worker-laid eggs) and β_s (the proportion of sexuals produced by workers in orphaned colonies). Dashed lines show the sex allocation (m) in queen-right colonies. Solid lines show the population sex ratio (M). The $\beta_s = 0$ line (the dotted line in figure b) holds both for queen-right colonies and for the whole population because under this circumstance there is no reproduction by orphaned colonies; (a) sex allocation under queen control; (b) sex allocation under worker control. Based on eqn. (5.6).

Mated workers occur in some bees and wasps which have little or no physical differentiation between queen and worker (Michener, 1974:85–6; 1990). Although it is uncertain how often these mated workers produce fertilized eggs, they are able to do so in orphaned colonies (in which case they often become replacement queens). In the bee *Halictus ligatus* workers seem to produce female offspring whereas the queen produces males (Richards *et al.*, 1995). Pamilo (1991a) analysed a model in which nests produce both queen-laid and worker-laid brood and found that reproduction by mated workers did not change the expected sex ratios from those given by eqn. (5.3). Although the diploid eggs laid by a worker are less related to both the queen and the worker nest mates, the reproductive success of females increases and these two trends cancel each other. This result, of course, depends on the specific assumptions of the model.

Many species of ponerine ants have mated workers called gamergates (Peeters, 1987b, 1991). Most species that have gamergates lack a morphologically distinct queen caste; in such cases gamergates may be regarded as queens for the purposes of sex allocation theory. But the ant *Rhytidoponera confusa* has two kinds of colonies, those headed by a single queen and those headed by a group of gamergates (Ward, 1983b). Ward (1983b) found that whereas the queen-right colonies produce both queens and males, gamergate-headed ones produce only males and workers, presumably otherwise reproducing by colony fragmentation ('colony fission'). In such cases, a clear distinction obviously has to be made between queens and gamergates and the continued existence of two different kinds of colonies in the population has to be recognized.

Another source of worker-produced females is thelytoky (the origin of females from unfertilized eggs). Thelytoky occurs sporadically in the workers of a great many social Hymenoptera (Table 4.4). Reports of a major role for thelytoky in some ant species (Ledoux, 1950, 1954; Soulié, 1960) remain controversial, but in the ant *Cataglyphis cursor* thelytoky appears to enable colonies to survive queenless periods (Cagniant, 1979; Lenoir *et al.*, 1987), which is also a reasonable explanation for the thelytokous workers of the Cape honey-bee, *Apis mellifera capensis* (Moritz and Hillesheim, 1985). If thelytoky occurs regularly, we might expect the population to adjust its sex allocation to take into account these worker-produced females. One completely thelytokous ant has been found, the Japanese *Pristomyrmex pungens* (Itow *et al.*, 1984); in such cases there are no questions to ask about sex ratios!

Worker-reproduction in all the models presented above affected not only genetic relatedness but also the sex-specific reproductive values. The reason for this is that worker-reproduction provides another route for the transmission of genes between generations. For example, when the workers produce males, the genes in the male offspring may have been descended from the males of the preceding generation via the workers. This increases the reproductive value of males. The model assumes non-overlapping generations which makes it possible to construct the transmission matrix **P** for gene transmission between generations. In many cases, of course, generations overlap; it is reasonable to suppose that the results from discrete models apply reasonably well to the overlapping-generations case, but this remains to be verified.

A final possible, although perhaps implausible, complication would be if the sex allocation pattern of a colony made up of several worker genotypes differing in their influences on sex allocation was a non-linear function of the worker genotype proportions, in other words that certain genotypes dominated the results out of proportion to their numbers. This situation of *behavioural dominance* has been explicitly modelled by Charnov (1978), Craig (1980),

Pamilo (1982b) and Bulmer (1983a). Bulmer (1983a) provides a synthesis of these analyses, but also argues that behavioural dominance is unlikely to arise.

5.2 Evolution of sex ratio conflict

5.2.1 *Compromise or arms race?*

We derived above the sex allocation ratios expected assuming that allocation is controlled either by the queen(s) or by the workers. The expectation is different for the two groups, creating a potential conflict of interest. This realization leads naturally to the question whether it is possible to predict what will happen when the queens and workers try to gain their preferred allocation ratio simultaneously. The first prediction is a conflict that leads to a kind of arms race (Dawkins and Krebs, 1979) between the two castes.

Alternatively, it has been suggested that the queens and workers can arrive at some kind of compromise leading to a sex ratio which is the 'best of the suboptimal ones'. The solution can be thought of as a set of individual strategies or as a problem of colony-level selection (Oster and Wilson, 1978:83–92). Oster and Wilson (see also Oster *et al.*, 1977) examined the effects of selection for the sex ratio at the colony level using a fitness function (V) essentially identical with those used here. They derived the functions separately for queens (V_Q) and workers (V_W) and noted that they cannot be maximized at the same time. They then introduced a third function as one possible solution to this conflict of interests, one based purely on selection at the colony level, as if the colony itself is an organism. Based on this fitness function, which includes no coefficients of genetic relatedness, they showed that the equilibrium investment ratios are $M = 0.5$ if all males are produced by the queen and $M = 0.67$ if the workers are the only source of the males. The problem with this approach is that, without a genetic basis, there is no theoretical justification for this particular fitness measure to be maximized, because the 'good-of-the-colony' is not defined in it. In fact, this 'good' *cannot* be defined, because the colony has no reality apart from the queens and workers of which it is composed, so that it is not suprising that the model breaks down when one tries to analyse it genetically. The compromise proposed would always be vulnerable to cheating, because both queen and workers would be selected to adopt a different investment ratio.

It seems best to approach the problem as an arms race between queens and workers. Pamilo (1982b) simulated a two-locus genetic model in which one locus was expressed in queens and the other in workers. Because the model did not include any costs or constraints, it led to complete queen control over the sex ratio. This control was achieved by the queens laying only haploid eggs in one half and diploid eggs in the other half of the colonies. Such a strongly bimodal sex ratio distribution agrees with observations made for many ant species (Pamilo and Rosengren, 1983; Nonacs, 1986a) but perhaps for a different reason. It will occur to readers that the option of producing only haploid eggs is unlikely if it also stops worker production in the colony. The option is in fact available for species in which the sexual brood is produced at a different time to the worker brood, as is often the case in *Formica* ants.

If we wish to examine the arms race at the colony level, we need to make some assumptions about the constraints and costs affecting queen–worker conflict. For a given behaviour to be selected for, these costs must be balanced by the genetic gain.

If there is no cost involved in sex ratio biasing by workers, they would in monogynous colonies raise broods leading to a population sex ratio of $M = 0.25$, namely an investment ratio of 3:1 females:males. This biasing is possible if the workers can selectively eliminate extra males or if they can increase the investment in females either by inducing additional diploid eggs to develop into queens, or by increasing the investment in individual queens.

Assume that the total number of haploid eggs in a colony is H and that of diploid eggs is D. If the colony has sufficient resources to raise all $H + D$ eggs into sexuals, elimination of males would mean a reduction of the total output, which is not to the benefit of the workers. Also if all diploid eggs are raised as new queens, no new workers are produced and this may reduce the probability that the colony will survive, reducing its future reproductive value. Clearly, there are limits within which the workers have to operate when they try to bias the sex ratio. This is the background for models developed by Bulmer (1981) for annual and by Bulmer and Taylor (1981) for perennial colonies, both models including specific colony growth functions. A more general treatment of the conflict situation was presented by Matessi and Eshel (1992) using a two-locus model, one locus active in queens and the other in workers.

5.2.2 *Conflicts in annual colonies*

Because the success of a colony, in a genetic and evolutionary sense, is measured by the sexuals it produces, it is reasonable that an annual colony has first a period of growth and then, at the end of the season, releases new sexuals. The life of an annual colony, such as those of many social wasps and bumble-bees, can be divided according to this scheme into founding, ergonomic and reproductive stages (Oster and Wilson, 1978:26–30). The reproductive stage is often coupled with conflicts between queen and workers or among workers and can lead to a complete breakdown of the social harmony of the colony. Cannibalism, in the form of egg-eating, can increase during this phase when the workers try to lay their own (haploid) eggs and to destroy the eggs laid by the queen or by other workers. The workers may also openly oppose the queen and even kill her (Fletcher and Ross, 1985; Bourke, 1988). This rebellion by the workers can take place because the queen is no longer required for colony growth (she is normally the only individual capable of laying diploid eggs) and because the resource allocation pattern optimal for workers is in general not optimal for the queen. Selection at the queen level is expected to adjust the timing of her production of haploid and diploid eggs so as to make it more expensive for the workers to alter the queen's allocation pattern.

If new sexuals are produced only at the very end of the season, the queen should have the ability to control completely the investment ratio by producing one-sex broods, all males in half the colonies and all-female in the other half (Bulmer, 1981; Pamilo, 1982b), or mixed broods of the maximal size the workers can raise (but not larger). This minimizes the opportunity of the workers to alter the sex ratio produced by the queen. The only avenue still open for conflict is over the source of the males: the workers could replace the queen-produced haploid eggs by unfertilized eggs of their own.

If the queen lays both haploid and diploid eggs at the end of the season, the allocation of resources within the colonies becomes crucial. As is well known, the workers should try to make the sex ratio female biased, the preferred bias depending on the number of matings. If there are more eggs laid than the colony can raise to adults, the workers should allocate the resources so as to obtain the optimal investment ratio, or a ratio as close to their optimum as possible. The queen can restrict this worker influence by limiting the supply of diploid eggs and laying a surplus of haploid eggs. The workers will then rear all of the diploid eggs as queens and spend all remaining resources in raising males. If the queens have managed to optimize the number of diploid eggs, they achieve the goal of a 1:1 ratio of investment in the population.

Many annual colonies of bees and wasps do not produce all sexuals as a single ultimate brood, but rather produce sexuals over a considerable portion of the second half of the season. This complicates the sex ratio conflict. Let us follow the model of Bulmer (1981) and assume that the colony lives through successive cohorts of brood and sexuals are produced in the last two cohorts of the colony's life. The penultimate cohort may also yield workers, whereas the ultimate cohort only yields new reproductives. If the queen lays diploid eggs in the penultimate cohort, 'hoping' that they will become workers, the old workers could feed them well enough to yield new queens. As a consequence, the colony gets less workers and the size of the ultimate cohort will be reduced. If the workers behave according to their own genetic interests, they should preferentially sacrifice males of the ultimate cohort (if there is an excess), making the sex ratio female biased. If the queen lays haploid eggs in the penultimate cohort, there will be no more workers and therefore no ultimate cohort.

To model the situation, let us assume that it costs c_F to produce a single female (a potential new queen) and c_M a single male, and that the cost of producing a new worker is $c_W = 1$. The costs c_F and c_M thus tell us how much more expensive it is to produce sexuals than new workers. The investment ratio in the population is then $F{:}M = (1 - r)c_F{:}rc_M$, where as usual r is the numerical proportion of sexuals which are males.

We focus our attention on one colony and examine how its members should behave according to selection imposed by the population investment ratio. The important limits, assuming single mating and that the workers lay no eggs, are as already given:

$rc_M > (1 - r)c_F$	there is an excess of males in the population from the points of view of queen and workers, and all individuals in our focal colony should invest in new queens;
$3rc_M > (1 - r)c_F > rc_M$	the queen should favour males and the workers should favour females;
$(1 - r)c_F > 3rc_M$	there are too many females from both standpoints, and the focal colony should invest in males.

When the queens and workers agree (i.e. selection on each favours the same bias to the sex ratio), as in the first and last of the above cases, the maximum production for each caste is achieved by raising workers in the penultimate cohort and sexuals (either males or females) in the ultimate one. The middle case is one of conflict over the colony's sex allocation.

Following Bulmer (1981), we examine the conflict situation and try to determine how the colony should allocate its resources in the last two cohorts.

We first give a brief intuitive overview. The queens should in the conflict situation favour males. This aim could be achieved by producing workers in the penultimate cohort and males in the ultimate cohort. But the workers prefer raising females instead of males, and they could raise the diploid eggs of the penultimate cohort as females instead of workers. This action would reduce the total sexual production of the colony and the critical question is whether or not workers benefit more from producing a smaller number of sisters than from a larger number of brothers. The queen could force the colony to produce males by laying only haploid eggs in the penultimate cohort and thus give the workers no option to affect sex allocation through caste determination. But this also reduces the sexual output of the colony, as well as leading to protandry. The abilities of the workers to bias the colony's sex allocation is constrained by the number of haploid (H) and diploid (D) eggs laid by the queen in each cohort, and by the total resources (R) available for raising a cohort. The amount of resources available for the ultimate cohort depends on the number of workers (n_W) raised in the penultimate cohort; following Bulmer we assume a linear relationship and that each worker adds an amount Δ to the total resources (with $\Delta > 1$). These resources are measured in the same units as the production costs c_M and c_F, so that the ultimate cohort could consist of $n_W\Delta/c_M$ males or $n_W\Delta/c_F$ females, or a combination of the two sexes.

The conflict situation can now be handled with the already familiar fitness functions, and we have to include not only the allocation component but also the actual numbers of sexuals, so we use the eqn. (2.6). The inclusive fitnesses for a party I are then given by

$$W_I = N_F g_{FI} v_F/(1 - r) + N_M g_{MI} v_M/r \tag{5.7}$$

where N_F and N_M are the total numbers of females and males produced by the end of the year by the colony (i.e. by the end of its existence). Substituting in the usual relatednesses and sex-specific reproductive values, we obtain the fitnesses of queens and workers. It is worth noting that, because we are examining the conflict case, the queen should lay only haploid eggs in the ultimate cohort. If the numbers of females and males produced in the penultimate cohort are n_F and n_M, the numbers at the end of the season will be n_F and $n_M + n_W\Delta/c_M$, respectively. The fitnesses can then be seen to be

$$W_Q = \frac{rn_F + (1-r)(n_M + n_W\Delta/c_M)}{r(1-r)} \quad (5.8a)$$

$$W_W = \frac{3rn_F + (1-r)(n_M + n_W\Delta/c_M)}{4r(1-r)} \quad (5.8b)$$

Because the denominators are not affected by the investment of an individual colony (except in small populations), the maximization of eqn. (5.8) can be conveniently examined using the numerators alone. The solution of the workers' optimization problem is taken from Bulmer's (1981) results and is presented in Table 5.1.

When the growth factor, Δ, is high, the workers would benefit by raising new workers in the penultimate cohort and males in the ultimate cohort, although the population sex ratio is already too male biased from their point of view (case (a) in Table 5.1). When the growth factor is low, they should

Table 5.1. The optimal strategies for workers in the penultimate cohort of an annual colony when there is conflict over the colony sex allocation due to the population sex ratio r being such that $rc_M < (1-r)c_F < 3rc_M$. The solutions show the proportions of resources workers would favour being spent raising the diploid eggs as females or as workers; if there are not enough diploid eggs to use all the resources available, the remaining resources should be used to raise males (subcases of case (b)).

	Proportion of investment in	
	Females	Workers
(a) $(1-r)(c_F + \Delta - 1) > 3rc_M$	None	All
(b) $(1-r)\Delta c_F > 3rc_M > (1-r)(c_F + \Delta - 1)$		
$D > R$	None	All
$R/c_F < D < R$	$\dfrac{R-D}{c_F-1}$	$\dfrac{c_F D - R}{c_F - 1}$
$R/c_F > D$	All	None
(c) $3rc_M > (1-r)\Delta c_F$	All	None

raise as many females as possible in the penultimate cohort (case (c) in Table 5.1). There is also a narrow region of intermediate values of Δ in which the workers should divide the resources in the penultimate cohort between female and worker larvae depending on the number of diploid eggs (D) laid by the queen and on the resources available (R) to raise these. As noted by Bulmer (1981), important limits are the workers raising the entire penultimate cohort as workers ($D > R$) or as females ($D < R/c_F$).

As can be verified from Table 5.1, the optimal solution from the workers' point of view depends on the growth factor Δ and on the availability of diploid eggs (D). This means that the decision the workers make depends on the strategy followed by the queen. If the growth factor is high, the queen would be selected to lay diploid eggs in the penultimate cohort because these will be raised as workers and the ultimate cohort will then consist of males (because she can then lay solely haploid eggs). But if the growth factor is sufficiently low workers should try to raise the diploid larvae as new queens instead of as workers. Because this outcome is suboptimal for the queen, she should limit the number of diploid eggs laid in the penultimate cohort in order to guarantee resources being spent on her sons. Thus, this model shows, as did the one-brood model, that the colony queen can control the colony sex ratio, and hence that queens can control the population sex ratio. But this transition to protandry involves a cost to the queen: the number of sons produced in the penultimate cohort is smaller than the maximum which could be raised in the ultimate cohort.

The results above were derived in the absence of worker reproduction. If the workers can lay haploid eggs and replace the queen's haploid eggs with their own, the scene alters substantially. To investigate this situation, we now turn to the case, also presented by Bulmer (1981), in which the workers produce a proportion ψ of all the males. Worker reproduction can also cause aggression in the colony, and we follow Bulmer in assuming that the workers have to kill the old queen in order to produce their own sons. If this event happens during the penultimate cohort, the colony will die before it can produce the ultimate cohort. (Strictly speaking of course, in the event of the workers killing the queen in the penultimate cohort, that cohort then becomes the ultimate one.)

Although we have followed Bulmer in suggesting that the workers win their contest with the queen by killing her, this assumption may not be correct. If the workers can either prevent the queen laying haploid eggs, or succeed in destroying all or most of these and in preventing her from destroying their haploid eggs, then the workers not only gain fitness returns through production of their sons but also preserve the colony's life into the phase of producing the ultimate cohort.

If all the sexuals are produced only in the ultimate cohort, the workers should replace all queen-laid haploid eggs with their own eggs, and this leads to an expectation of a 1:1 allocation ratio. This result relies on single mating by the queen, otherwise worker policing is expected to evolve (Section 4.2.2).

The decisions concerning the penultimate cohort form a more complex and interesting situation when there is worker reproduction. The fitnesses can be obtained using eqn. (5.7) by inserting the appropriate relatednesses and noting that the ratio of the sex-specific reproductive values is $\nu_F/\nu_M = 2 - \psi$ (see eqn. 4.1), and by denoting the total number of queen-produced males as N_M and of worker-produced males as $N_M{}^*$:

$$W_Q = \frac{r(2-\psi)N_F + (1-r)(2N_M + N_M{}^*)}{2r(1-r)/\nu_M} \tag{5.9a}$$

$$W_W = \frac{3r(2-\psi)N_F + (1-r)(2N_M + 3N_M{}^*)}{4r(1-r)/\nu_M}\,. \tag{5.9b}$$

If the workers produce males in the penultimate cohort, there will be a struggle between the queens and the workers resulting in the death of the queen, and the colony produces no ultimate cohort. In that case, we need not pay attention to the growth factor, Δ. If the workers refrain from laying eggs in the penultimate cohort, the ultimate cohort should consist of either females or worker-produced males, because the workers will replace any queen-produced haploid eggs with their own. It is interesting to note that in this situation selection will favour recognition of the origin (workers or queen) of brood, and its sex, as soon as possible, thus reducing the loss of resources liable to incur as the doomed individuals are cannibalized.

Which sex should the colony produce in the ultimate cohort? When the males are produced by the workers, the female and male offspring are equally related both to the queen (by $g = 0.5$) and to the workers ($g = 0.75$, assuming monandry). Therefore, the queen and workers have no conflict over the sex ratio in the ultimate cohort and we can neglect relatedness when comparing the fitness returns from male and female offspring, i.e. we compare the returns from males $\nu_M(R/c_M)/r$ and from females $\nu_F(R/c_F)/(1-r)$. If we take the ratio of these values, we obtain a measure of the *relative male advantage*, (RMA), namely the relative advantage in producing males as against new queens, which is

$$RMA = \frac{(1-r)c_F}{(2-\psi)rc_M}\,. \tag{5.10}$$

When the returns from females are the greater, i.e. RMA < 1, the colony should produce females, otherwise it should raise worker-produced males. There are, however, some additional constraints as shown by Bulmer (1981).

RMA < 1. When the fitness returns from females are higher than those from males, the last cohort is expected to consist only of females. The workers are expected to use all the resources in the penultimate cohort to raise workers, to lead to the female production in the ultimate cohort being maximal. If there are not enough diploid eggs, the workers have to raise some males, and if the growth factor Δ is small enough also some queens, in the penultimate cohort.

Table 5.2. Optimal investment from the workers' viewpoint in new queens (n_F), new queen-produced males (n_M), new worker-produced males (n_M*) and new workers in the penultimate cohort of an annual colony's existence when there is conflict between queen and workers. The results are expressed in terms of the numbers of each group produced. A worker costs one resource unit, whereas new queens and males cost c_F and c_M resource units, respectively. Thus investment of R resources in workers raises R of them, but investment of R resources in males would rear R/c_M of them.

		n_F	n_M	n_M^*	n_W
(a) RMA < 1					
$D \geq R$		None	None	None	R
$\Delta > \Delta_L$,	$D \geq D_0$	None	$\frac{R-D}{c_M}$	None	D
	$D < D_0$	D	None	$\frac{R-Dc_F}{c_M}$	None
$\Delta < \Delta_L$,	$D > R/c_F$	$\frac{R-D}{c_F-1}$	None	None	$\frac{Dc_F-R}{c_F-1}$
	$D < R/c_F$	D	None	$\frac{R-Dc_F}{c_M}$	None
(b) RMA > 1					
$D \geq R$		None	None	None	R
$\Delta > \Delta_L$,	$D > D_0$	None	$\frac{R-D}{c_M}$	None	D
	$D < D_0$	None	None	R/c_M	None
$\Delta < \Delta_L$,	$D > \frac{R}{c_F}$ & $D > D_1$	$\frac{R-D}{c_F-1}$	None	None	$\frac{Dc_F-R}{c_F-1}$
	$D < \frac{R}{c_F}$ or $D < D_1$	None	None	R/c_M	None

RMA is the relative male advantage as defined in eqn. (5.10).
Δ_L is the lower limit of the colony growth factor, and is given for case (a) by eqn. (5.11) and for case (b) by eqn. (5.13).
D_0 is the optimal number of diploid eggs from the queen's point of view, and is given for case (a) by eqn. (5.12) and for case (b) by eqn. (5.14).

$$D_1 = \frac{[(c_F + \Delta - 1)\,RMA - c_F]R}{c_F \Delta RMA - c_F}.$$

The optimal worker strategies were solved by Bulmer (1981) and are presented in Table 5.2.

Whether the queen should produce diploid or haploid eggs in the penultimate cohort depends on the growth factor Δ. If Δ lies outside the range

$$c_F - 2\text{RMA}(c_F - 1)/3 < \Delta < 2\text{RMA} \tag{5.11}$$

then the queen should lay diploid eggs which are raised as workers. If Δ is within these limits, the queen should lay both diploid and haploid eggs. The optimal number of diploid eggs from the point of view of the queen is (Bulmer, 1981)

$$D_0 = \frac{\text{RMA}\ R}{3(\Delta - c_F) + (3c_F - 2)\text{RMA}} \tag{5.12}$$

and it would at the same time be optimal for the workers to raise them all as workers (Table 5.2(a)).

RMA > 1. When the fitness returns from males exceed those from females, the workers should replace all queen-laid eggs in the ultimate cohort with their own haploid eggs. In the penultimate cohort, the optimal worker strategies are very much as in the previous case (Table 5.2). Once again the strategy of the queen depends on the growth factor Δ. If Δ lies outside the range

$$\frac{c_F - 2\text{RMA}(c_F - 1)/3}{\text{RMA}} < \Delta < 2 \tag{5.13}$$

then the queen should lay only diploid eggs in the penultimate cohort and these will be raised as workers. When Δ is within the range specified by eqn. (5.13) then the optimal number of diploid eggs for the queen is

$$D_0 = \frac{R}{3\Delta - 2}\,. \tag{5.14}$$

It is optimal for the workers to raise all of these diploid eggs as workers, and to spend any remaining resources on rearing queen-produced males (Table 5.2(b)).

The above analyses indicate that the queen has a very considerable capacity to dominate the sex ratio in annual colonies, even though she may lose control of the ability to produce the males. A major assumption, however, was that the costs of the sexuals, c_F and c_M, are fixed. If workers can increase their investment in new queens, i.e. increase c_F, and the fitness of new queens increases with increased investment in them, then these conclusions should be modified, as we discuss below (Section 5.2.3).

The results in this section show that there is not only a conflict over the investment ratio (*f*:*m*) but also over the allocation of resources to the production of workers. According to Bulmer (1981), the outcome in both parts of the conflict depends critically on the costs of individual females (c_F) and on the colony growth factor, Δ.

5.2.3 *Conflicts in perennial colonies*

The conflict in perennial colonies differs from that of annual nests in that the queen normally outlives many generations of workers. In annual colonies, although there will be a significant attrition rate of the first workers, we could assume that the queen dies, or is even killed by the workers, before the end of the season and that the workers can then profit from laying haploid eggs. The conflict in perennial colonies is strongly asymmetric. If the workers harm the queen, that will be the end of the colony, whereas the queen can harm a worker without any significant loss to the colony's ability to function. For this reason several authors have assumed that the queen should be the main source of haploid eggs (e.g. Trivers and Hare, 1976; Bulmer and Taylor, 1981), because she can simply destroy worker-laid eggs. In large colonies, however, it becomes impossible for the queen to control all the workers in different parts of the nest.

We discuss the model of Bulmer and Taylor (1981) in which it is assumed that all eggs are laid by the queen. This model addresses the question of the compromise between the queen and workers of a monogynous colony. The strategy of the queen is defined by the proportions of haploid and diploid eggs produced. The strategy of the workers is defined by the proportions of males, new queens, and new workers reared from these eggs. If the queen manages to lay only haploid *or* diploid eggs, she can control the allocation ratio as in annual colonies. This has been suggested for ants of the *Formica rufa* group which produce sexuals in the early spring without any overlap with worker production (Pamilo and Rosengren, 1983). Single-sex colonies are common in other ant species (Elmes, 1987b). But in many species the queens cannot lay broods consisting entirely of haploid eggs because that stops worker production and weakens the colony. The queens must therefore lay both types of egg, which gives the workers the possibility to adjust the investment ratios.

Bulmer and Taylor (1981) used simulation based on linear programming to approach the problem, which we outline here. In year t a colony has a worker force consisting of $n_{W,t}$ workers. The resources available next year, R_{t+1}, in this model are given by

$$R_{t+1} = [1 - \exp(n_{w,t}\, u/K)]\, K \tag{5.15}$$

where R_{t+1} refers to the amount of resource at time $t + 1$ (meaning that it takes one resource unit to raise one new worker), u is the maximal growth rate of the colony, and K is the maximum colony size. The queen's preferred investment ratios may be determined using eqns. (5.7) and (5.8a).

The lifetime production, over T years, of a colony is given by

$$N_F = \sum_{t=1}^{T} n_{F,t}\, s^t \tag{5.16a}$$

$$N_M = \sum_{t=1}^{T} n_{M,t}\, s^t \tag{5.16b}$$

where s is the chance of the colony surviving from one season to the next, and $n_{F,t}$ and $n_{M,t}$ are the numbers of females and males produced at time t.

The procedure used by Bulmer and Taylor (1981) to find the evolutionarily stable strategy (ESS) value of r, the proportion of males, was to work backwards from an arbitrary time t. Assume arbitrarily a population sex ratio of r_t. Let $n_{F,t}$, $n_{M,t}$ and R_t for our colony have arbitrary values. The value of R_{t-1} is now known. Select an arbitrary value of D_{t-1}. Given R_{t-1} and the starting value D_{t-1}, determine the workers' investment ratio in queens and males from eqns. (5.8b) and (5.16). Given the workers' strategy, work out the best strategy (in terms of D) for the queen, and take this back to the workers again. When these strategies for time $t-1$ have been computed, proceed back to year $t-2$. Keep proceeding backwards until the proportions of queens and males stop changing (or, in practice, change by some arbitrarily small amount): these values then give the compromise for the chosen population value r. Then repeat for another r until the compromise value of the colony is the same as that for the population, which is then the ESS value. The population investment, M, can be obtained from

$$M = \frac{rc_M}{rc_M + (1-r)c_F} \tag{5.17}$$

Bulmer and Taylor followed this procedure assuming that $s = 0.8$, $u = 10$ and $R_t = K/100$ (the value of the scalar K chosen has no effect on sex ratio). The values obtained by this approach are as follows:

relative cost of a queen (c_F)	1	2	5	10
investment ratio (M)	0.499	0.478	0.443	0.385

If these assumptions are at least moderately realistic, then the 'compromise' is much closer to the queen's optimum ($M = 0.5$) than to that of the workers ($M = 0.25$), although at extreme values ($c_F = 10$) the compromise investment is close to intermediate between them. The increasing power of the workers as c_F increases stems from their increasing power to invest in queens in the face of their mother's limitation of D, because they need convert fewer and fewer of the female larvae from workers to queens to absorb their investment.

Bulmer and Taylor note that deriving the values above required (then) a lot of computer time. Nevertheless, further work is needed because they assumed colonies with potentially infinite lifetimes (despite the setting of $s = 0.8$), so that the finite lifetimes of monogynous social insect colonies require a somewhat different model, intermediate between the eternal colony assumed and the annual model of the previous section. For example, the best answer from an initial run with arbitrary D, n_F, and n_M values could be used as the starting values for another run. Furthermore, computer time is increasingly cheaper, making further such studies more feasible.

Whereas the relative costs of queens and males have often been discussed in empirical studies of sex-ratios, that of workers has not, and is shown by the model above to be important, because both c_F and c_M are given relative to the cost of a worker.

5.2.4 *A general two-locus model*

The conflict over the sex allocation can be interpreted in terms of a queen–worker game in which each party tries to maximize a different payment function. When the genes affecting the behaviour of one party are also carried by members of the other, it is not evident that these payment functions would be the same as the inclusive fitnesses. Matessi and Eshel (1992) examined a two-locus model, one locus expressed in queens and the other in workers, trying to identify genetic equilibria that cannot be invaded by any possible mutation at either locus. Such equilibria will show long-term stability.

The model considered by Matessi and Eshel assumes that two separate loci affect the sex ratio phenotypes of the queen (α) and of the workers (β). These phenotypes determine the colony sex ratio, namely the proportion of investment in females as a function $F(\alpha,\beta)$. The difference between the phenotypes affects also the reproductive success of the colony as a function $S(\alpha,\beta)$. The greater the difference, $\alpha - \beta$, between the queen and worker phenotypes, the greater is the conflict and the lower is the reproductive output. Matessi and Eshel used the model to find values of α and β that form an ESS (see Section 2.1). They further assumed the simplest colony type, namely each colony has a single monandrous queen and all the workers are sterile.

The expected outcome depends on the relative competitive powers of the queen and workers, which are determined according to the functions F and S, and on the behaviour of the reproductive success function at equilibrium. If colony success is affected only slightly by small conflicts between the queen and her workers (mathematically, S is a function with continuous derivatives at its maximum, the case of smooth punishment), the resulting sex ratio will be intermediate betwen the queen and worker optima and there will be a reduction in the reproductive success of the colony. The location of the equilibrium point between the two optima depends on the biological conditions, such as physiological or behavioural possibilities open to the queen and to the workers.

If even a small disagreement between queen and workers strongly decreases reproductive output (i.e. function S has discontinuous derivatives at its maximum point, the case of strong punishment), the sex ratio conflict may be settled at no cost to the colony, provided that the decrease of S is steep enough on each side of its maximum. The resulting sex ratio is likely to lie between the queen's and the workers' optima, although it may also lie outside this range (although only in the case of extreme sensitivity to conflict).

This genetic model, which we have not presented in detail here, indicates that a number of equilibria between the two optima may occur as an evolutionary compromise and that the observed variation in actual sex ratios recorded in natural populations could be explained as reflecting different values of the competitive abilities of queens and workers, and of the sensitivity of colony reproductive success to conflict.

5.3 Conflict over sexual production and colony maintenance

5.3.1 *Sexual production versus maintenance in monogynous colonies*

Individuals in perennial colonies should also decide how to divide the available resources between sexual production and worker production for colony maintenance. We consider here a model of Pamilo (1991a) for a species with nest founding by independent females. To make the treatment simple, we further assume monogyny and monandry. It is intuitively clear that a successful queen wants to continue reproduction also in the future. She can do that if resources are invested in the maintenance and growth of the colony. Workers, instead, should be interested in seeing their highly related sisters as new reproductives. One option, which we do not discuss here (but see Section 4.4.3 and eqns (4.29)), is queen supersedure in the colony. Alternatively, workers might simply benefit from investing a large fraction of resources in new highly related sexuals and care less for the maintenance of the old mother queen.

The optimization problem is to divide the colony's total resources (R) between the production of workers (w), new alate females (f) and males (m), with $w + f + m = 1$, in a way which gives the best representation in the asymptotic gene pool. When neither local mate competition nor local resource competition occurs, the fitness returns through sons and daughters can be considered as linear functions of the investment. The relative fitness through male function of colony i is m_i/M and through females is f_i/F, where F and M refer to the population means.

Being perennial, the colony can survive to the next season with a probability which can be given as a function of w. We denote the probability of surviving over the season of sexual production as $s(w)$, where w is the amount of resources invested in colony maintenance simultaneously with the production of new sexuals. If a constant, α, is used to denote the success of colony foundresses in such a way that the daughters produce α new nests when all resources are invested in daughters (i.e. when $f = 1$), then the proportions of old and new nests in the population are in the next generation

$$\text{old nests} \quad \frac{s(W)}{s(W) + \alpha F}$$

$$\text{new nests} \quad \frac{\alpha F}{s(W) + \alpha F}. \tag{5.18}$$

Taking into account the genetic relatednesses, reproductive values, relative successes of old and new nests and the relative fitnesses within each class of individuals, the inclusive fitness of the queen is

$$V_Q = \frac{g_{QQ}\nu_F s(w_i)}{s(W)+\alpha F} + \frac{g_{DQ}\nu_F \alpha f_i}{s(W)+\alpha F} + \left[\frac{\alpha F}{s(W)+\alpha F}\right]\left[\frac{g_{MQ}\nu_M m_i}{M}\right] \tag{5.19}$$

where g_{QQ} is the relatedness of the queen to herself, equal to 1, g_{DQ} is the relatedness of daughters to the queen, and g_{MQ} is the relatedness of males to the queen.

Because the sexual offspring contribute only through new nests, the terms related to the production of sons and daughters in eqn. (5.19) are weighted by the relative success of new nests, as obtained from eqn. (5.18).

The population values αF and $s(W)$ can be considered fixed, and the fitnesses given by eqn. (5.19) are maximized by maximizing

$$\frac{g_{QQ}\nu_F s(w_i)}{\alpha F} + \frac{g_{DQ}\nu_F f_i}{F} + \frac{g_{MQ}\nu_M m_i}{M}. \tag{5.20}$$

At equilibrium, the fitness returns through all three components (old queen, daughters, sons) should be equal in the sense that a small change in investment ratios does not change the mean reproductive returns. This equilibrium point can be found by taking derivatives of eqn. (5.20) and making pairwise comparisons between the three terms.

Sons versus daughters. The last two terms are the same as in eqn. (2.8), and the solution is $F = M$.

Maintenance versus sexual brood. By taking derivatives of the first two terms of eqn. (5.20), we obtain the condition

$$\frac{g_{QQ}\nu_F s'(W)}{\alpha F} = \frac{g_{DQ}\nu_F}{F}$$

or

$$s'(W) = \frac{\alpha g_{DQ}}{g_{QQ}}. \tag{5.21}$$

Under queen control of investment, this becomes $s'(W) = \alpha/2$. Note that this result is unaffected by worker reproduction.

Under worker control of investment we insert the relatednesses of the sexual offspring to the workers into eqn. (5.20). In the first term, however, we have to replace the product $g_{QQ}\nu_F$ by the sum $g_{DW}\nu_F + g_{MW}\nu_M$, because not only is the old queen related to the workers but so is the sperm in her spermatheca. We therefore have to use the relatednesses and reproductive values of her offspring. Inserting the appropriate numerical values in eqns. (5.20) and (5.21), we obtain

$$M = \frac{F}{3}$$

$$s'(W) = \frac{3\alpha}{4} \tag{5.22a}$$

when $\psi = 0$, and

$$M = F$$

$$s'(W) = \frac{\alpha}{2} \tag{5.22b}$$

when $\psi = 1$. The conditions with $\psi = 1$ are the same as under queen control, which is the general result for most cases of queen–worker conflict. We later assume that $\psi = 0$, i.e. the queens produce all males in the population.

To examine the results given by eqns. (5.21) and (5.22), we introduce a specific survivorship function

$$s(W) = \frac{1 - e^{-cW}}{1 - e^{-c}} \tag{5.23}$$

where W is the proportion of resources invested in colony maintenance, and c is a constant. This model assumes that all colonies have the same amount of resources (R). The function $s(W)$ is shown for various values of c in Fig. 5.4. As c increases survival becomes close to a threshold function: a certain level of resource is sufficient to guarantee survival and additional resources are optimally assigned to reproduction. With $c = 1$, on the other hand, there is an almost linear relationship between the probability of survival and the resources allocated to maintenance over the whole range of possible resource allocation. With these notations, we obtain the equilibrium values, W^*, for the proportion

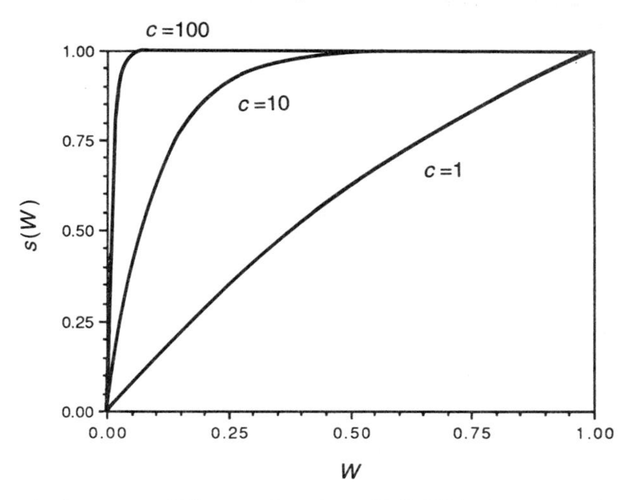

Fig. 5.4 Survival curves, $s(W)$, as a function of W (W is the proportion of resources invested in colony maintenance) for three different values for the constant c in eqn. (5.23).

of resources invested in workers from eqns. (5.21) and (5.23) under queen control

$$W^* = -\frac{1}{c} \ln\left[\frac{(\alpha/2)\,(1 - e^{-c})}{c}\right] \tag{5.24a}$$

and from eqns. (5.22) and (5.23) under worker control

$$W^* = -\frac{1}{c} \ln\left[\frac{(3\alpha/4)\,(1 - e^{-c})}{c}\right] \tag{5.24b}$$

Some numerical values are given in Table 5.3.

We have already discussed at some length, and will do so again, that there is an expected conflict between queen and workers over the allocation ratio between new queens and males. What is new about the model in this section is that it demonstrates that there may also be conflict between queen and workers over the splitting of colony resources between maintenance and reproduction. Equations (5.24) and the values in Table 5.3 indicate that the queen always prefers a greater allocation of resources to colony maintenance than is selected for at the worker level. The differences in the preferred allocations to males and new queens follow the expectations established elsewhere in this book of a 1:1 allocation being selected for at the queen level but a 3:1 allocation at the worker level.

Table 5.3. Expected allocation of resources in colony maintenance and sexual production under queen (Q) or worker (W) control in perennial monogynous colonies with monandrous queens according to the model of Section 5.3. c is the coefficient in the survivorship function (see eqn. (5.23) and Fig. 5.4), and α is the number of new nests successfully founded by daughter queens if all resources are invested in daughters (from Pamilo, 1991a).

α	c		Maintenance	Daughters	Sons	% new nests
1	1	Q	1.000	—	—	0
		W	0.746	0.191	0.063	19
	10	Q	0.300	0.350	0.350	27
		W	0.259	0.556	0.185	38
	100	Q	0.053	0.478	0.478	32
		W	0.049	0.713	0.238	42
2	1	Q	0.459	0.270	0.270	48
		W	0.053	0.710	0.237	95
	10	Q	0.230	0.385	0.385	46
		W	0.190	0.608	0.202	59
	100	Q	0.046	0.477	0.477	49
		W	0.042	0.718	0.240	59

The conflict between queens and workers affects the turnover rate of colonies, in that the greater allocation of resources to reproduction under worker control than under queen control increases the proportion of colonies which are new each year. The course of this conflict would be expected to be mediated by the timing of production of reproductives as against workers, and by caste determination, and also to affect these life history factors. For example, in species where the reproductives are produced as a separate brood the development of which overlaps with the development of worker brood (i.e. the two castes compete to some extent for the same resources), the queen in a monogynous colony has the option of limiting the number of eggs entering the sexual brood. In a polygynous colony, on the other hand, a queen producing fewer eggs for the reproductive brood could be cheated against by other queens producing more at that time.

The results of the model suggest that colony turnover, proportion of resources allocated to maintenance, and the sex allocation ratio, are all connected to the life-cycle parameters affecting the outcome of queen–worker conflict over resource allocation. It could therefore be predicted for example that the longevity of colonies with separate sexual broods would be greater than those of related species with simultaneous worker and sexual production. Data on the longevity of monogynous colonies in natural populations are scarce (Wilson, 1971:440): *Solenopsis invicta* queens (and their colonies in the monogyne form) have an effective lifespan of about 6 years (Tschinkel, 1987), *Pogonomyrmex owhyeei* colonies have a mean lifespan of 17 years (Porter and Jorgensen, 1988), and those of *Formica exsecta* have one over 20 years (Pamilo, 1991d). There are more data from laboratory cultures (Hölldobler and Wilson, 1990) but the maximum lifespans in captivity cannot be safely extrapolated to the field.

A further, complex, case is analysed by Bourke *et al.* (1988). In *Harpagoxenus sublaevis* queens live a maximum of 10 years and workers 3 years, leading Bourke *et al.* to predict a proportion of orphaned colonies of 3/13 = 23.1%. Of 47 colonies collected in 1985, a queen was not found for 14 (29.8%), which Bourke *et al.* note is not significantly different from 23.1%. Moreover, some of the colonies for which queens were not found would in fact have had one which eluded capture. A second estimate comes from rearing the brood taken with the colonies. In the 30 cases where this could be done (for 20 colonies captured with a queen and 10 without), the brood from all 20 colonies captured with a queen yielded both female and male offspring as did that from six of the colonies captured without, indicating the presence of a queen in the field (the remaining four yielded only male brood). This yields an estimate of 13%, which is also not significantly different from 23.1% but in better agreement with an orphaned period of under 2 years if the estimate of the queen-right period is correct. But it seems likely that orphaned colonies usually survive for at least a year. Taking 1 year as the survival period for orphaned colonies yields a mean longevity of queenright colonies of 6.7 years. A further complication is that *H. sublaevis* is a slave-maker, so that the efforts of both queen and workers to modify the colony's allocation to maintenance have to be channelled through the slave workers.

5.3.2 *Resource allocation when colonies accept new queens*

In the previous section we examined a case in which it was easy to distinguish between resource allocation to sexual production and resource allocation to colony maintenance or growth. The distinction was clear because it was assumed that all the sexuals leave the colony to disperse and hence the production of workers is completely devoted to colony maintenance. The situation changes when the workers not only maintain the old queen but also aid the new queens. This is the case when the new queens are assisted in colony foundation by a worker force derived from their natal nest (colony fission: discussed below in Section 5.4) or when the colonies are polygynous and accept new queens. Trivers and Hare (1976) realized that when nests recruit their own daughters as new reproductives these daughter queens require a share of the colony's workers (which should therefore be regarded as investment in female function), leading to an expectation of male-biased sex ratios when only the sexuals are included in the calculation of investments. The effects of such recruitment were analysed by Pamilo (1990b), and we follow his presentation here.

As above, assume that the resources in a colony are divided three ways, w in workers, f in new females, and m in males ($w + f + m = 1$). The sexual allocation, when measured only from the point of view of the sexuals, is then given as $m/(f + m)$. We use this measure, which is independent of the proportion of resources allocated to the workers, because it can be difficult in practice to estimate what fraction of w is used to maintain the old queens and what fraction is used to support the new ones. Pamilo's model also has the restrictive assumption that the investments m and f refer only to those sexuals which disperse from the natal colony. If the new females that are recruited back to their original colony form a small part of all females, the model holds approximately for all sexuals.

The expected sex allocation depends on various factors: on whether the queens or the workers control allocation, on the genetic heterogeneity of colonies, and on the relative success of the new queens that either disperse or stay in the natal colony after they have mated.

It is convenient to start by examining the situation where there is no recruitment at all, and in which all new females leave the colony. As shown already, in deriving eqn. (2.10), the optimal sex allocation under this circumstance will be

$$M^* = \frac{g_{MI}v_M}{g_{MI}v_M + g_{FI}v_F} \tag{5.25}$$

where g_{MI} and g_{FI} are the relatednesses of new males and females to the controlling individuals I (either queens or workers), and v_M and v_F are the sex-specific reproductive values of males and females respectively.

Under queen control, the predicted investment ratio is 0.5. If the workers produce none of the males, the ratio of the sex-specific reproductive values is $v_F/v_M = 2$, and the relatednesses of the offspring to the workers in the nest are

$$g_{FW} = \frac{3 + (n-1)g_{QQ}}{4n}$$

$$g_{MW} = \frac{1 + (n-1)g_{QQ}}{2n}$$

where n is the number of coexisting old queens and g_{QQ} is the relatedness among them. In deriving these values we also assume that the queens are monandrous and that they share the egg-laying evenly. With these assumptions, the investment ratio under worker control is expected to be

$$M^* = \frac{1 + (n-1)g_{QQ}}{4 + 2(n-1)g_{QQ}} . \tag{5.26}$$

If the queens are not related ($g_{QQ} = 0$), the sex allocation is expected to be 0.25 (Frank, 1987c). If the queens are highly related to each other ($g_{QQ} >> 1/n$), the sex ratio should approach 0.5 with increasing n (Trivers and Hare, 1976). If the relatedness of the queens is an inverse ratio of their number ($g_{QQ} \approx 1/(n-1)$), the sex allocation is expected to be about 0.33 (Pamilo, 1990b). But these predictions do not hold when some of the new queens are recruited back into their natal colony. The positive estimates of g_{QQ} found in genetic studies (Table 4.8) indicate that such recruitment does occur, and we have to take into account its effects on resource allocation.

When the queens recruited to a colony include its own daughters, the optimal allocation depends on the relative successes of dispersing and stay-at-home queens. The investments m and f now refer to the dispersing sexuals. We also assume that no local mate competition or local resource competition occurs among related same-sex offspring.

Let the probability that an old nest survives to the following year be S, and let the overall reproductive value of such a nest be v_o. This reproductive value is proportional to the probability that a randomly chosen gene in a remote future generation originated in this nest. Assume further that the dispersing daughters establish new nests, the reproductive value of each new nest being $v_n = C_v v_o$. If, on average, C_n new nests are established when all resources are lavished on daughters (i.e. $f = 1$), then the combined success of new nests is fCv_o, where $C = C_v C_n$. The proportional representation in the population of nests is therefore $Cfv_o/(CFv_o + Sv_o)$, with F being the mean investment in females in the whole population. The contribution to the inclusive fitness of the old colony members through dispersing daughters is thus

$$\frac{g_{FI}\nu_F C_f}{CF + S}. \tag{5.27a}$$

Males contribute to future generations by inseminating either dispersing queens or the queens staying at home as new reproductives. Let p be the proportion of such new recruits among queens of an old colony. The relative contribution of new queens to the whole population is $(CF\nu_o + pS\nu_o)/(CF\nu_o + S\nu_o)$. The contribution to the inclusive fitness of individual I through male function of a colony allocating a proportion m of its resources to the production of sons is

$$\frac{g_{MI}\nu_M(m/M)\,(CF + pS)}{CF + S} \tag{5.27b}$$

where M is the mean investment in males in the whole population.

It is convenient to combine the parameters C, p, and S into a single one (Pamilo, 1990b):

$$d = \frac{CF}{CF + pS} \tag{5.28}$$

where the numerator and the denominator give the contributions by new queens in the gene pool; the numerator gives the contribution of dispersing queens and the denominator that of all the new queens. The expected equilibrial sex ratio can then be found by taking derivatives of eqns. (5.27a) and (5.27b) with respect to f and m. Setting these derivatives equal yields

$$M = \frac{g_{MI}\nu_M(CF + pS)}{g_{FI}\nu_F C}$$

and the expected proportions of males among all sexuals is

$$\frac{M}{M + F} = \frac{g_{MI}\nu_M}{g_{MI}\nu_M + dg_{FI}\nu_F}. \tag{5.29}$$

When the parameter $d = 1$, this ratio equals that given by eqn. (5.25), and when $d = 0$, this ratio approaches the value of one. This latter result means that when dispersing females are unsuccessful, nests should only produce a small number of new queens (which are all recruited back) and use remaining resources to produce workers and males.

Inserting the relatednesses of sons and daughters to queens in eqn. (5.29) gives us the expected allocation ratio under queen control:

$$\frac{M}{M + F} = \frac{1}{1 + d} \tag{5.30}$$

which is independent of the proportion of worker-produced males (Pamilo, 1990b).

The relatednesses of the sexual offspring to a randomly chosen worker also depend on the number of matings of each queen, i.e. on how many patrilines she produces (actually, the effective promiscuity given by eqn. (4.15)), and on the proportion of worker-produced males. We denote these quantities k and ψ. Inserting the appropriate relatedness values into eqn. (5.29) gives, under worker control (Pamilo, 1990b)

$$\frac{1 + 2/k + (n-1)g_{QQ} - (4/k)(1-\psi)(2-\psi)}{(1+d)[1 + 2/k + (n-1)g_{QQ}] - (4/k)(1-\psi)(2-\psi)}. \qquad (5.31a)$$

If we set $d = 1$ in eqn. (5.31a) we obtain the prediction under worker control for the case where no daughters are recruited, i.e. all new queens disperse. If we set both $d = 1$ and $k = 1$ (i.e. monandry), the formula devolves to that derived by Benford (1978:eqn. 29).

The real aim of deriving such a model, however, is to extend theory to cases when some of the daughter queens are recruited back into their natal colony, i.e. $d < 1$ (Pamilo, 1990b). If the workers produce all of the males ($\psi = 1$), or if the number of matings is large, then (5.31a) can be simplified to

$$\frac{M}{M+F} = \frac{1}{1+d} \qquad (5.31b)$$

which is independent of n and of g_{QQ}. This is, of course, the same result as obtained under queen control in eqn. (5.30).

When the queens produce all of the sexual offspring ($\psi = 0$), the expected investment ratio under worker control depends on the relatedness among the old queens. The results can be roughly generalized as follows (Pamilo, 1990b):

$$g_{QQ} >> \frac{1}{n} \text{ yields } \frac{M}{M+F} = \frac{1}{1+d} \qquad (5.31c)$$

$$g_{QQ} \approx \frac{1}{n} \text{ yields } \frac{M}{M+F} = \frac{1}{1+(1+1/k)d} \qquad (5.31d)$$

$$g_{QQ} << \frac{1}{n} \text{ yields } \frac{M}{M+F} = \frac{1}{1+(1+2/k)d}. \qquad (5.31e)$$

Of these equations, (5.31d) is especially important because the recruitment of daughters is expected to lead to a relatedness between queens of $g_{QQ} = 3/(3n+1)$ when there is no reproductive dominance between the queens (Pamilo and Varvio-Aho, 1979). With single-mating of queens, the predictions from eqns. (5.31d) and (5.31e) reduce to $1/(1+2d)$ and $1/(1+3d)$ respectively.

These predictions are pleasingly simple. Unfortunately we lack data so far on the values taken by the parameter d in nature.

The higher is the proportion of successfully dispersing new queens (i.e. d is large), the closer becomes the expected ratio from eqn. (5.29) to that given by eqn. (5.25). When dispersal is risky, a high proportion of new queens will be recruited back into their natal colonies, and in that case (d is small) it is profitable to produce both workers maintaining the old nest and males inseminating the queens recruited, but it does not pay to produce surplus females to those which can be added to the colony if their fate will be to perish on dispersal.

All the above equations predict a male-biased sex allocation ratio among the sexual offspring. One factor that would reduce this male bias is the recruitment of *unrelated* queens to colonies. Pamilo (1990b) showed that eqn. (5.29) holds in this case if we redefine the dispersal parameter, d, as

$$d = \frac{cF + p_2 S}{cF + p_1 S + p_2 S} \tag{5.32}$$

where p_1 and p_2 are the frequencies of new queens recruited by a colony that are its own daughters or unrelated females, respectively. If p_2 is not negligible, dispersal becomes profitable. Recruitment of unrelated new queens also lowers the average relatedness among queens which, under worker control, should further reduce male bias.

5.4 Sex allocation in species with colony fission

The previous section, 5.3, showed that there are cases in which the allocation problem includes not only sexuals but also workers, and that the workers can even form part of the allocation into sexual function. In that section we dealt with models in which the workers maintained the old colony and also helped new queens that were recruited by it. Workers can also take part in establishing new nests, because eusocial insects often found new nests by fission or by budding from established ones. Fission is a process whereby a colony splits to form two or more daughter colonies that become functionally independent (Franks and Hölldobler, 1987). This is the situation, for example, in both stingless and honey-bees (Michener, 1974:131–7; Seeley, 1985:36–8), in the Polybiini and Ropalidiini wasps (Jeanne, 1980), and in army ants (Schneirla, 1971), as well as in some primitively social bees and wasps in which several females establish a new nest in which some of the females remain in the role of helpers (Schwarz, 1986). A different situation is found in many polygynous species of ants, wasps, and termites, which can produce new colonies by budding (Rosengren and Pamilo, 1983; Roisin and Pasteels, 1986). In this process the nests normally remain in close contact with each other and form polydomous colonies (Section 4.5.3).

What kind of resource allocation is expected in such populations? The worker force and the nest structures inherited by the daughter colonies are big investments and should be taken into account. We will next examine some models of colony fission, following the approaches of Bulmer (1983b) and Pamilo (1991a).

The species which reproduce by colony fission (swarming) generally produce a much greater weight of new males as against new queens. This is the case for example in army ants, in which several new queens are produced at one time and the colony divides between the old queen and one of the newcomers (the rest are excluded and die). Hamilton (1975) noted that the workers in fissioning colonies should be counted as an investment in females when they support the new queens heading the daughter colonies. This idea was formalized by Macevicz (1979). Craig (1980) offered a different explanation, suggesting that the colonies should produce only a few new queens to head the daughter colonies and produce as many males as possible. This view has been interpreted as local resource competition (Bulmer, 1983b). The main difference between the two hypotheses is that the Hamilton–Macevicz hypothesis implies that the optimal allocation ratios are not affected by the mode of colony foundation, whereas under the Craig–Bulmer hypothesis, no single optimum exists but the sex allocation depends on the colony's size. As emphasized by Pamilo (1991a), it is important to distinguish between these hypotheses, because they actually refer to different situations, colonial fission and budding.

5.4.1 *Colonies fissioning in a regular manner*

Let us first view an example with a very simple colony growth pattern. Assume that a population has colonies, each of size R at the beginning of the growing season (R refers to worker resources). Each nest then grows deterministically to size kR and reproduces. Reproduction takes place in the way that a fixed fraction, M, of resources is devoted to new males and the rest evenly to $(1 - M)k$ daughter colonies each headed by a single daughter queen from the maternal nest. Each daughter colony thus receives the amount R of resources. Under the hypothesis of optimal investment in two sexes, we should expect that under queen control $M = 0.5$.

Let us assume that the resources are indeed divided with $M = 0.5$, and examine what happens to queens which depart from this investment ratio. We use a simplified version of Frank's (1987b) approach. The inclusive fitness of a queen with a variant investment allocation, m, will be

$$g_{MQ}\left[\frac{mkR}{MkR}\right] + 2g_{DQ}\left[\frac{(1-m)kR}{(1-M)kR}\right]$$

where g_{MQ} and g_{DQ} are the relatednesses of the queen's sons and daughters to her, and the sex-specific reproductive values have been calculated in the absence of worker reproduction. Substituting in the usual values for the relatednesses yields the queen's inclusive fitness under these 'standard' conditions:

$$\frac{m}{M} + \frac{1-m}{1-M}\ .$$

The ESS sex allocation is, as usual, that which has a higher fitness than any variant. This formulation is the same as that considered previously (Section 2.1), and yields the expected result that $M = 0.5$ is an ESS. Substituting the relatednesses to the workers into the equation yields $M = 0.25$ as the ESS.

However, not all colonies are equally large, and the resources are not always easily divided between the daughter colonies. For example, if the minimum viable size of a colony requires R resources and the size of a colony at the time of division is, say, $3R$, we have a problem. Equal investment would require allocating $1.5R$ in sons and $1.5R$ in one daughter colony. One alternative would be that the colony divides into two daughter colonies, R each, and spends the remaining R in male production. The best answer depends not only on the investment ratio but also on the survival and growth rates of the daughter colonies of different sizes. If a larger daughter colony has a considerably higher survival probability, it might be good to invest in one large daughter colony. On the other hand, if a small colony has a much higher growth rate (exponential growth), it might be selectively advantageous to produce two small daughter colonies. The main point is that a unit investment in sons gives a fixed fitness return, but the return from a unit investment in daughters is a variable depending on the size of the daughter colonies.

We can examine separately two cases depending on how the colony divides.

Case 1. In this case, a colony of size R_T divides into two daughter colonies headed by new queens which are daughters of the old mother queen. The old mother queen dies at the time of nest division. The investment ratio of a colony with a rare mutant allele, a, is examined. Assume that the queens of genotype AA invest a proportion M of the available resource in males ($R_M = MR_T$) and the remainder, $(1-M)R_T = R_d$ in workers for the daughter colonies, whereas the queen genotype Aa invests m in males. When the allele a is a rare mutant, we do not have to bother about the aa queens. Let p be the frequency of type of $AA \times a$ colonies and q the frequency of $AA \times a$ colonies; both p and q are very small. The linearized recursion relations for the frequencies of the rare colony types are given (approximately) by

$$\begin{bmatrix} p' \\ q' \end{bmatrix} = \mathbf{A} \begin{bmatrix} p \\ q \end{bmatrix}$$

where p' and q' are the new frequencies and the matrix $\mathbf{A}$ is of form

$$\mathbf{A} = \begin{bmatrix} 0 & \dfrac{m}{2M} \\ 1 & \dfrac{s(R_T(1-m)/2)}{2s(R_T(1-M)/2} \end{bmatrix}$$

and $s(.)$ is the value of the survival function. The equilibrium value of M (M^*) is found to satisfy the equation

$$R_T M^* = R^*_M = \frac{2s}{s'} \tag{5.33}$$

where both the survival function s and its derivative s' are evaluated at $(R_T - R_M{}^*)/2$ (Bulmer, 1983b).

Alternatively, we could use the approach used when deriving the eqn. (2.18). The fitness returns from sons are directly proportional to their number and the fitness function is linear, $h(M) = \gamma R_M$, where γ is a constant. Each colony, when it divides, is assumed to divide into two with initial sizes of $R_O = \frac{1}{2}R_T(1-M)$.

We then apply eqn. (2.18), and because there are two daughter colonies, the equilibrium is solved from

$$\frac{s'(R_0)}{s(R_0} = \frac{2h'(M)}{h(M)} = \frac{2}{R_{\mathrm{M}}}$$

which gives the same solution as in eqn. (5.33) obtained using Bulmer's (1983b) approach.

Under worker control and assuming that the queen produces all males, we get from eqn. (2.18) the solution for optimal investment

$$R^*_{\mathrm{M}} = \frac{2s(R_0)}{3s'(R_0)} \,. \tag{5.34}$$

More generally, if a proportion ψ of males is produced by workers, we get from eqn. (2.18) and from the values of g and ν derived in Section 5.1,

$$R^*_{\mathrm{M}} = 2\,\frac{(0.5 + \psi/4)\,[1 + \psi/(3 - \psi)]}{0.75[2 - \psi/(3 - \psi)]}\;\frac{s}{s'} \tag{5.35}$$

under worker control. The reader can verify that, under queen control, eqn. (5.33) holds for all values of ψ.

We need to know the survival function, s, in order to get numerical solutions to the resource allocation. We use for this the model suggested by Bulmer (1983b). The colonies must be larger than a minimum size, R_{min}, before they can survive and above it the fate of a colony depends on a survival function $s(R_0)$, where R_0 refers to the initial size of the worker force in a nest. Bulmer used

$$s(R_0) = 1 - e^{-\alpha(R_0 - R_{\mathrm{min}})} \quad (\text{with } R_0 > R_{\mathrm{min}}) \,. \tag{5.36}$$

The survival function is shown in Fig. 5.5 using the numerical values $\alpha = 0.001$ and $R_{\mathrm{min}} = 1000$ also used by Bulmer.

Some numerical results for resource allocation are given in Fig. 5.6 using the above survival function (5.36) with the above parameter values. As Bulmer (1983b) noted, and as can be verified from Fig. 5.6(a), there is, under queen control, an equal investment in sons and daughters at the point where $s(R_0) = s'(R_0)R_0$. This result is easily seen from eqn. (5.33). The increase in the probability of survival declines with size.

This model assumes that the survival of a colony is a specific function of the colony size. It then assumes that the colony always divides in two and the equilibrium is solved for a situation where each colony has at the time of fission the same size R_{T}. These assumptions are easily violated in nature and the numerical solutions will be different. But Bulmer's point is important: the success of the daughter colonies is a variable depending on their initial size. Therefore we cannot apply the approach of Section 2.2 based on linear fitness functions, but rather we should determine the fitness functions of the daughter colonies and use eqn. (2.18).

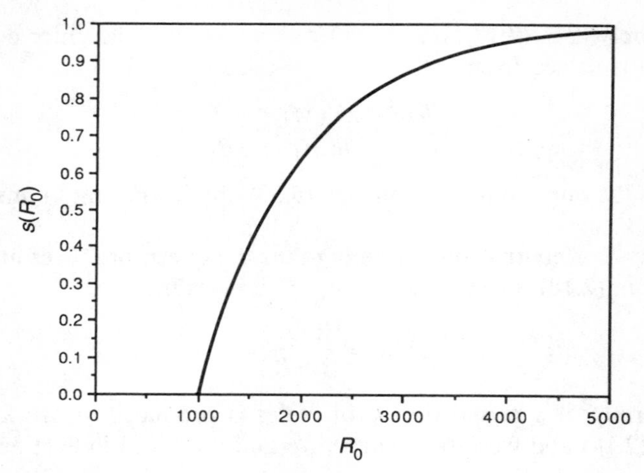

Fig. 5.5 Relationship between survival probability $s(R_0)$ and R_0, as given by eqn (5.36), with $\alpha = 0.001$ and $R_{\min} = 1000$.

Case 2. We next examine the same model in the case where the mother colony continues to function but produces one new daughter colony by fission. The question now concerns not only the proportion of resources invested in sons, but also the division of the worker force between the mother colony and the daughter colony. Bulmer (1983b) used the recursion method for this problem, while Pamilo (1991a) applied the inclusive fitness approach.

Let us assume that a colony always divides in two: one colony has the old queen and the daughter colony has a queen which is a daughter of the old queen. Each colony is strictly monogynous and the queens are monandrous. The problem of allocation is between males, workers associating with the old mother, and workers following the new queen to the daughter colony. We have to remember in this case that workers of the initial nest are related not only to the queens of the two colonies but also to the sperm in the old queen's spermatheca. The situation is the same as that presented in Section 5.3. Therefore, we have, under worker control, to use the relatednesses of the offspring produced in the mother colony. If the workers are divided R_Q with the old queen and R_D with the daughter queen, we can first solve the distribution of the worker force between the two colonies. This can be solved under worker control from

$$(g_F v_F + g_M v_M) \frac{s'(R_Q)}{\bar{s}} = g_F v_F \frac{s'(R_D)}{\bar{s}}$$

where g_F and g_M are the relatednesses to the workers of the female and male offspring produced by the old queen. Assuming that the queen lays all male eggs, worker control gives a relationship

$$R_D = R_Q - \frac{1}{\alpha} \ln(2)$$

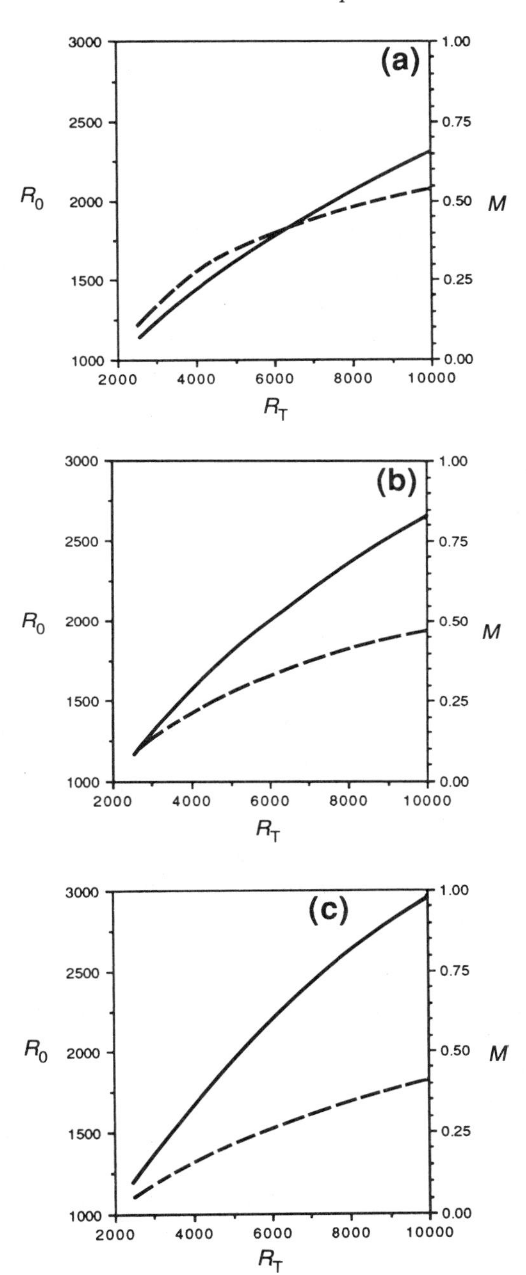

Fig. 5.6 Optimal values of R_0 (solid curves, allocation of resource units to each of two daughter colonies) and M (dashed curve, proportion of resources invested in males) for various colony sizes, R_T, under (a) queen control, (b) worker control with half the males produced from worker-laid eggs, and (c) worker control with all reproduction by the queen. Based on eqns. (5.33), (5.34) and (5.35), and on survival function (5.36) with $\alpha = 0.001$ and $R_{min} = 1000$ (Fig. 5.5).

where α is the growth parameter of eqn. (5.36). Under queen control, we can use the relatedness of the queen to herself, and the relationship becomes

$$R_D = R_Q - \frac{1}{\alpha}\ln(4/3)$$

as also given by Bulmer (1983b).

The equilibrium of sex allocation can be now solved by equalizing the returns from the daughter colony and from the males:

$$g_F v_F \frac{s'(R_D)}{s(R_D)} = g_M v_M \frac{1}{R_M}$$

and

$$R_M = \frac{s(R_D)}{s'(R_D)}$$

under queen control, and

$$R_M = \frac{1}{3}\frac{s(R_D)}{s'(R_D)}$$

under worker control (Pamilo, 1991a).

Examples of equilibrium investments for various colony sizes R_T ($R_T = R_Q + R_D + R_M$) are given in Table 5.4.

The models in Table 5.4 differ slightly from those given by Bulmer (1983b) who examined only a mixed model where workers decide how to divide themselves between the two colonies, whereas the old queen controls the proportion invested in males.

The total investment in males in Case 2 is slightly less than in Case 1 where two daughter colonies were formed. But if the parental investment is calculated using only the males and the daughter colony, investment in males is larger than in Case 1.

Bulmer (1983b) suggested that it is not possible to predict the population ratios because these are overruled by the optima of individual colonies. However, Pamilo (1991a) showed that this conclusion must be altered if we relax the limitations of the models. First, it was assumed that all colonies are of same size, i.e. both s' and s were evaluated at the same point R_D. Second, the treatment forced each colony always to divide in two, independently of the colony size. In real life, the size of the colonies varies and, depending on the colony size, it may be better not to divide at all or sometimes to make more than two daughter colonies.

5.4.2 *Evolution of colony fission patterns*

The above models, based on those of Bulmer (1983b) and Pamilo (1991a), assumed that all colonies fission in the same pattern, independently of the colony size. This may not be the case in reality, but small colonies may not

Table 5.4. The equilibrium pattern of colony fission when the colonies of size R_T always divide in two: one new daughter colony and a colony headed by the old queen. The fitness function is given in Fig. 5.5, and it is assumed that the queen produces all males (i.e. $\psi = 0$). R_Q is the size of worker force remaining with the old queen, R_D is the worker force associating with the new queen, and M is the amount of resources invested in male production, so that the total resource invested in males is $R_M = MR_T$.

Controller of		R_T	R_Q	R_D	M
Sex allocation	Division of workers				
Queen	Queen	2500	1542	849	0.04
		4000	2025	1332	0.16
		8000	2955	2262	0.35
		12000	3523	2850	0.47
Queen	Workers	2500	1328	1040	0.05
		4000	1816	1528	0.16
		8000	2756	2468	0.35
		12000	3350	3062	0.47
Workers	Workers	2500	1360	1072	0.03
		4000	1962	1674	0.09
		8000	3182	2894	0.24
		12000	3936	3648	0.37

fission at all whereas large colonies may produce more than two daughter colonies.

Let us go back to the model where colonies always produce two daughter colonies and the survival function is that given by eqn. (5.36). The optimal fission patterns were shown in Fig. 5.6. We note that under queen control, a colony of size $R_T = 6000$ will produce two daughter colonies of size $R_0 = 1792$ each, and invests $R_M = 2416$ in males. This means that an investment of 2416 units in sons gives the same fitness return as an investment of $2 \times 1792 = 3584$ in daughter colonies. If there is a rare colony type which does not fission but invests all resources in male production, this new type would spread in the population until the increased male production will devalue the males to the extent that a balance is achieved.

We can see that the critical point under queen control occurs where the resources are evenly divided between male production and daughter colonies. In the model used to calculate Fig. 5.6(a), this point is given at $R_T = 8584$. At that point, individual colonies can either produce new daughter colonies of size $R_0 = 2146$ or invest in males, or do both without affecting fitness. If the population is monomorphic for a fission pattern giving two daughter colonies and the colony size at fission is $R_T = 10\,000$, we see from Fig. 5.6(a) that an investment of $2 \times 2307 = 4614$ in daughter colonies gives the same genetic return as an investment of 5386 units in sons. Clearly, any mutant colony type splitting into three or four daughter colonies will have an evolutionary advantage, provided that there are no other costs in producing many daughter colonies.

The formal presentation can be based on eqn. (2.16). Namely, the controlling individuals should maximize the function

$$V_{\mathrm{I}} = \frac{g_{\mathrm{FI}} \nu_{\mathrm{F}} s_{\mathrm{I}}}{\bar{s}} + \frac{g_{\mathrm{MI}} \nu_{\mathrm{M}} h_{\mathrm{I}}}{\bar{h}} \tag{5.37}$$

where g_{FI} and g_{MI} are the values of genetic relatedness of daughters and sons to the controlling individual, ν_{F} and ν_{M} are the reproductive values, s_{I} and h_{I} are the values of the fitness functions of daughters and sons of I evaluated at the point determined by the fission strategy, and $\bar{s}$ and $\bar{h}$ are the mean fitnesses of females and males in the population. The maximum of function V_{I} is found by setting the derivative equal to zero, i.e.

$$\frac{g_{\mathrm{FI}} \nu_{\mathrm{F}} s'}{\bar{s}} = \frac{g_{\mathrm{MI}} \nu_{\mathrm{M}} h'}{\bar{h}} \tag{5.38}$$

where the derivatives are taken with respect to the amounts invested in males and females, respectively. This is of the same form as eqn. (2.18) except that the derivative is now for an individual strategy and the population mean is considered to be constant.

For a concrete example, let us take the fitness function

$$s(R_0) = 1 - e^{-\alpha(R_0 - R_{\min})} \tag{5.39}$$

also given above as eqn. (5.36). The argument (R_0) is the initial size of a daughter colony. If we assume that the population is monomorphic for the colony size $R_{\mathrm{T}} = 6000$ and each colony always splits into two daughter colonies, the optimal strategy is $R_0 = 1792$ and $R_{\mathrm{M}} = 2416$ under queen control (Fig. 5.6(a)). The mean fitness resulting from the daughter colonies is $2[1 - e^{-\alpha(1792 - R_{\min})}]$. Because the combined success of sons can be considered as a linear function of the number of them, the mean fitness resulting from the matings of the sons is proportional to 2416, the number of sons. Under queen control, the product of relatedness and the reproductive value is the same for sons and daughters and cancels out from eqn. (5.38). Using the parameter values $\alpha = 0.001$ and $R_{\min} = 1000$ (as for Fig. 5.6), the mean fitness from two daughter colonies is $\bar{s} = 1.094$. If a colony of size R_{T} produces k daughter colonies of size R_0, we therefore have the fitness proportional to

$$V = \frac{k\,\{1 - e^{-\alpha(R_0 - 1000)}\}}{1.094} + \frac{R_{\mathrm{T}} - kR_0}{2416} \; .$$

This is maximized when

$$\frac{k\alpha e^{-\alpha(R_0 - 1000)}}{1.094} = \frac{k}{2416}$$

which gives the solution $R_0 = 1792$ independently of k. Whatever the pattern of fission, it is optimal to produce daughter colonies of the same size. When $R_T = 6000$, the inclusive fitness takes the values (Pamilo, 1991a)

$$V_0 = 2.483 \text{ for } k = 0,$$

$$V_1 = 2.242 \text{ for } k = 1,$$

$$V_2 = 2.000 \text{ for } k = 2,$$

$$V_3 = 1.758 \text{ for } k = 3.$$

The fitness is highest when the colony invests all available resources in producing sons.

It is evident that, if the fission pattern in a monomorphic population is such that the equilibrium investment in males is below 50%, colonies investing exclusively in males are favoured. When the investment pattern at equilibrium is such that investment in males is over 50%, colonies investing more in daughter colonies are favoured. As a result, we can expect that the investment ratio, males versus daughter colonies, approaches the 1:1 ratio under queen control. Under worker control, the critical values also equal those derived earlier in this chapter. Let us examine the fitness function (5.37), but allowing colonies to vary with regard to the number of daughter colonies they produce. The results for several representative cases are given in Table 5.5.

If the colony size at the time of fission is $R_T = 4000$, all the equilibrium points have M less than 0.5. This result indicates that it would be beneficial for both the queen and workers to produce only males and invest nothing in a daughter colony. We will next examine only queen control. If the colony size at the time of fission is $R_T = 6000$, a fission type producing two daughter colonies has $M = 0.4$. Such a monomorphic population would be invaded by colony types which produce only one daughter colony and more males, or by a type producing only males. A monomorphic population producing just one daughter colony has an equilibrium of $M = 0.58$. Such a population would be therefore invaded by a colony type fissioning in two daughter colonies. Actually, it would also be invaded by a colony type splitting in three colonies of size $R_0 = 2000$ each. We would thus expect that an equilibrium population would be polymorphic with some colonies producing one daughter colony and some producing two. If the colony size $R_T = 10\,000$, the equilibrium would be polymorphic for colony types giving two or three daughter colonies. At equilibrium, there could also be other colony types, some nests producing only males and some producing more daughter colonies.

Allowing the population to be polymorphic shows that the investment ratio in the case of colony fission should evolve to the same values as in other types of reproduction.

Table 5.5. The equilibrial pattern of colony fission in monomorphic populations where each colony produces either (1) one, (2) two, or (3) three daughter colonies of size R_0 each.

	Queen control			Worker control					
				$\psi = 0.5$			$\psi = 0$		
	R_M	R_0	M	R_M	R_0	M	R_M	R_0	M
$R_T = 4000$									
(1)	1926	2074	0.48	1630	2370	0.41	1369	2631	0.34
(2)	1114	1443	0.28	856	1572	0.21	646	1677	0.16
(3)	520	1160	0.13	382	1206	0.10	274	1242	0.09
$R_T = 6000$									
(1)	3497	2503	0.58	3113	2887	0.52	2769	3231	0.46
(2)	2416	1792	0.40	1964	2018	0.33	1574	2213	0.26
(3)	1671	1443	0.28	1284	1572	0.21	969	1677	0.16
$R_T = 10\,000$									
(1)	6929	3071	0.69	6464	3536	0.65	6048	3952	0.61
(2)	5386	2307	0.54	4694	2653	0.47	4076	2962	0.41
(3)	4324	1892	0.43	3568	2144	0.35	2911	2363	0.29

5.4.3 *The effect of the fitness functions*

The above section (5.4.2) gave the general approach which can be used when examining the equilibrium patterns of colony fission. In this section we extend the topic by considering several different fitness functions of the daughter colonies in order to see how they affect the fission pattern.

The model we use assumes a three-step colony cycle. First, there is a critical period determining whether the colony survives or not. The survival probability is a function of the initial colony size. Second, there is a period of colony growth and the growth is also a function of the colony size. Third, at the end of the season the colony gives rise to k daughter colonies headed by daughters of the old queen and it also produces males. There is random mating in the population. It might be better to cover the survival and growth with a single function but this is not essential for general patterns of the model.

For colony survival, we use two different functions:

$$s_1(R_0) = 1 - e^{-\alpha(R_0 - R_{min})}$$

$$s_2(R_0) = 1 - e^{-\alpha R_0} \qquad (5.40)$$

$$\text{and } s_1 = s_2 = 0 \text{ for } R_0 < R_{min}\ .$$

The function s_1 is the same as that used by Bulmer (1983b) and given as eqns. (5.36) and (5.39) above. s_2 is essentially the same function with a discontinuity point at R_{min}.

The colonies are assumed to grow logistically, and the colony size at the end of the growing period is

$$R_T = \frac{R_{max}}{1 + (R_{max}/R_0 - 1)e^{-ut}} \tag{5.41}$$

where R_{max} is the maximal colony size supported by the environment and u is the intrinsic growth rate (commonly denoted by r, but we have reserved that symbol for the numerical sex ratio).

Production of new colonies takes place only by fission. When the colonies enter the reproductive size, the colony resources equivalent to R_T are divided in the production of sons which enter the mating pool and workers which will be evenly divided by k daughter colonies. Each daughter colony receives a queen which is a sister of the workers, and the old queen dies. The queen is inseminated by one male only, and there is population-wide random mating.

We combine the survival functions and colony growth functions to make two models: model I has the survival function s_1 and model II the function s_2. As a third model we consider the case in which both survival and growth are linear functions of the initial colony size ($s_3 = \gamma R_0$ and $R_T = \tau R_0$). The overall fitness of the daughter colonies in all three models is determined as the product of the survival function and the final colony size. The fitness curves for the models are shown in Fig. 5.7 as functions of the initial colony size R_0. Because the fitness functions are not linear, we need to use the same methods as in Section 5.4.2.

Concerning investment in males, we can assume that the contribution in the asymptotic gene pool is directly proportional to the number of sons produced. In other words, the fitness return through sons is a linear function. This may not be true in small populations, because an additional investment in males will decrease the average mating probability per male. But for large populations and for small differences in the colony-level sex allocation, linearity is a fair assumption.

We can now find the equilibrium investments under queen control and under worker control. We can also include the effect of ψ, the proportion of males produced by the workers.

The optimal investment ratio is a function of the final colony size. The colony size can vary for a number of reasons, but in a deterministic model it is a simple function of the initial size and growth. Therefore it is possible to obtain an equilibrium colony size. At equilibrium there is a specific final colony size R_T^* which produces k daughter colonies exactly of the size which in a deterministic growth model will grow back to this R_T^* before they fission again. In other words there is an initial size x for which

$$x = [R_T^*(x) - R_M(x)]/k \tag{5.42}$$

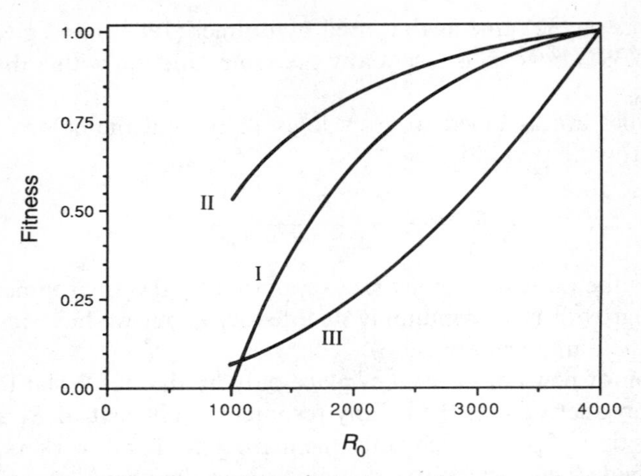

Fig. 5.7 Relationship of fitness (defined as the product of survival function and final colony size) under three models discussed in Section 5.4.3. Models I and II use survival functions s_1 and s_2 from eqn. (5.40) and R_T from eqn. (5.41), and model III uses the linear functions $s_3 = \gamma R_0$ and $R_T = \tau R_0$. The fitnesses are set relative to that of a colony of size 4000. The values of the other parameters used are $R_{min} = 1000$, $R_{max} = 8000$, $e^{-ut} = 0.04$, $\alpha = 0.001$, $\tau = 3.5$ and $\gamma = 0.0005$.

where $R_M = MR_T$ is the amount of resource invested in males.

The equilibrium colony sizes and the sex allocation ratios associated with them are listed in Table 5.6 for fixed values of k (ranging from 1 to 4). When k is taken as a variable, selection should lead to equilibrium with $M = 0.5$ under queen control, $M = 0.37$ under worker control with $\psi = 0.5$, and $M = 0.25$ under worker control with $\psi = 0$. We note that in model II, the equilibrium fission pattern is obtained with a larger k value than in model I. This means that the colonies tend to produce a larger number of small-sized daughter colonies in that model. The reason is that the fitness curve of model I has a steeper slope and therefore the individuals get a better fitness return by investing more per single daughter colony. Note that the entries with $R_0 = 1000$ in Table 5.6 represent the minimum viable colony size. The fission pattern in those cases is not optimal in the sense of satisfying eqn. (5.42) because the daughter colonies cannot become smaller than 1000. Both models show, as expected, that there would be a larger number of daughter colonies under worker control, but the equilibrium colony size is approximately the same under queen and worker control.

Model III, based on linear functions for both survivorship and colony growth, has a concave overall fitness function (Fig. 5.7). For this reason, we do not get the same equilibrium points as in the other models. Instead, the analysis suggests that the colonies should always invest a fixed fraction of resources in males, that fraction being 1/7 under queen control.

Table 5.6. Patterns of colony fission under queen control (QC) or worker control (WC) under the three models discussed in Section 5.4.3, with k the number of daughter colonies. The population equilibria are expected to occur at the points $M = 0.5$ under queen control, and under worker control at $M = 0.357$ with a proportion of worker-produced males of $\psi = 0.5$, and $M = 0.25$ with $\psi = 0$. Under the assumptions of the models, colonies with $R_0 < 1000$ die out, so that no colonies below this size are included in this table.

	$k = 1$		$k = 2$		$k = 3$		$k = 4$	
	R_0	M	R_0	M	R_0	M	R_0	M
Model I								
QC	2882	0.61	1995	0.43	1547	0.30		
WC								
$\psi = 0.5$	3378	0.55	2301	0.35	1728	0.23		
$\psi = 0$	3829	0.50	2569	0.30	1880	0.18		
Model II								
QC	2166	0.69	1336	0.58	1000	0.49		
WC								
$\psi = 0.5$	2707	0.63	1717	0.49	1213	0.42	1000	0.33
$\psi = 0$	3204	0.57	2050	0.41	1453	0.33	1084	0.16
Model III								
QC	2286	0.71	2286	0.43	1219	0.14		
WC								
$\psi = 0.5$	2286	0.71	2286	0.43	1496	0.14		
$\psi = 0$	2286	0.71	2286	0.43	2147	0.14		

The models presented here have been selected to represent different possible alternatives rather than some known examples. Our main emphasis has been to show that the allocation associated with colony fission can be treated within the same general framework as the other sex allocation problems. The crucial point is that the answer depends on the shapes of the fitness functions. Finding the shapes of fitness functions and estimating the investment proportions are important tasks of empirical research.

We should also point out some complicating factors which have not been included in our models. The models assumed a monomorphic population, all colonies obeying the same strategy of fission. In reality, colony size varies and populations are age structured. These factors are likely to cause individual variation, and it is also possible that they affect the average strategy of the colonies in the population.

The models clearly show that when the colonies fission and the daughter colonies compete neither with each other nor with the mother colony, the investment in male and female functions can be predicted by the relatednesses and reproductive values alone. We must just remember that the workers supporting the daughter queens must be counted as an investment in female function. However, if the daughter colonies remain in the close vicinity of the mother colony, it is likely that their survivorship and growth are affected by the number of daughter colonies. Such daughter colonies compete with each other, not only for the worker force which is part of the female function, but for the space and environmental resources. That would devalue extra daughter colonies and favour investing more in males (local resource competition, LRC). Similarly, the fitness function in our third model was exceptional and affected the true investment ratios. Such concave functions have been called local resource enhancement, LRE (Schwarz, 1988a; Section 2.8). The more one invests in daughter colonies, the higher are the fitness returns per unit invested. This situation leads to more female-biased investment ratios than otherwise expected.

5.4.4 *How to estimate the investment: the honey-bee case*

In the above hypothetical examples it was easy to decide what proportion of investments went into sons and what proportion into daughter colonies. The allocation proportions were simply determined by the division of individuals at the time of the reproductive stage of the colony cycle. The estimation becomes more difficult when there is no separate reproductive stage but sons are produced throughout the colony life, and when the old colonies recruit new queens and are polygynous. The functioning honey-bee colony has a number of successive worker generations within a season (mean worker longevity in summer can fall to about 15 days, depending on the locality, Michener, 1974:359). During colony life, drones are regularly produced and at the end of the cycle, the colony fissions by swarming. The old queen leaves the nest with a swarm of workers and some drones. As expected from the results presented in Table 5.4, the old queen has the larger share of the worker force (66% on average, Martin, 1963) One daughter queen inherits the old nest with the remaining workers, brood and combs. Two lines of argument have been used in estimating the allocation of resources in males and females.

According to the first estimation method, introduced by Macevicz (1979), one counts the final outcomes from the colony as sexual investments. The investment in sons equals the total number (or biomass) of drones produced. The investment in daughters includes the workers, brood and cell combs remaining in the old nest and supporting the life of the daughter queen. This approach does not consider how the colony functions when making these investments. The internal colony life, the succession of worker generations during its history, is not counted.

Page and Metcalf (1984), on the other hand, argued that one should also take into account the worker force which maintains the colony. From an evolutionary point of view the only function of the colony is to help the queen to produce sexual offspring. With this in mind, we should divide the worker force between different reproductive activities. The investment in sons includes the drones produced and also that part of the worker force which was required to raise the drones. The rest of the worker force raises new workers which then swarm with the old queen or stay with the daughter queen. (Only the latter fraction of them should be counted as an investment in daughters.) If the total amount of drone brood during the life of the colony is denoted by H (haploid) and the amount of worker brood by D (diploid), the colony investment in males is (Page and Metcalf, 1984)

$$M = \frac{H + HD/(H + D)}{H + D}. \tag{5.43}$$

Pamilo (1991a) doubted the validity of this estimator on the following grounds. First, the investment in those workers swarming with the old queen should not be counted as an investment in daughters. Our second remark concerns the time depth when dividing the worker force. The approach of Page and Metcalf goes back only one step and counts the workers which raise the drones as an investment in males. But surely one should not go back further steps and count also the workers which were required to raise the drone-raising workers etc.? This leads to a complicated counting procedure.

We tend to favour Macevicz's approach, because it argues that the allocation should be that made at the time of decision. Workers generated during colony growth through their activities increase the colony's ability to produce both queens and males. How these resources are allocated is finally determined at the times of swarming and of drone production.

Using the first approach, Macevicz (1979) estimated the sex allocation in four colonies studied by Weiss (1962). In each of these he found a clear female bias, the mean values for the investment in sons being 0.08 when the combs were included in the investment in daughters and 0.32 when the combs were not included but the estimates were based on the adults and brood only. The analysis of Page and Metcalf (1984) gave the estimates for the proportion invested in males as 0.31 when using the wet weights and 0.37 when using the dry weights.

Both theoretical and empirical questions remain for the honey-bee case. Page and Metcalf (1984) note that the models of Macevicz (1979) and Bulmer (1983b) assume that sex allocation occurs at discrete episodes, namely at fission time, whereas colonies produce some drones even when not fissioning. For example, relatively small colonies are likely to invest more in colony growth and maintenance, so that counting all workers as female investment would be a significant overestimate. In fact, Page and Metcalf (1984) try to cut the Gordian knot with respect to estimating swarming propensity with colony size—an important variable in the models we have discussed — by concentrating solely on worker and nest resources as the estimator of female investment. The difficulty here can be assessed in terms of the models we have discussed: if a colony never fissioned, the likely relationship between survival probability and colony size (see Fig. 5.5) would mean that its fitness would be less than of a colony which fissioned at the optimum rate. So we are left with a continuing need for estimates in natural populations of fission rates and associated colony sizes, production rates of drones at different colony sizes, and the proportion of naturally queenless colonies. With regard to the last quantity it should be remembered that managed honey-bee populations lack such colonies (due to pre-emptive requeening by apiarists), hindering their study.

5.4.5 *Population investments in ants with colony fission*

Ants present a variety of life-patterns involving fission. We noted earlier the distinction made by Franks and Hölldobler (1987) between budding and fission. Budding as defined by Franks and Hölldobler (1987) is not a means of colony reproduction but a strategy of resource use: a colony may establish many nests each with several to many queens. The factors governing budding therefore pertain to foraging efficiency and resource holding power (Hölldobler and Lumsden, 1980). Of course, if a polydomous colony fragments, then the queen-right fragments may succeed as independent colonies, but this then fits the definition of fission anyway.

The difficulty seen in many species is that the crucial distinction between fission and budding — the independence or not of daughter nests — is not clear-cut or, to put it another way, the notion of an absolute distinction is not valid. This difficulty is seen particularly strongly in *Formica* species, many of which have two types of populations, some consisting mainly of monodomous and monogynous colonies and others consisting of polydomous and polygynous colonies. For example, *F. exsecta* and *F. truncorum* present differences in production rates of males and queens between monodomous and polydomous populations (Pamilo and Rosengren, 1983), but these differences are also associated with the colonies being monogynous or polygynous, respectively. It is therefore difficult to separate the effects of queen-number from those of nest-number per colony.

The models of this section fit better the army ant life-patterns than those of honey-bees and *Formica* ants. After several broods producing only workers, the army ant colony produces a brood that is almost all (ca. 99%) males, the remaining biomass being six or fewer new queens. The colony divides into two or more parts, one containing the old queen. The colonies are very large, making them relatively easy to find but hindering the collection of data from many colonies. Hence the data tend to be tantalizing rather than voluminous despite great efforts by researchers on these insects. For example, Macevicz (1979) was forced to use the data on just one colony division, of *Aenictus gracilis* (Schneirla, 1971), to estimate an allocation ratio of $M = 0.27$. This estimate came from assuming that the sexual brood consumed the same resources to produce as one made up exclusively of workers (usually 25 500), and 70 000 workers departed with the daughter colonies.

Franks (1985) found a curvilinear relationship between mass of brood produced and colony size in *Eciton burchelli*, and that fission only occurred among the largest colonies (Fig. 5.8), as would be predicted by the models of this chapter. Unfortunately, information on the number of daughter colonies produced during fission by *E. burchelli* colonies is not known.

Clearly, much empirical work needs to be done on army ants before reliable sex allocation estimates are possible. Such work would determine the energetic costs of producing males and workers, the relationships between colony size, survival and fission probability, and the number and sizes of daughter colonies. Although in *Dorylus* army ants worker egg-production does not yield adults (Raignier, 1972), this needs to be checked for other species, especially given the widespread suspicion among myrmecologists that army ants form a polyphyletic group. In fact, phylogenetic work is also much needed in order to assess the hypothesis of polyphyly and, if this is supported to identify extant ponerine relatives of the various army ant groups; if army ants are polyphyletic, findings based on studying one group would not be as predictive with regard to the others as would be the case if they are found to be monophyletic.

Several ponerine groups have adopted fission as the obligatory mode of colony production, namely those species with gamergates. Gamergates have been reported from three ponerine tribes, in two of which species with queen-right colonies have also been reported (Peeters, 1987b). Consistent with the diversity of groups in which this life-pattern occurs, there are significant differences in genetic structure of the colonies. In *Rhytidoponera* species, mature colonies with gamergates average 5–20 of them, depending on species (Ward, 1983a; Pamilo et al., 1985). Up to 100 gamergates occur in nests of *Ophthalmopone berthoudi* (Peeters and Crewe, 1985). By contrast, the number of gamergates is regulated at one in colonies of *Diacamma* (Peeters and Higashi, 1989).

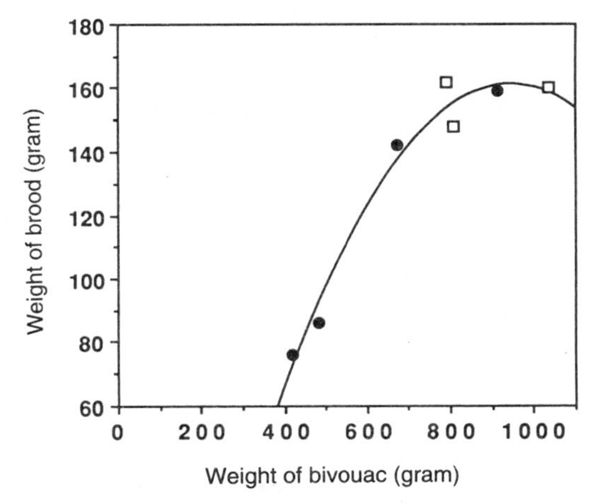

Fig. 5.8 Weight of brood as a function of total colony size in colonies of *Eciton burchelli*. Solid dots show all-worker broods, open squares show sexual broods. (Data courtesy of N. Franks after Fig. 6 in Franks (1985).

The relatedness structure of colonies has been studied in various *Rhytidoponera* species (Ward, 1983a,b; Crozier *et al.*, 1984; Crozier and Pamilo, 1986). Relatedness declines with the number of gamergates per colony. In species of the *R. impressa* group it appears as though gamergates are generally full sisters. This sisterliness of gamergates need not indicate that they are always derived from a queen, but may reflect habitual founding of new colonies by a single gamergate accompanied by some unmated workers (Ward, 1981).

The models of this section are therefore applicable to species with gamergates, given the appropriate data. Unfortunately, there are still major gaps in our knowledge, the same as mentioned above for army ants. For example, the most complete study of *Rhytidoponera* biology comes from the insightful series of papers by Ward, yet the lack of an estimate of fission probability prevents use of his data to estimate sex allocation patterns in gamergate-containing colonies.

5.5 Mechanisms for biasing the final sex ratio

As noted previously (Section 2.6), it is conventional in studies on vertebrates to distinguish between *primary, secondary*, and *tertiary* sex ratios, but these distinctions are less appropriate for male-haploids than the simple distinction between the *initial* and *final* sex ratios. The initial sex ratio is that at the time the eggs are laid, and the final sex ratio is that occurring at the time individuals become adult.

For social Hymenoptera, males can be discounted as effective participants in the struggle to control the sex ratio, because their sperm are identical with respect to sex determination and females have total control over the release of sperm from their spermatheca. (Trivers and Hare (1976) suggested that sperm could be selected for their ability to wrestle their way out of the spermatheca and enter eggs, but doubted whether this has been a significant factor in evolution.) Of course, although males are probably powerless to determine sex allocation patterns, they can compete with each other to inseminate females, and in multiply mated species competition may occur within the spermatheca (Starr, 1984).

The contestants in the sex ratio game are therefore queens and workers. On *a priori* grounds we can determine the likely weapons of the two sides.

Queens can control the number and sex ratio (Section 1.5) of the eggs they entrust to the workers, as well as possibly determining some diploid eggs to develop as workers while providing others with the potential to develop as queens.

Through differential feeding and cannibalism, workers can determine the relative survival of diploid and haploid larvae and the proportion that become queens among the diploid larvae. Workers also have the ability to invest in queen larvae in excess of the limit required to reach the queen threshold.

As reviewed by Brian (1983:194–6) and Wilson (1971:146–54), queen-laid ant eggs have been shown to vary in the potential for development as queens in several species (*Formica polyctena, Myrmica ruginodis, Monomorium pharaonis*). Where genetic variation affects the capacity to develop as a queen, this would add a further dimension additional to queen age. It might be expected that increased heterozygosity might predispose female larvae to develop as queens, but this effect is negligible in several species tested (Craig and Crozier, 1978; Bourke *et al.*, 1988). There does appear, however, to be a significant influence of genetic factors on the capacity to develop as morphologically complete queens in some leptothoracine species. In these cases, reproduction is by a fully developed queen or by either a worker-like individual (*Harpagoxenus sublaevis*) or a morphologically distinct queen with a reduced alitrunk (*Leptothorax* sp. A). In each case, laboratory breeding experiments show a strong influence of strain on ability to produce winged queens. The variation can be explained by a single-locus model in which homozygosity for a recessive allele is necessary for a female larva to have the capacity to develop as a queen (Winter and Buschinger, 1986; Heinze and Buschinger, 1989), although it would be desirable for further analyses to be made to test whether this simple model is fully adequate.

Cannibalism of males has been reported for *Monomorium pharaonis* (Peacock *et al.*, 1954) and for *Formica polyctena* (Schmidt, 1974). In *Myrmica* (species unstated), workers in the presence of queens may prevent large larvae from developing as queens by biting them; under this circumstance they also underfeed male larvae leading to the production of smaller adult males (Brian, 1983:192). Crespi (1992a) and Kukuk (1992) review the occurrences of cannibalism in a variety of social insects, and Nonacs (1993) discusses the consequences of the cannibalism of male eggs. Hasegawa (1992) observed that haploid eggs (identified by karyotypes) were fed to other colony members in the ant *Colobopsis nipponicus*, and recent work of Aron *et al.* (1994) showed that the haploid and diploid eggs are differentially removed from Argentine ant (*Iridomyrmex humilis*) colonies, and the final sex ratio produced differed from that initially present among the queen-laid eggs.

Crespi (1992) notes that differential cannibalism, such as that expected to be selected for under sex allocation pressure, is likely to select for deception with regard to apparent sex of the brood. For example, male larvae would be selected for their resemblance to female larvae. Of course, the workers would be selected for their ability to see through such deception; Peacock *et al.* (1954) reported that workers can distinguish male larvae in *Monomorium pharaonis*, but Edwards (1991) implies that the workers of this species can only distinguish sexual from worker brood, not queen from male brood. Nonacs and Carlin (1990) found that sex in the carpenter ant *Camponotus floridanus* becomes detectable only at pupation, which is consistent with sexual deception as an aspect of queen control. Fisher and Pomeroy (1990) showed that a bumble-bee queen can react to the presence of male larvae and start destroying her own eggs. Cannibalism could be connected to conflicts between the castes or other kin groups within a colony, but it can also serve to regulate the colony size in order to guarantee sufficient resources to all of the developing larvae. Underfed sexuals are likely to have low fitness, and sufficient size is important both for nest-founding queens (Keller and Passera, 1989) and for males in sexual competition (Lloyd, 1981; Davidson, 1982).

At this point it is worth noting that various authors refer to 'queen control' over workers as being evidenced by the differential rearing conditions imposed by workers on brood depending on whether queens are present or not. This inference is unwarranted. The alternative view is more useful, namely that workers respond adaptively to the presence or absence of queens by changing their brood-rearing patterns (Crozier, 1992, 1994; Keller and Nonacs, 1993). Of course, queens may influence and manipulate worker behaviour, for example by adjusting the number and sex ratio of their eggs, as we have stressed above. Keller and Nonacs (1993) make a distinction between the terms queen control, queen signal and queen effect, and point out that true pheromonal queen control has not been demonstrated, nor can it be easily justified evolutionarily (but see also Heinze *et al.*, 1994b).

In honey-bees, workers have strong control over sex allocation, because they make the drone cells prior to oviposition and can make queen-producing cells either prior to oviposition or later by modification of worker cells (Michener, 1974:102–3). The honey-bee queen thus lacks the power of ant queens to control the sex allocation ratio; although she could refuse to oviposit in drone cells, this power is negated by the fact that she is selected to produce a larger number of drones than is optimal for the workers.

In those cell-making species with differentiated queens but with equal-sized cells, the situation would seem to be more like that of ants than that of honey-bees, in that workers lack the ability to force the queen to lay particular types of eggs.

Kerr (e.g. 1974) has reported that, in various *Melipona* species, queen potential depends on two, or in some cases three, caste-determining loci: only females heterozygous for all of the caste-determining loci have the potential to develop as queens. However, the ratios giving rise to this inference, of 25% or 12.5% of diploids becoming queens, are seen only under the right feeding conditions. Thus, as in ants, workers have the ability to modify the caste ratios in *Melipona*.

Of course, in many species workers have the ability to modify the investment ratio by producing male-destined eggs. The conditions selecting for or against this behaviour have been examined earlier (e.g. Section 4.2).

While polyandry affects the payoffs for workers laying male-destined eggs, polygyny changes the ability of individual queens to control the allocation ratio by restricting the volume of eggs available for larval rearing and, especially, by restricting the number of diploid as against haploid eggs. This loss of control capacity under polygyny occurs because a restriction in production by one queen simply means that the eggs of more prolific queens are reared instead. However, relatedness patterns in such cases mean that workers may still fail to produce a 3:1 allocation ratio of siblings unless they can distribute aid differentially to their own matrilines. There is no strong evidence that the workers would have the capability to distinguish between kin groups within a colony and direct their aid preferentially to closer kin. While there is some positive evidence for such nepotism in queen rearing in the honey-bee (Noonan, 1986; Visscher, 1986; Page *et al.*, 1989a), negative observations have also been reported (Woyciechowski, 1990b). Honey-bee workers do not seem to favour their full sisters over half-sisters as egg-layers in orphaned colonies (Hogendoorn and Velthuis, 1988; van der Blom and Verkade, 1991), nor do *Polistes annularis* females preferentially join full sisters when founding new nests (Queller *et al.*, 1990). (See also the discussion 'Nepotism in the honey bee' in *Nature* 346:706–9 in 1990). Theoretically, the evolution of nepotism is expected to depend on its costs and on the level of polyandry in monogynous societies (Ratnieks and Reeve, 1991).

In termites, the role of king and queen pheromones in mediating the production of replacement sexuals, and in the production of new ones, has long been known (reviews by Wilson, 1971:188–96; Sewell and Watson, 1981; Watson and Sewell, 1981, 1985; Brian, 1983:206–16). The action of workers in attacking and cannibalizing 'surplus' reproductives has also long been known. Zimmerman (1983) found that attacks on the wingpads of potential winged sexuals in *Pterotermes occidentis* occur in mature colonies but not in immature ones (see also Roisin, 1994).

5.6 Who wins?

The meaning of the question 'Who wins?' is best given by asking 'On which group of colony members has selection for sex allocation been most effective?'. As first noted by Trivers and Hare (1976), the question should be asked because of the different relatednesses of the males and females in the sexual brood to the workers and queens. To see which party has the strongest effect on sex allocation, we need to predict the expected sex allocation ratio paying attention not only to genetic but also to ecological determinants of sex ratio. It is not always easy to test these predictions, because the estimation of investment can itself be a problem (Crozier and Pamilo, 1993).

5.6.1 *Estimation*

The quantity of interest is the relative investment in queens and males, or in their equivalents. Even in the apparently simplest case of species which have differentiated queens, and which found new colonies by propagules composed solely of these queens, there are cautions to heed when estimating the investment ratio.

Trivers and Hare (1976) argued that using the live ('wet') weights of queens and males is inappropriate, because of the relative inexpensiveness of water.

They consequently used dry weight. However ratios determined from wet weights are highly correlated with ratios determined from dry weights ($r = 0.97$), with a slope of the regression line between them statistically indistinguishable from unity (Crozier and Pamilo, 1993). However, weight is a biased estimate because the materials making up queens and males differ in their relative energetic costs (Boomsma and Isaaks, 1985; Boomsma 1987, 1989). Boomsma (1989) provides a first approximation to correct the ratio of individual dry weights (female/male) to a ratio of costs:

$$C = D^{0.7}$$

where D is the ratio of dry weights, and C is the ratio of costs, taken using ants ready to leave the nest.

The difficulty of estimating cost ratios from weight ratios may be greater with ants than with other Hymenoptera. Danforth (1990) found that the provision cost and adult weight ratios are very similar for the solitary mass-provisioning bee *Calliopsis (Hypomacrotera) persimilis.*

When should the insects be sampled? It is tempting to use pupae, because colonies collected at that stage are less likely to have already emitted new reproductives, and it is true that the ratios of individuals may be well estimated at that time. But there are two interconnected problems. First, females and males may differ in their rates of surviving the pupal/adult transition, as noted for *Monomorium pharaonis* (Peacock, 1951).

Secondly, and more seriously, considerable changes in dry weight may take place after eclosion. Boomsma and Isaaks (1985) found that queens of *Lasius niger* increased in dry weight from about 4 mg at eclosion to about 15.5 mg at flight time, whereas males remained approximately constant at 0.9 mg. Of course, both males and queens incur maintenance costs over this period.

Thirdly, differences between the sexes in rates of respiration mean that accurate estimates of the investment in the various classes of individuals require studies of metabolism until the reproductives leave the nest. For mass-provisioning insects, such estimates could be made using the provision store, otherwise more detailed studies are required (MacKay, 1985). Studies on *Pogonomyrmex* species (MacKay, 1985) and various other ant species (Keller and Passera, 1989) suggest that the exponent in Boomsma's cost-ratio estimator may vary significantly between species. There are thus significant problems to estimating the relative production costs, as also further discussed by Crozier and Pamilo (1993).

Boomsma (1989) also brings attention to the need for an unbiased sample of sexual production in each population, reinforcing earlier observations by Owen *et al.* (1980) and Forsyth (1981). The need for this stems from the fact that colonies of different sizes often have different production rates of males and queens. Not only do smaller colonies tend to produce more males (Section 6.5) while queen-right but also, in species with potentially fertile workers, the death of the queen releases worker reproduction via the laying of male-destined eggs (Section 4.2). Earlier papers show an increasingly female-biased investment ratio with smaller data sets, which Boomsma (1989) attributes to researchers tending to sample preferentially from larger colonies!

The testing of observations against expectations naturally relies on a correct assessment of the expectations.

As emphasized earlier (Section 2.8), female- or male-biased sex ratios may arise from local mate competition (LMC) or local resource competition (LRC), respectively. Adequate population genetic data are needed to determine the magnitude of LMC and LRC. The possibility of interdemic selection between populations has also been raised as an explanation of female-biased sex ratios in social spiders (Aviles, 1986, 1993; see Table 1.6).

The number of times a queen mates (Section 4.3), and the proportion of male-destined eggs laid by the workers (Sections 4.2, 5.1), also affect the expected optima for both parties, and need to be estimated. Such estimation can be done by various means, including genetic analyses of colony structures. Naturally, whether or not there is one functional queen or many must also be known!

Where worker laying of male-destined eggs may occur, sex ratio compensation is expected (Section 5.1), so that a complete survey of the population is needed to determine the expected sex ratio in queen-right colonies. This is another example showing the need for a balanced sampling of all colony sizes in a population, and the need for assessing their relative contributions to the pools of queens and males.

One complication in estimating investment ratios as well as in predicting the expected optima can be caused by morphological polymorphism within a sex. This affects not only the amount of energy invested in the sexuals but also their dispersal strategies and individual reproductive values. Polymorphism of females can involve various intermorphs and winged and wingless queens (e.g. Bolton, 1986; Yamauchi *et al.*, 1991; Buschinger and Heinze, 1992; Heinze *et al.* 1992), or queens and mated workers (gamergates) (Ward, 1983a; Sommer and Hölldobler, 1992). Males can also show both size dimorphism (Scherba, 1961; Fortelius *et al.*, 1987) and dispersal polymorphism (Kinomura and Yamauchi, 1987; Yamauchi *et al.*, 1991; Yamauchi and Kawase, 1992).

Once these various factors have been determined or reasonably inferred, then other data are useful in determining the strategies used by the two parties in modifying the sex ratio. Seldom estimated is the weight of workers, yet we have seen earlier that this affects the ability of workers to counter the control of the queen. Increasing the investment in new queens is one way workers have to counter the power of the queen to restrict the supply of diploid eggs. If this effect is significant, then larger queens would show an increase in personal fitness less than that justified by the extra investment in them.

As noted in models discussed earlier, the queen may control the workers by laying only male (or female) eggs at crucial times. These life history characteristics should be noted and fitted to the models.

5.6.2 The findings

The current interest in the sex ratios of social insects stems from the work of Trivers and Hare (1976), who used observed sex ratios and dry weights in various social insect species to argue that the workers win the contest for control over sex allocation. We have seen that Trivers and Hare (1976) were correct that there will often be a conflict between queens and workers over the sex ratio, the question remains as to how well their analyses have stood the course of time.

Figure 5.9 gives the chief results of Trivers and Hare. According to Trivers and Hare, the results show that in monogynous ants there is a tendency towards a sex allocation pattern of $M = 0.25$, reflecting the differences in relatedness of new queens and males to the workers. The contrast of the results between monogynous ants and various other categories, in which expectations are different, led Trivers and Hare (1976) to adduce these differences as further support for their conclusion.

Trivers and Hare (1976) were swiftly attacked by Alexander and Sherman (1977), whose criticisms can be reduced to two claims.

1. The analysis of Trivers and Hare (1976) is statistically flawed.
2. Other explanations, especially queen control under local mate competition, explain the data better.

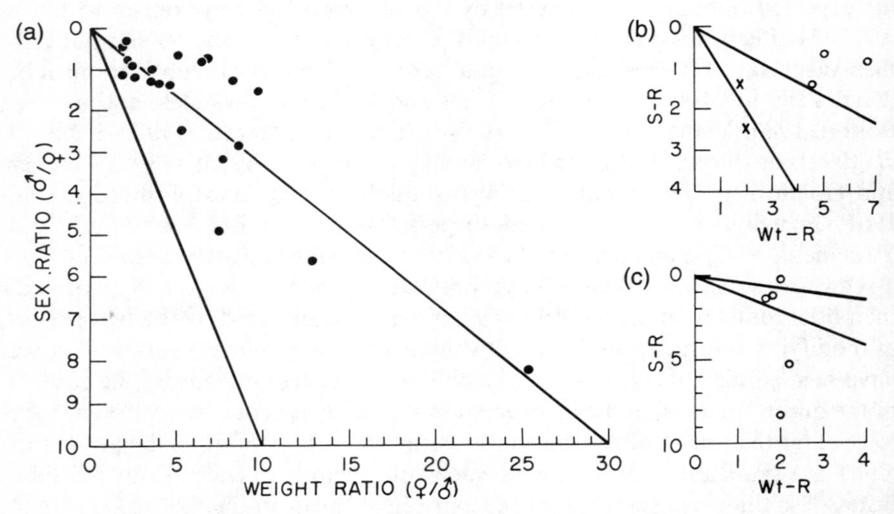

Fig. 5.9 The ratio of the numbers of winged individuals (male/female) plotted against adult dry weights of the same species (female/male) for various ant species. Lines showing $M = 0.25$ (upper diagonal lines in each part) and $M = 0.5$ are drawn for comparison. (A) Monogynous species. (B) Two slave-making *Leptothorax* (x) and three non-slaver congeners (o). (C) Five species of *Myrmica* varying in their propensity to polygyny. (This is Fig. 4 of Trivers and Hare (1976), reprinted by permission.)

Inspection of the data (Fig. 5.9) shows a considerable scatter of the points, including the crucial case of monogynous ants. Alexander and Sherman (1977) reanalysed the data after normalizing them using logarithmic transformation, and found that 95% confidence intervals of the investment ratios of 'monogynous ants, termites, bumblebees, and nonsocial trapnested bees and wasps all overlap'.

Alexander and Sherman also seem correct in some of their theoretical criticisms, but not in others. In particular, some of the species in the sample collected by Trivers and Hare showed female-biased sex ratios markedly in excess of $F = 0.75$. Alexander and Sherman claimed that such extreme investments must reflect local mate competition (LMC) but, as noted by Nonacs (1986a), this claim seems most unlikely. As discussed in Section 2.8, the optimal sex ratio under LMC with n mothers per patch is $r^* = (n - 1)/2n$ if inbreeding is allowed (eqn. (2.19)) and $r^* = (n - 2)/(2n - 3)$ if it is not (eqn. 2.20)). Under LMC a patch size no larger than two would be required to yield a sex ratio of 0.75 if inbreeding is allowed and this sex ratio is unobtainable if inbreeding is avoided. The evidence (Crozier, 1980) is that LMC is undetectable in most social insect populations. The combination of LMC and queen control is therefore insufficient as an explanation of highly female-biased sex ratios.

Trivers and Hare (1976) assume that their monogynous ants also have once-mated queens, otherwise a 3:1 investment ratio would not be expected under selection at the worker level. Alexander and Sherman (1977) correctly note that multiple mating by queens is widespread (see Table 4.1), and that this alters the expected sex ratio preference of workers but not of queens. Trivers and Hare also assume that male-production by workers is rare, whereas Alexander and Sherman correctly regard it as common (see Table 4.3), but their suggestion that this factor be controlled for by eliminating from the analysis '. . . all nests that produced only males . . .' would be a mistake because of the interaction of all nests in a population in determining the overall sex ratios. We should also note that both polyandry and worker reproduction are expected to reduce the worker-preferred female bias in sex allocation. Therefore the occurrence of these complications does not diminish the significance of the female bias observed by Trivers and Hare.

Nonacs (1986a) performed an extensive reanalysis of the data collected by Trivers and Hare (1976), adding some additional records, and concluded that the hypothesis of queen control is excluded by various tests. In particular, Nonacs (1986a) observed that the numerical proportion of males in monogynous ants is in excess of 0.5, even though the investment ratio is female biased; under queen control combined with LMC a deficiency of males is expected. However, Nonacs's (1986a) claim that the existence of specialist colonies (male- or female-producers) is inconsistent with queen control under LMC can now be doubted to some extent (Section 6.5).

Boomsma (1988) concludes that Nonacs (1986a) is correct in his arguments against queen control and LMC being the best explanation for ant sex allocation patterns, but found that the data used were too heterogeneous to support a conclusion of any particular sex ratio being general. The difficulties concerned (a) heterogeneity in the cost estimates for males and females, and inappropriate measures used to convert weights to costs, and (b) strong bias in sampling. The bias in sampling comes from an increasing excess of investment in females as the number of nests sampled decreases. Boomsma plausibly attributes this trend to the tendency of smaller colonies to produce a higher proportion of males than larger colonies, and for researchers to be prone to collect preferentially from larger colonies. Nonacs (1986a) noted the trend to higher male production in smaller colonies, but did not extrapolate from this to the effect on the overall conclusions.

Boomsma (1988) also notes that Nonacs's (1986a) data yield a strong correlation between sex ratio and the extent of male/queen dimorphism, but concludes that this is an artefact resulting from the difficulty of converting weights to costs. Helms (1994) finds a similar correlation in bees and wasps, and notes that this needs not be artifactual if species with small males (i.e. strong dimorphism) tend to have more local mate competition.

These considerations lead to the conclusion that it is difficult to combine large and inevitably heterogeneous data sets to derive conclusions about the extent of worker or queen control in social insects. It seems preferable to examine particular species in depth, attempting to estimate as many of the parameters listed in Section 5.6.1 above in the effort to determine not only the sex allocation pattern but also the expectations under either queen or worker control.

Various of the papers cited above have described the hypotheses under test as being those of worker control versus local mate competition. However, these are not opposed phenomena, and the most useful hypotheses are those of worker versus queen control. LMC, LRC (local resource competition), and the relatednesses of the various parties will enter into determination of expectations under both worker and queen control.

What is the current state of the data? We present compilations from the literature of the data from eusocial (Table 5.7) and non-eusocial species (Table 5.8). Noteworthy is the spread of the data and the high variability between species even of the same genus.

Table 5.7. Observed sex ratios in eusocial Hymenoptera (those in ants are based largely on calculations of Pamilo, 1990b). C is the number of colonies studied, log N gives the approximate number of individuals examined (2, $100 < N < 1000$; 3, $1000 \leq N < 10\,000$; 4, $10\,000 \leq N < 100\,000$; 5, $N > 100\,000$), r is the numerical proportion of males in the population and SE is its standard error over colonies; m is the proportion of resources invested in males estimated using a dry weights, b fresh weights, c caloric contents, d total energetic costs including respiration. Occasional colonies in ant species classified below as 'monogynous' have more than one queen, and occasional colonies of 'polygynous' ants have one (or no) queen.

					m				
	C	log*N*	*r*	SE	a	b	c	d	Notes
Eusocial bees									
Augochlorella									
striata									92
parasocial	12	2				0.59			
eusocial	12	2				0.33			
Bombus									
affinis	22	3	0.79			0.51			1
americanorum	25	3	0.60		0.27				2
auricomus	12	2	0.53		0.32				2
fraternus	4	2	0.57		0.45				2
griseocollis	20	2	0.63		0.46				2
impatiens	5	2	0.59		0.24				2
melanopygus	17	3	0.87	0.04		0.73			3
ruderatus	5	3	0.83						4
terrestris	26	3	0.77		0.62				88
terricola	32	3	0.83	0.03	0.74	0.74			5
Exoneura bicolor									6
NFN	152	3	0.11						
OWN	74	2	0.25						
Halictus									
ligatus	> 130	2	0.34				0.75		11
ligatus	39	2	0.46		0.39				90
Lasioglossum									
imitatum			0.50						8
marginatum	9	3	0.70	0.13					9
rhytidophorum			0.41						12
rohweri			0.48						7
umbripenne			0.20						7
versatum			0.27						7
zephyrum		2	0.29						10
Parasitic bees									
Psithyrus									
ashtoni	39	3	0.64			0.41			1
variabilis	4	2	0.48		0.31				2

Table 5.7 *Continued*

					m				
	C	logN	r	SE	a	b	c	d	Notes
Eusocial wasps									
Dolichovespula									
arenaria	6	2	0.50	0.15					13
Microstigmus									
comes		2	0.43		0.36				89
Mischocyttarus									
drewseni		2	0.67			0.67			14
flavitarsis	7	2	0.39			0.38			15
Parapolybia									
indica	18	3	0.06			0.03			18
varia	?	?	?		0.19				16
Polistes									
bischoffi	20	2	0.56			0.39			17
chinensis	12	3	0.54	0.03		0.40			18
exclamans	78	3	0.45			0.50			19
fuscatus	17	2	0.52			0.49			20
fuscatus	24	3	0.48		0.50				21
gallicus	17	3	0.29			0.51			17
jadwigae	8	2	0.51			0.18			18
japonicus	8	2	0.24			0.35			18
mandarinus	3	1	0.42			0.50			18
metricus	228	4	0.45		0.49	0.18			21
nimpha	?	?	0.22			0.15			16
snelleni	12	2	0.22			0.49			18
variatus	24	3	0.48		0.50				21
Vespula									
atropilosa	21	3	0.70	0.04					22
consobrina	11	3	0.65	0.04					23
maculifrons		4	0.65						24
pensylvanica	31	3	0.83	0.03					22
vidua	6	3	0.53	0.04					25
Parasitic wasps									
Dolichovespula									
arctica	5	2	0.60	0.13					25
Monogynous ants									
Acromyrmex									
lundi	?	4	0.61						26
octospinosus	10	3	0.46		0.25				27
Aphaenogaster									
rudis	14	2	0.85	0.06	0.30				28
treatae	12	3	0.61		0.13				29

Table 5.7 *Continued*

	C	logN	*r*	SE	*m* a	b	c	d	Notes
Apterostigma									
dentigerum	53	2	0.51		0.51				30
Atta									
bisphaerica	5	4	0.76	0.05	0.28				31
laevigata	6	4	0.74	0.05	0.26				31
sexdens	7	5	0.83	0.06	0.39				31
Camponotus									
ferrugineus	6	3	0.56	0.11	0.17				32
herculeanus	1	3	0.71		0.32				33
pennsylvanicus	12	3	0.56	0.04	0.16				32,34
Carebara vidua	7	2	0.30	0.17	0.08	0.11			35
Colobopsis									
nipponicus	20	2			0.25				91
Formica									
exsecta	15	2	0.43	0.11	0.41				36
exsecta	15	3	0.75	0.09	0.73				36
fusca	29	3	0.60	0.08	0.59				36
fusca	?	2	0.72			0.48			37
nitidiventris	19	3	0.23	0.09	0.14				38
pratensis	35	3	0.72	0.07	0.71				36
rufa	32	3	0.50		0.46				39
truncorum	63	3			0.47				40
Lasius									
alienus									
(Ditch)	?	4	0.95		—	0.54			41
(Knoll)	?	4	0.78		—	0.17			41
alienus	17	?	?		0.03	0.05			42
alienus-niger									
hybrids	6	?	?		0.11	0.18			42
flavus	12	3	0.69	0.10	0.10				43
niger	?	5	0.79		—	0.24			44
niger	9	?	?		0.61	0.72			42
Leptothorax									
ambiguus	12	2	0.45		0.12				45
1986	27	2	0.72		0.46				93
1990	16	2	0.98		0.94				93
1992	13	2	0.75		0.49				93
curvispinosus	97	3	0.57		0.23				46
provancheri	13	2	0.42	0.06					47
Myrmecina									
americana	10	2	0.54		0.31				48
Myrmica									
ruginodis	10	3	0.53	0.10	—	0.42			49
ruginodis	5	3	0.33	0.06	—	0.24			50
schencki	10	2	0.24	0.07	0.13				51

Table 5.7 *Continued*

					m				
	C	logN	*r*	SE	a	b	c	d	Notes
Pheidole									
desertorum	40	?	?			0.42			52
desertorum	115	3	0.42						53
xerophila	19	2	0.68						53
Pogonomyrmex									
desertorum	6	2	0.36			0.17			54
montanus	35	4	0.51	0.05	0.40		0.34	0.51	55
rugosus	4	3	0.73	0.15	0.48		0.41	0.94	55
subnitidus	7	3	0.70	0.08	0.40		0.41	0.46	55
Prenolepis									
imparis	12	3	0.89	0.04	0.24				56
Pseudomyrmex									
belti	1	3	0.55		0.40				57
ferruginea	1	3	0.31		0.21				57
Rhytidoponera									
purpurea	10	3	0.47		0.26				58
Sericomyrmex									
urichi	1	4	0.94						59
Solenopsis									
invicta	?	?	0.48		0.21	0.39			60
Stenamma									
brevicorne	8	2	0.42	0.12	0.23				61
diecki	10	2	0.56	0.09	0.27				61
Tetramorium									
caespitum									
(Ditch)	?	4	0.47	0.18		0.36			62
(Knoll)	?	4	0.65	0.31		0.52			62
Trachymyrmex									
septentrionalis	40	3	0.31	0.05		0.14			63
Veromessor									
pergandei	8	3	0.35	0.09		0.14			64
Polygynous ants									
Atta									
texana	2	3	0.54	0.01					65
volleweideri	2	3	0.92	0.01		0.58			66
Crematogaster									
mimosae	?	3	0.92	0.01		0.53			67
nigriceps	?	3	0.72	0.05		0.38			67
Formica									
aquilonia	20	3	0.24	0.06		0.21			36
aquilonia	32	3	0.30	0.05		0.26			36
aquilonia	10	2	0.30	0.13		0.28			36
aquilonia	116	3	0.28						39
cinerea	9	2	0.37	0.14		0.35			36

Table 5.7 *Continued*

					m				
	C	logN	*r*	SE	a	b	c	d	Notes
exsecta	18	2	0.93	0.05	0.93				36
exsecta	10	2	1.00		1.00				36
exsecta	10	2	0.83	0.07	0.81				36
incerta	12	2	0.45	0.13	0.31				38
lugubris	11	3	0.47	0.15	0.46				36
obscuripes	46	4	0.58		0.54				68
opaciventris	152	3	0.95						69
polyctena	8	2	0.44	0.17	0.43				36
polyctena	19	2	0.19	0.08	0.18				36
pressilabris	16	2	0.97	0.02	0.97				36
pressilabris	20	3	0.98	0.01	0.98				36
rufibarbis	11	2	0.47	0.12	0.46				36
sanguinea	12	2	0.22	0.11	0.22				36
truncorum	158	3	0.74						40
yessensis	26	4	0.73	0.06					70
yessensis	11	2	0.49	0.09					71
Iridomyrmex									
humilis	?	?				0.94			72
humilis	?	4	0.98	0.00	0.97				73
Leptothorax									
diversipilosus	9	2	0.55	0.12					76
longispinosus									74, 75
New York	381	3	0.59		0.31				
Vermont	488	3	0.87		0.66				
Monomorium									
pharaonis	24	3	0.28						77
Myrmica									
rubra	4	3	0.88	0.08					78
ruginodis									
(microgyne)	5	2	0.89	0.09		0.84			50
sabuleti	35	3	0.84	0.04	0.70	0.77			79
sabuleti	95	3	0.74		0.57	0.64			80
scabrinodis	12	2	0.50			0.40			80
sulcinodis (S)	73	3	0.62		0.35	0.47			81
sulcinodis (W)	69	3	0.86		0.67	0.77			81
Pheidole									
pallidula	23	3	0.86			0.53			82
Pseudomyrmex									
nigrocincta	1	3	0.58		0.41				57
nigropilosa	1	2	0.64		0.49				57
venefica	2	3	0.51	0.06					83
Tetraponera									
penzegi	?	2	0.57	0.07	0.41				67

Table 5.7 *Continued*

	C	logN	r	SE	m				Notes
					a	b	c	d	
Slave-making ants									
Epimyrma									
ravouxi	30	3	0.60	0.06	0.48				84
Harpagoxenus									
sublaevis	36	3	0.56	0.05	0.56				85
Parasitic ants									
Chalepoxenus									
brunneus	9	?	0.30						86
Epimyrma									
adlerzia									
field	5		0.21						87
laboratory	19		0.15						87
kraussei	17	2	0.24	0.07	0.16				84

1. Fisher (1987).
2. Webb (1961), in Trivers and Hare (1976).
3. Owen and Plowright (1982).
4. Pomeroy (1979), in Owen *et al.* (1980).
5. Owen *et al.* (1980).
6. Schwarz (1988a). NFN is newly founded nests, OWN is overwintered nests. This species is quasi-social, not eusocial, and shows a pattern of brood-size dependent sex allocation (see Fig. 2.5) that is attributable to local resource enhancement (see Section 2.6)
7. Breed (1976). The figures shown are the maximum production rates of males.
8. Michener and Wille (1961); wing-length data combined with r yield an estimate of $m = 0.51$ (Strassmann, 1984).
9. Michener (1974:298).
10. Batra (1966); maximum production of males.
11. Packer (1986a, 1987) used pollen ball weights to estimate investment in queens, workers, and males in an Ontario population; information on a Florida population of the same species is given by Packer and Knerer (1986).
12. Michener and Lange (1958b); the figure shown is the maximum production of the brood to be male.
13. Greene *et al.* (1976).
14. Jeanne (1972), in Strassmann (1984).
15. Litte (1979).
16. Yamane (1980), in Suzuki (1986). According to Suzuki (1986), Yamane's (1980) identification of this population as *Polistes biglumis* was in error.
17. Pardi (1943), in Strassmann (1984).
18. Suzuki (1986).
19. Strassmann (1984).
20. Noonan (1978).

21. Metcalf (1980).
22. MacDonald *et al.* (1974).
23. Akre *et al.* (1982).
24. MacDonald and Matthews (1981).
25. MacDonald and Matthews (1976).
26. Kusnezov (1962). Taken from mating flights over three years.
27. Lewis (1975). Data combined from two years.
28. Headley (1949) and Talbot (1951).
29. Talbot (1954).
30. Forsyth (1981). ca. 42% of the 285 males in the 53 colonies studied were produced by workers in orphaned colonies; 269 queens were produced in queen-right colonies; β_s is therefore estimated as 0.22; no worker-produced males were evident in queen-right colonies; the new queen dry weight was 1.02 times that of a male. Queen-right colonies had ca. 35 individuals, orphaned colonies about 9.
31. Autuori (1950).
32. Pricer (1908). Samples were taken from large colonies.
33. Hölldobler and Maschwitz (1965). (The species should be classified as polygynous (oligogynous) although considered monogynous by Trivers and Hare (1976)).
34. Fowler and Roberts (1982). Data include at least one queenless colony.
35. Lepage and Darlington (1984). The means are calculated by weighting the colonies equally, because total counts of sexuals are not available from all colonies.
36. Pamilo and Rosengren (1983). The means are calculated by weighting the colonies equally, because total counts of sexuals are not available.
37. Darling (1978), in Nonacs (1986a).
38. Talbot (1948). Data combined from two years.
39. Rosengren and Pamilo (1986).
40. Pamilo and Rosengren (1983) and Rosengren *et al.* (1986).
41. Brian (1979b). Data combined from five years. 'Ditch' and 'Knoll' refer to two different sites.
42. Pearson (1987).
43. Pickles (1940). Data combined from five years.
44. Boomsma *et al.* (1982).
45. Talbot unpublished (taken from Trivers and Hare (1976), some colonies are polygynous (Alloway *et al.*, 1982)).
46. Data combined from Headley (1943) (10 colonies) and Talbot (unpublished data cited by Trivers and Hare, 1976)(87 colonies).
47. Figures shown were calculated from the pupal production; Buschinger *et al.* (1980) note that mated egg-laying females of this species include both normal queens and relatively worker-like 'intermorphs'.
48. Talbot unpublished (in Trivers and Hare (1976).
49. Brian and Brian (1951). The macrogyne form of the species is monogynous.
50. Mizutani (1981). One third of the macrogyne colonies are polygynous.
51. Talbot (1945a). Data combined from three years.
52. Droual (1982).
53. Helms and Rissing (1990). Colonies of *Pheidole desertorum* and *P. xerophila tucsonica* have almost complete specialization into male- and female-producers.
54. Davidson (1982).
55. MacKay (1981, 1985). Data for *P. montanus* were combined from three years and include some experimental colonies.

56. Numerical data were combined from nests dug up over two years at Tiffin, OH, and St. Charles, MO (Talbot, 1943). Dry weights from Trivers and Hare (1976).
57. Janzen (1975). Only one colony per species was examined, but the author states that the sex ratios were very similar in other colonies.
58. Ward (1983b).
59. Weber (1972:52). Recorded as deaths in a laboratory colony over four years.
60. Markin and Dillier (1971) and Morrill (1974). Colonies were followed throughout a year.
61. Talbot (1975) provides the numerical ratio; data combined from different years and including pupae and adults. Trivers and Hare (1976) provide the dry weight ratio.
62. Brian (1979b). Data combined from the same 28 colonies from two sites over seven years.
63. Cole (1939).
64. Pollock and Rissing (1985).
65. Moser (1967).
66. Jonkman (1980). Only one good colony (may be monogyne).
67. Hocking (1970). Data combined from two years. Trivers and Hare (1976) and Nonacs (1986a) differ widely in their interpretation of the data. Our interpretations agree with those of Nonacs.
68. Herbers (1979).
69. Scherba (1961). The very high proportion of males probably results from the mode of colony formation (fission) in this species.
70. Kim and Murakami (1980).
71. Ito and Imamura (1974). Ten pupae were randomly sampled from each nest, and the total counts of sexuals were not done.
72. Markin (1970). Colonies were followed throughout a year.
73. Bartels (unpublished). Data combined from two years. 90% of the nests produced single-sex broods.
74. Herbers (1984). The number of queens in nests varies.
75. Herbers (1989, 1990). As reflected in the figures in the table, the Vermont site showed a consistently higher numerical and investment ratio than did the New York site, but there was considerable year-to-year variation over a ten year study.
76. Alpert and Akre (1973). Data combined from three years.
77. Peacock (1951). Laboratory 'colonies' were used for this unicolonial ant.
78. Brian (1972), data combined over 4–6 years.
79. Brian (1972). Data combined from six years.
80. Elmes and Wardlaw (1982). Both *M. sabuleti* and *M. scabrinodis* may be mainly monogynous.
81. Elmes (1987b). Data combined from seven years. The degree of polygny varies temporally.
82. Passera unpublished (in Trivers and Hare (1976)).
83. Janzen (1973).
84. Winter and Buschinger (1983). *Epimyrma kraussei* is a workerless parasite with intra-nest mating, hence the female-biased sex allocation is explicable as due to inbreeding. *E. ravouxi* is an outbreeding species whose workers raid *Leptothorax* spp. See also Bourke (1989).
85. Bourke *et al.* (1988).
86. Buschinger *et al.* (1988). This is a workerless species.

87. Douwes *et al.* (1988). *Epimyrma adlerzia* is a workerless parasite in which the new queens mate within the host nest; the female-biased allocation is therefore expected as due to inbreeding.
88. Duchateau and Velthuis (1988)
89. Ross and Matthews (1989b)
90. Boomsma and Eickwort (1993). The sex ratio is for the summer brood.
91. Hasegawa (1994)
92. Mueller *et al.* (1994)
93. Herbers and Greco (1994). About half of the colonies are polygynous and half are monogynous.

Table 5.8. Sex ratios in non-eusocial bees (superfamily Apoidea) and aculeate wasps. *r* is the numerical proportion of males, WR is the weight ratio (females/males), and *M* is the proportion invested in males assuming that WR refers to the ratio of investment. All weight ratios were determined from dry weights save those of *Osmia bruneri*, for which fresh weights (FW), and *Sceliphron assimile* and *Trypoxylon politum*, for which provision fresh weights (PW) were used. When two sex ratios are given on the same line, they refer to summer and autumn generations of a bivoltine population.

Species	*r*	WR	*M*	Notes
BEES				
Halictidae				
Agapostemon nasutus	0.68	1.52	0.58	1, 25
Augochlora pura	0.27			2
Augochloropsis sparsilis	0.43			3
Neocorynura fumipennis	0.38			4
Nomia melanderi	0.50	0.80	0.56	25
Pseudagapostemon divaricatus	0.62			5
Pseudaugochloropsis				
graminea	0.20			6
nigerrima	0.43			6
Megachilidae				
Anthidium maculosum	0.23	0.93	0.24	7
Ashmeadiella				
meliloti	0.39	2.31	0.22	7
occipitalis	0.24	2.10	0.13	7
Chelostoma minutum	0.51	1.21	0.42	8
Hoplitus anthocopoides	0.66	0.94	0.67	25
Megachile				
gentilis	0.83	1.46	0.77	7
inermis	0.17, 0.88			9
mendica	0.70, 0.54			9
mendica	0.73	1.99	0.58	7
relativa	0.32, 0.54			9
Osmia				
excavata	0.63	1.53	0.53	10, 25
bruneri	0.57	1.29 (FW)	0.51	11

Table 5.8 *Continued*

Species	*r*	WR	*M*	Notes
lignaria lignaria	0.68	2.40	0.47	7
lignaria propinqua	0.67	1.93	0.51	12
pumila	0.28	1.84	0.17	7
Prochelostoma philadelphi	0.20	1.11	0.18	7
Anthophoridae				
Allodape				
ceratinoides, and *panurgoides*	0.29			13
exoloma	0.13			13
friesei	0.41			13
mucronata	0.55			13
rufogastra	0.29			13
Allodapula				
acutigera	0.37			13
dichroa	0.42			13
melanopus, and *turneri*	0.38			13
variegata	0.31			13
Anthophora				
abrupta	0.62	1.58	0.51	25
edwardsii	0.60	1.34	0.52	25
flexipes	0.60	1.15	0.57	14, 25
occidentalis	0.51	1.76	0.38	25
peritomae	0.50	2.36	0.30	25
Braunsapis				
bouyssoui	0.24			13
draconis	0.56			13
facialis	0.30			13
foveata	0.06			13
foveata, and *leptozonia*	0.23			13
leptozonia	0.14			13
luapulana	0.41			13
simplicipes	0.40			13
stuckenbergorum	0.16			13
Ceratina smaragdula	0.25, 0.47			15
Exoneura				
hamulata	0.42			13
variabilis	0.40			13
Exoneurella lawsoni	0.36, 0.66			13
Halterapis nigrinervis	0.11			13
Xylocopa				
caffra	0.36			16
capitata	0.48			16
erythrina	0.33			16
flavorufa	0.34			16
hottentotta	0.47			16
inconstans	0.34			16
lugubris	0.12			16

Table 5.8 *Continued*

Species	r	WR	M	Notes
rufitarsis	0.44			16
sicheli	0.44			16
somalica	0.31			16
Apidae				
Euplusia surinamensis	0.59	1.20	0.54	17
Andrenidae				
Calliopsis persimilis	0.47	1.47	0.38	26
Colletidae				
Chilicola ashmeadi	0.74	1.60	0.64	25
WASPS				
Eumenidae				
Ancistrocerus				
adiabatus	0.24	1.95	0.14	27
antilope	0.33, 0.84			18
antilope	0.09, 0.67			9
antilope	0.65	1.90	0.49	7
campestris	0.64	2.10	0.46	7
catskill albophaleratus	0.30, 0.85			18
catskill	0.49	2.28	0.30	7
tigris	0.21	2.37	0.10	7
Antodynerus flavescens	0.61	1.52	0.51	19, 25
Eumenes				
campaniformis	0.45			19
emarginatus	0.31			19
Euodynerus				
foraminatus	0.50, 0.52			9
foraminatus	0.47, 0.52			20
foraminatus	0.47			21
foraminatus				
apopkensis	0.70	2.07	0.53	7
foraminatus	0.72	1.49	0.63	7
leucomelas	0.32, 0.54			18
megaera	0.40	2.19	0.23	7
Monobia quadridens	0.47	1.93	0.31	7
Pachodynerus erynnis	0.42	2.16	0.25	7
Stenodynerus				
ineatiformis	0.32	1.95	0.19	7
krombeini	0.46	1.07	0.44	7
saecularis	0.41	1.39	0.33	7
toltecus	0.42			7
Symmorphus cristatus	0.51	1.92	0.35	7
Sphecidae				
Cemonus lethifer 1966	0.72, 0.19			22
1967	0.68, 0.39			22
1968	0.45, n.a.			22

Table 5.8 *Continued*

Species	r	WR	M	Notes
Chalybion bengalense	0.60	2.25	0.40	19
Ectemnius paucimaculatus	0.65	1.33	0.58	25
Passaloecus				
eremita	0.41	2.13	0.25	25
ithacae	0.29, 0.31			18
Sceliphron				
assimile	0.45		0.36 (PW)	23
spirifex	0.49			25
Trypoxylon				
clavatum	0.47	1.50	0.37	7
frigidum	0.42	1.40	0.34	7
johannis	0.54	1.26	0.48	7
pileatum	0.38			19
politum N,NJ	0.52	1.2	0.48	24
N,PA	0.54	1.2	0.50	24
S,FL	0.52, 0.63	1.2 (PW)	0.47, 0.59	24
striatum	0.62	1.30	0.56	7
tridentatum	0.44	0.90	0.47	7
Pompilidae				
Dipogon				
sayi	0.36, 0.55			18
sayi	0.29	3.46	0.11	7

1. Eickwort and Eickwort (1969); figure determined from pupal numbers from two excavated nests.
2. Stockhammer (1966)
3. Michener and Lange (1959).
4. Michener *et al.* (1966)
5. Michener and Lange (1958a).
6. Michener and Kerfoot (1967)
7. Krombein (1967), taken from Trivers and Hare (1976).
8. Information from several sites (Parker, 1988); method of determining weights not stated.
9. Longair (1981).
10. Hirashima (1958), taken from Michener (1974).
11. Frohlich and Tepedino (1986).
12. Tepedino and Torchio (1982).
13. Michener (1971).
14. Torchio and Youssef (1968).
15. Kapil and Kumar (1969).
16. Watmough (1983).
17. Janzen unpublished, taken from Michener (1974:75).
18. Fye (1965).
19. Jayakar and Spurway (1966).
20. Stubblefield unpublished, taken from Seger (1983).
21. Cowan (1979) stresses that many females mate with their brothers in this species.

22. Danks (1983).
23. Freeman (1981) used the fresh weight of provisioned prey to determine investment ratios by females of different sizes.
24. Brockmann and Grafen (1989); the northern populations (NJ New Jersey, PA Pennsylvania) are univoltine and the southern one (FL Florida) is bivoltine. Figures shown are averages over several years. See also Brockmann and Grafen (1992).
25. Various sources taken from Trivers and Hare (1976).
26. Danforth (1990).
27. Cowan (1981).

The period of broad surveys has thus been equivocal in giving a single answer to the seemingly simple question 'Who wins?'. It is worth asking how future research should proceed in order to obtain answers. Broadly following Pamilo (1991a) and Crozier and Pamilo (1993), we suggest following the approaches below.

1. Comparisons between many species. This, of course, continues the approach begun by Trivers and Hare (1976) which, we have just seen, has not yielded definitive results. Nevertheless, as noted by Trivers and Hare (1976), such surveys can pick out broad patterns. However, more detailed information should be sought for each species in future, especially on the levels of polyandry and polygyny.

2. Comparisons between closely related populations differing in some important feature believed to affect sex allocation. The test of the effect of polyandry in *Lasius niger* by Van der Have *et al.* (1988) is a case in point.

3. Studies of variation between colonies of the same population. Are trends as predicted by theory? Can the various possible causes of variation be untangled? The study by Herbers (1990) discussed below is an exemplar of this genre. We will return to the colony-level sex ratio variation in the next chapter.

4. Behavioural studies. Can queen–worker conflict be observed, namely, can a physical basis be seen? A comparison between initial (eggs) and final (adults) sex ratios reveals any changes that may have taken place while the brood was raised in the colony (Aron *et al.*, 1994).

5. Precise prediction of sex allocation patterns. A wealth of parameters is needed for this boldest of all courses: the relatedness structure of colonies, differences in relatedness structure between colonies, population structures (inbreeding and interaction structures (Wade, 1980), proportions of colonies of different types, effective population size), reproductive values of the various classes of individuals, production costs of queens, males, and

workers, and the abilities of individuals able to affect sex allocation to detect the kind of colony they are in.

6. Interference from other levels of selection than the population, colony and individual levels we have discussed. Are there intragenomic agents affecting the sex allocation of social insects in a manner analogous to the various infectious agents and selfish genetic elements that occur in the parasitoid wasp *Nasonia vitripennis* (Werren, 1987; Nur *et al.*, 1988)? Such agents are more likely in polygynous species than in monogynous ones because in the latter marked derangement of the sex ratio would often sterilize the colony preventing establishment of the element.

After the pessimism engendered by our survey of surveys, we will end this section on an optimistic note by briefly discussing some case histories auguring well for the future.

Apterostigma dentigerum. As noted earlier, Forsyth's (1981) pathbreaking study on this ant species brought the phenomenon of sex ratio compensation to attention, although more through inference than through experiment. However, concentrating on the question of this section, we repeat our earlier observation that according to the predictions of theory the sex allocation patterns of queen-right colonies depart significantly from the predictions of both queen and worker control, falling between them. This is evidence suggestive of queen–worker conflict, with neither 'winning' — or 'losing' completely either.

Lasius niger. Dutch populations of this ant vary in the proportion of nests headed by twice-mated as against once-mated queens. The populations vary in their average sex allocation patterns in the direction predicted by theory (van der Have *et al.*, 1988), and Boomsma and Grafen (1990) adduce evidence from many ant species that such variation is widespread and due to worker detection of the relatedness structure of the colonies in which they are present. The ability to respond is evidence of a considerable level of worker control over the sex ratio.

Augochlorella striata. Mueller (1991) went beyond inference and manipulated colonies of this bee. He set up pairs of colonies, one colony with a queen (eusocial) and one without (parasocial, see Table 1.1), manipulating other attributes to be the same. Loss of the queen leads to her replacement, and the colony lacks the relatedness asymmetry characteristic of colonies with a simple family structure. Theory predicts that, under 'worker' control, the sex allocation of eusocial colonies should be more female biased than in parasocial colonies (Boomsma, 1991), and this is what Mueller found. The ability to manipulate other factors to be the same between the two classes of colonies gives confidence that it was the relatedness structure alone that led to the change in allocation pattern.

Leptothorax longispinosus. Herbers (1990) carried out a ten-year study of populations of this ant at two locations, and analysed the various possible factors influencing allocation patterns using path analysis. She was able to eliminate LMC and LRC as significant determinants of sex allocation in these populations, and interpreted significant effects of queen and worker numbers as evidence of queen–worker conflict. The occurrence of a large unexplained variance remaining led Herbers to suggest that unexamined ecological factors might be important. Significant year-to-year variation was a feature of the data, leading Herbers also to stress the importance of long-term studies.

As suggested in point 5 of the above list, we could estimate the necessary parameters (relatednesses, sex-specific reproductive values) and calculate the predicted sex ratio under both queen and worker control. Enzyme electrophoretic data have been used for estimating relatednesses (Table 4.7) and such estimates have been further used to predict sex ratios in the ants *Formica sanguinea* and *F. exsecta* (Pamilo and Rosengren, 1983) and in the sphecid wasp *Microstigmus comes* (Ross and Matthews, 1989b). The observed sex ratios in these species agreed well with the expectations under worker control, except in *F. exsecta* in which the observed sex ratio was intermediate between those expected under queen and worker control. Packer and Owen (1994) could also take into account worker reproduction both in queen-right and in orphaned colonies of the bee *Lasioglossum laevissimum*, and they calculated the predictions from eqn. (5.6). The observed female bias exceeded that expected under worker control (and of course that expected under queen control), and the results indicate that it is the workers which are in control of sexual investment in this bee.

5.7 Summary

Reproductive value is the expected contribution to the asymptotic gene pool, and is affected by such factors as whether the workers produce some or all of the males and, if worker male-production occurs, the proportion of orphaned colonies (those whose queens have died). In many species workers reproduce only in orphaned colonies, and the male-production of such colonies devalues males for the queen-right colonies, which are expected to invest more in females than they would otherwise do, the phenomenon called 'sex ratio compensation'. The proportion of orphaned colonies is small in some species examined, but appears to be large enough in *Apterostigma* to affect the expected allocation patterns, and is clearly implicated in affecting allocation patterns in species of *Rhytidoponera* with substantial numbers of colonies headed by gamergates (mated workers producing both haploid and diploid eggs).

A further possible factor in determining sex allocation patterns is whether or not there is a linear relationship between the preferences of genetically different types of workers and their proportion in the colony. If a worker genotype's preferences determine that of the colony, even when other genotypes are present, then we describe that genotype as showing 'behavioural dominance' over the others. If behavioural dominance exists it could lead to a much broader range of sex allocation patterns than otherwise expected. Behavioural work is needed to ascertain the likelihood of behavioural dominance occurring at all.

Joint models of queen and worker control show that the queen is expected to be able to exert considerable influence on sex allocation patterns in perennial (usually large) and, especially, in annual (often small) colonies. In perennial colonies the leverage of the queen is reduced as the production costs of queen and worker diverge, because when queens become extremely expensive it is easier for workers to pour investment into a small number of female larvae.

In the conflict over the allocation of resources between reproductive and maintenance functions, the workers are expected to favour reproduction more than the queen. Because queens can cheat against each other, and will be selected to do so, queen efforts to maintain colony longevity are expected to be undermined in polygynous colonies. In monogynous species, queens are expected to have more leverage when the sexual and worker broods occur at separate times, leading to an expected difference in longevities between related species depending on whether or not they have separate sexual broods. Too few data are yet available to test this prediction.

Eusocial insects often found new nests by budding, and sometimes by fission of old ones. Fission is a process where a colony forms two or more daughter colonies which become functionally independent. In the honey-bee, a fissioning species, the colony usually divides in two when the old queen leaves the hive with a group of workers and starts a new nest. The old hive is inherited by one of the colony's daughters. There can also be afterswarms in the same season, where the colony divides again. Similar colony fission occurs in army ants, which have monogynous colonies usually dividing in two.

A somewhat different situation is found in many polygynous species of ants, especially those that are also polydomous. These colonies normally reproduce by budding. In this process, the workers build new nests and then transfer queens to them. These queens may be either old or new. An old colony may divide into a number of daughter nests during a single season, but it is problematic to say how independent these units are from each other. Therefore the situation is different from that of honey-bees or army ants, whose daughter colonies are functionally separate units. In polydomous ants the nests often (if not always) remain functionally connected.

The question is: what kind of investment is expected? In addition to new queens and males, the worker force and the nest structures inherited by the daughter colony are also big investments.

When the daughter colonies arise by fission and do not compete with each other for resources such as space, food sources, or nest sites, there is no real local resource competition, and we expect the same investment ratios as in species with independent colony founding. If the daughter colonies are formed by budding, remain in the vicinity of the natal nest, and compete for resources, the number of successful daughter colonies can be greatly restricted. In this case of limited dispersal of daughter queens, LRC is produced between them, and we expect male-biased investment ratios even when the workers are counted as an investment in females. It is worthwhile to keep a clear distinction between fission and budding, because the latter normally includes complicating fitness effects through interactions between daughter colonies. In field studies it may be difficult to distinguish between the alternatives of fission and budding, but this analysis shows that it is worth the effort. It should be noted that the present results are based on the assumption that colonies always grow to the same size before fission. When they do not, we can expect some differences of allocation between colonies of different size.

It is appropriate to review the means by which the sex ratio can be biased by the actors in the sex allocation drama. Queens control the number and sex ratio of the eggs; they may also sometimes be able to produce eggs biased to develop as workers. Workers can cannibalize eggs, and the evidence suggests that they can do so differentially by sex. Workers also can affect the developmental track of female larvae, so that they are more or less likely to become new queens.

The concept of 'queen control', frequently encountered in the literature, can now be seen to be extremely problematic. We suggest that, although the queen can attempt to manipulate the workers through such means as controlling the supply of eggs, and perhaps through an effect on the relatedness structure of the colony through the number of times she mates, her production of pheromones is best seen not as queen control but rather as queen signalling. The workers then behave adaptively according to the presence or absence of a queen, and are not 'controlled' by her in any other sense.

The problem of deciding 'who wins' has itself two parts. First, there is the need to predict the allocation patterns, in both the male/female and reproduction/maintenance dichotomies, of the two parties. Prediction requires a better knowledge of the biology of the population than has usually been acquired. It may generally be more practical to look for either natural or experimental perturbations of specific factors in order to determine their relative importance, rather than attempting a complete prediction. The second problem is estimating the allocation pattern itself, in order to compare it with the prediction. The ease with which this can be done varies: in ants it is often difficult, whereas in mass-provisioning bees one can collect the entire mass of provisions supplied to the developing larva. Weight ratios have often been used; in ants these depart from the true cost ratios, requiring correction factors.

The confident finding of Trivers and Hare of worker control was tarnished by statistical flaws and lacunae in the then available knowledge about social insect biology. Recently, it has been often argued that, in broad terms, Trivers and Hare were probably right. However, the broad sweep of the data show a broad range of values, lacking in consistency, making such a judgement equivocal. We present a programme for future work, concentrating more on in depth studies of particular cases than the continuation of the survey approach, and note that a handful of well-studied cases suggest that queen–worker conflict really does occur, as predicted, and that the workers are perhaps getting the best of things.

Appendix: Transition matrix for joint effects of sex-ratio compensation and worker male-production

This is the situation presented in Section 5.1.3. The proportion of sexuals produced in orphaned colonies is β_s and the proportion of worker-produced males in queenright colonies is ψ, as used in the text. If we denote $\beta = \beta_s$, the transition matrix $\mathbf{P}$ can be written as

$$\mathbf{P} = \begin{bmatrix} 0.5 & 1 \\ \dfrac{(1-\beta)\,M(1/2-\psi/4)+\beta/4}{\beta+(1-\beta)M} & \dfrac{(1-\beta)\,M\psi/2+\beta/2}{\beta+(1-\beta)M} \end{bmatrix}$$

This yields

$$\mathbf{P}_\infty = \frac{\beta+(1-\beta)M(2-\psi)}{2\beta+(1-\beta)M(3-\psi)} \begin{bmatrix} 1 & \dfrac{2(\beta+(1-\beta)M)}{\beta+(1-\beta)M(2-\psi)} \\ 0.5 & \dfrac{\beta+(1-\beta)M}{\beta+(1-\beta)M(2-\psi)} \end{bmatrix}.$$

The sex-specific reproductive values can be calculated from these matrices and they are given in the text eqn. (5.5).

6

Colony-level variation of sex ratios

We have so far seen that the intrinsic asymmetries of relatedness in most social insect groups have probably been involved in the evolution of eusociality itself, and certainly have led to a fundamental conflict of interest between queens and workers. In the development of the theory dealing with these phenomena we have touched on the importance of the interaction between colonies in populations: such interactions between colonies of different characteristics affect the sex allocation pattern of each. We close this book by considering this topic directly.

Studies on numerous social insects reveal considerable variation between the sex allocation patterns of different colonies of the same population. As shown previously (Section 2.1), when all individuals have the same optimal sex ratio and the population sex ratio is at equilibrium (and fitness functions are linear), any individual strategy is equally advantageous. In any equilibrium situation, we can expect wide variation in individual, i.e. colony-level, sex ratios. This variation can be considered as random noise around the equilibrium. Alternatively, sex ratio variation can be adaptive in the sense that the ratio produced by a given colony is somehow correlated with characteristics of that colony. These characteristics could include both intrinsic features of the colony, i.e. its genetic structure, and extrinsic, environmental factors determining the level of resources available in sexual production.

We can identify several general principles which can affect the colony-level variation in sex ratios:

- variation uncorrelated with colony characteristics,
- variation due to the amount of resources available,
- variation due to differences in colony structure,
- variation in the outcome of queen–worker conflict.

We will first explain how these factors affect the sex ratios and then examine how they fit with the observational data.

6.1 Variation uncorrelated with colony characteristics

Colony sex ratios can vary either *stochastically* around the expected mean, or *deterministically* if they are strictly genetically determined. In both cases, the sex allocation pattern within a colony is not correlated with colony type (the strength of the colony, the number of reproductives etc.).

When the population sex ratio is at equilibrium, any individual strategy should be equally favoured when the fitness returns through daughters and sons are linear (Chapter 2). This is clearly seen, for example, in the simulation results of Pamilo (1982b) based on deterministic genetic models. Under queen control, each initial set of allele frequencies gave different equilibrium frequencies with a 1:1 population sex ratio. These results were obtained in deterministic models assuming an infinite population size. In a population of finite size the situation can be different. If the population sex ratio varies stochastically from generation to generation, the losses due to producing offspring of the common sex become greater than the extra gains when producing offspring of the rare sex (Verner, 1965; Williams, 1979; Taylor and Sauer, 1980). The best long term strategy would be to produce brood sex ratios close to the population mean. Such a strategy would also reduce the variation in the population.

Deterministic simulations show that there will be less genetic variation for the sex ratio under worker control than under queen control, usually none at all (Pamilo, 1982b). While it could be argued that this result stems from the particular model examined, it was also found in an analysis of a two-allele male-haploid model (Oster *et al.*, 1977). Recalling that the equilibrium selected at the queen level is 1:1 and that at the worker level often departs from this, it therefore seems that there will be less genetic variation for the sex ratio when the population equilibrium departs from 1:1.

6.2 Sex ratios affected by resource availability

6.2.1 *Environment affects sex or caste determination*

It has been suggested in a number of species that environmental factors can influence the colony sex ratio by affecting directly either sex determination or caste determination. These effects may be non-adaptive. It has been known for a long time that ant colonies producing fewer sexuals tend to produce a higher proportion of these as males (Nonacs 1986a; Boomsma 1989), a phenomenon sometimes interpreted as being linked to the nests with a smaller production of sexuals being in poor condition (Nonacs, 1986a). Gösswald and Bier (1957) suggested that the insemination of eggs in *Formica* ants depends on temperature proposing a physiological mechanism ('sperm pump') by which fertilization is prevented in low temperatures. That would lead to male production in nests where temperature control does not work properly. This phenomenon would affect sex determination and the proportions of haploid and diploid eggs laid.

Another possibility would be that nest condition affects caste determination. Pamilo and Rosengren (1984) speculated that the highly male-biased sex ratios in polydomous *Formica exsecta* colonies could, at least partly, result from poor nutritional status of nests in a dense population. This effect would lead to diploid eggs developing mainly into workers and haploid eggs into small-sized 'micraner' males. The true nutritional status of the *F. exsecta* nests is not known, and other explanations (particularly local resource competition) lead to similar predictions. Nonacs (1986a) also considered the possibility that nutritional status affects the diploid caste determination and therefore the sex ratio, finding in a literature survey that in *Leptothorax* ants the total sexual biomass was more strongly correlated with sex ratio than was the total worker number (a point supported by Herbers (1990)). Such observations do not adequately answer the question whether the male-biased sex ratio in poor nests is a mechanistic (and even non-adaptive) effect following caste determination or whether it represents an adaptive response to the availability of resources.

The opposite trend was observed in *Pogonomyrmex* colonies where extra feeding of colonies led to increased male production (Mackay, 1985). This trend was interpreted as reflecting increased worker reproduction when the nutritional status of nests improves.

6.2.2 *Allocation between colony growth and sexual production*

The queens of highly evolved eusocial insects are normally much larger than males and workers, and the males are often larger than workers. Consequently the development times of larvae can vary and the developing larvae of different sexes or castes may vary in how well they can compete for the available food resources. In *Myrmica* ants, the queens lay eggs predominantly during the first half of summer (Brian, 1979a). Some of these eggs develop into workers and some continue growing through the latter half of summer and then hibernate. After hibernation many of them develop into new queens, but some of them still become workers. The rate of egg-laying by the queens decreases towards the end of the summer and, instead, the workers start producing haploid eggs in late summer. These eggs develop into males. As a result, in early summer the nests have developing diploid larvae and the problem of allocation is whether they become new queens or workers. In late summer, the nests have mainly queen and male larvae and the question of allocation concerns directly the sex ratio. If the nest invests heavily in queen production, such investment will diminish the future worker force (competition in early summer), whereas by this stage investment in males does not much affect the growth of the colony. A numerical model (Brian *et al.*, 1981) predicts that a colony with limited resources should guarantee the colony survival and therefore restrict queen production. This predicts male-biased investment in colonies with limited resources.

The *Myrmica* model also predicts a year-to-year variation of sex ratios for two reasons. First, in bad years, the colonies should invest a high proportion of food resources in colony maintenance which yields a male-biased population sex ratio. Second, the numerical model of Brian *et al.* (1981) suggests that a natural cycle can evolve. If a colony invests much in queen production, next year this colony has to put more resources in colony maintenance and the sex ratio becomes male dominated. Because the total productivity is also affected by the environment, the colonies may be in phase. That would predict cycles in population level of sexual production and sex ratio. The observations from *Myrmica sabuleti* (Brian, 1972), *M. sulcinodis* (Elmes, 1987b) and *Tetramorium caespitum* (Brian and Elmes, 1974) hint that such cycles may exist.

6.2.3 *Non-linear fitness functions*

We noted above (Section 5.4) that when colonies reproduce by colony fission, the established daughter colonies are expected to be of the same size. This means that the resources available for male production vary depending on the size of the colony before fission. This principle can be generalized.

When the fitness functions of sons and daughters are not linear, it may be optimal to the colony to invest a fixed amount either in sons or in daughters, depending on the fitness functions, and use the remaining resources for producing offspring of the other sex. This is easiest to examine if we assume that the fitness function is linear in one sex. Yamaguchi (1985) and Frank (1987c) showed this in the case of local mate competition. In LMC the fitness return through sons is a diminishing function of the amount of investment, whereas the overall fitness of the daughters increases linearly with the investment. The best strategy would then be that each colony invests the same amount in males and the remaining resources in females. This means that the colony sex ratio should become more female biased with increasing colony size.

Local resource competition, for example in the case of colony budding, leads to an opposite prediction. There is, however, one difference. When a colony divides, the resources invested in daughter colonies form a package. If a colony is big enough, it can produce more daughter colonies of the same size than the smaller colonies do. This means that the theoretically optimal investment ratio has a stepwise form (Fig. 6.1). We can predict that the investment becomes more male biased with increasing colony size but this trend holds only within a given category defined by the number of daughter colonies formed.

6.3 Variation due to differences in colony types

6.3.1 *Orphaned colonies produce only males*

A natural source of sex ratio variation among colonies is caused by orphaned colonies producing only males. This is limited to species whose workers have the ability to lay haploid eggs giving rise to males. The queen-right nests are

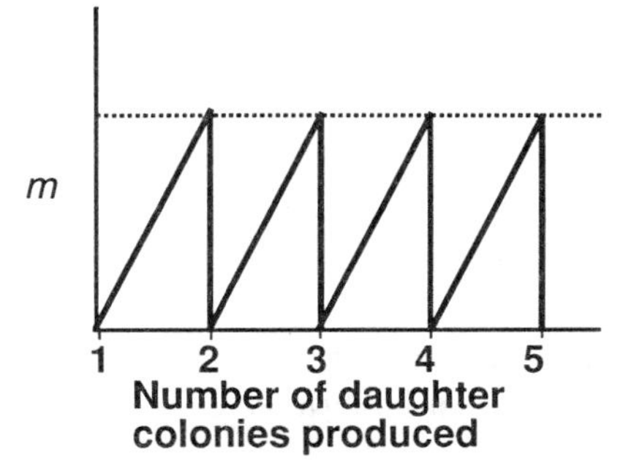

Fig. 6.1 The stepwise form of the function for investment in males in a species with colony fission. As usual, *m* denotes the proportion of the total investment in male and female functions devoted to males. The dotted line is the level of resources required for the production of a daughter colony. At fission time, the colony produces daughter colonies, with any additional resources going into males.

expected to compensate for this male bias by producing female-biased brood sex ratios (Section 5.1).

6.3.2 *Colony sex ratio may be associated with the colony type*

In addition to orphanage, there can be other types of variation in colony types which are reflected in their sex allocation. Some species of the ant genus *Rhytidoponera* have two types of reproducing females: morphologically differentiated **queens** and mated workers (**gamergates**). Moreover, there are two types of colonies, type A colonies headed by queens and type B colonies headed by gamergates (Section 5.1.4). In the *Rhytidoponera impressa* group species, Ward (1983a) found that the nests with queens are monogynous and the queen is the only egg-layer. In the absence of a queen, there are several gamergates which share the egg-laying. The gamergates produce new workers and males but no queens. There are thus two important differences between these colony types: the genetic structure is different depending on the number of egg-layers and the type B colonies are, like orphaned colonies, destined to be male producers.

The *Rhytidoponera* case is a combination of two separate problems: the restricted source of queens, and the difference in the genetic structure of colonies. The latter aspect alone could be enough to cause colony-level sex ratio variation. Relatednesses are affected by the number of matings, the number of queens per colony and by worker reproduction (Chapter 4). If the colonies differ with respect to such factors, the optimal sex ratios will also differ. This principle was generalized by Nonacs (1986a), who proposed that the workers should bias the colony's sex ratio in such a way as to maximize their inclusive fitness. If they can optimize their behaviour, this would lead to split sex ratios, i.e. some nests specializing in females and some in males. Boomsma and Grafen (1990, 1991) further quantified this model and its effects on the population as a whole. If we assume that the fitnesses of sons and daughters are linearly related to the investment, we can write the inclusive fitness as

$$V_i = \frac{g_{FI}v_F f}{F} + \frac{g_{MI}v_M m}{M} \tag{6.1}$$

where m is the proportion of sexuals which are sons in colony i and M is the population proportion of males, and f and F are the corresponding quantities for daughters. This function can be reorganized

$$V_i = \frac{g_{FI}v_F}{F} - m\left[\frac{g_{FI}v_F}{F} - \frac{g_{MI}v_M}{M}\right]. \tag{6.2}$$

This function is maximized by having

$$m = 1 \text{ when } \frac{g_{MI}v_M}{M} > \frac{g_{FI}v_F}{F}$$

$$m = 0 \text{ when } \frac{g_{MI}v_M}{M} < \frac{g_{FI}v_F}{F}.$$

If the daughters are highly related to the controlling individuals (I), the latter inequality is likely to be true, whereas low relatedness of the daughters leads to the first inequality. Let us briefly examine the colony characteristics expected to affect m.

Under queen control, the expected M at the population level is 0.5. Even under polygyny, the values of g_{MI} and g_{FI} decrease to the same extent and we do not predict any selection for sex ratio variation at the colony level. The situation may become different if workers produce males in part of the colonies. The expected population sex ratio under queen control is $M = 0.5$ even though workers produce males. If worker reproduction is unevenly distributed among the colonies, queens in the colonies where they are the only source of males should favour a male-biased colony sex ratio and the queens in the colonies with reproductive workers should favour a female-biased colony sex ratio. This simply means that there is selection on queens to prevent worker reproduction. Under worker control, the relatednesses vary as functions of polyandry and polygyny. If the effective polyandry of a single queen is k_E (eqn. 4.15), the bracketed term in the inclusive fitness relation (6.2), from the viewpoint of the workers in monogynous colonies, takes the form

$$\frac{3/4k_E + (k_E - 1)/4k_E}{F} - \frac{1}{2M} = \frac{0.5 + 1/k_E}{F} - \frac{1}{2M}.$$

For given M, and varying k_E, this function is likely to be positive when k_E is small and negative for large k_E. So, in monogynous colonies, the workers should favour a female-biased sex ratio in colonies with singly inseminated queens and a male-biased sex ratio in colonies with a multiply inseminated queen (Nonacs, 1986a; van der Have *et al.*, 1988). We would therefore expect colony-level sex ratio variation, provided that the workers can recognize the type of colony they live in and that they 'know' the population investment ratio. The latter knowledge can be built in by evolutionary history, and the recent studies on kin recognition indicate that the workers may be able to recognize the level of genetic diversity within a colony. Similar split sex ratios are expected in other cases where the relatedness asymmetries vary among the colonies, for example in primitively social bees that have both eusocial and parasocial colonies in the same population (Boomsma, 1991). Observations from both natural populations (Yanega, 1988; Boomsma, 1991) and experimentally manipulated nests (Mueller, 1991) show split sex ratios as predicted by the theory.

It should be noted that colony-level variation also affects the population sex ratio (Boomsma and Grafen, 1991). To take a simple case, let us assume that there are only two colony types in the population, monogynous nests with either singly or doubly inseminated queens. Assume further that the queen produces all the offspring but the workers control the sex ratio. The optimal population sex ratio for the workers with a singly inseminated queen is $M = 0.25$, whereas the optimum in the other colonies is $M = 0.33$. If the queens are largely monandrous, the population sex ratio will be 0.25. The workers of the rare nests with double mating will maximize their inclusive fitness by spending all available resources in raising males. If double mating occurs with a frequency less than 25%, the monandrous nests can adjust their sex ratio so as to keep the population ratio at $M = 0.25$. When the frequency of double mating is within the limits 25%–33%, both colony types should produce extreme sex ratios and the population allocation pattern shifts from $M = 0.25$ to 0.33. When double mating is more frequent than 33%, the monandrous colonies produce only females and the workers in the double-mating colonies adjust their sex ratio so as to keep the population ratio at $M = 0.33$. As noted above, this prediction holds only if the workers can recognize which colony types they are in. If they cannot, the population sex ratio is predicted by the mean relatednesses to the workers $g_{MW} = 0.5$ and $g_{FW} = (2 + k_E)/4k_E$ (Section 4.3).

One stimulus to the modelling by Boomsma and Grafen was the observation that two populations of the Dutch ant *Lasius niger* differed in M in the direction predicted by the proportion of queens that mated twice (van der Have *et al.*, 1988). These populations, however, are of recent age (one is on a dune only about 25 years old). Consequently, it seems likely that the populations have not achieved their M values through gene frequency change but that the differences in M are a direct consequence of differences in the queen mating frequencies between the populations. The suggestion is that the workers can determine the class of colony they are in by observing the variety of colony-recognition cues (Section 4.5.2).

Two complicating factors are immediately apparent. If the number of different colony types is large, there will be more thresholds where the sex ratios (both in the population and within the colonies) shift from one value to another (Boomsma and Grafen 1990, 1991). Another interesting point, also raised by Boomsma and Grafen, concerns the effects of population size and of colony size. In the above derivation we assumed that the fitness returns through both sons and daughters are linear functions of the investment. This is true in a large population, where the sexual production of an individual colony affects neither the mating success of males nor the success of colony founding by females. In a small population this is no longer true. If a single colony produces many males, it concurrently devalues the males and the fitness function is a convex curve. Similarly, heavy investment in females can lead to local resource competition among colony-founding sister queens. If a single colony produces a substantial fraction of all sexuals in a population, the investment pattern of such a colony should approach the population mean ratio. This leads to a prediction that in local populations, small colonies should produce extreme sex ratios of their optimal type, and the large colonies should produce colony-level sex ratios closer to the population mean (Boomsma and Grafen, 1991).

6.4 Connections to worker–queen conflict

It has been stressed throughout this book that queens and workers prefer different sex allocation ratios under a wide variety of conditions. This conflict can lead to a kind of arms race between queens and workers. If the workers try to bias the sex ratio, the queens are expected to respond by making the biasing less profitable to the workers. An extreme case would be that the queens lay eggs of only one ploidy level in any one nest (Williams, 1979; Bulmer, 1981). Such a course of action could guarantee the queens' complete control of the population sex ratio and simultaneously it leads to a maximal colony-level sex ratio variance. Simulations of genetically deterministic conflict models show that evolution may indeed lead to such an extreme specialization of colonies (Pamilo, 1982b). This is, of course, possible only if the production of haploid eggs in half of the colonies does not affect the colony growth.

Specialization of colonies is ruled out if the queens must lay diploid eggs in order to guarantee worker production and colony growth. In such cases queen–worker conflict over sex allocation becomes possible. Herbers (1984) argued that this conflict can have differing outcomes within colonies of a single species and that the workers could win more control with the increasing worker/queen ratio. The prediction in that case would be that large nests invest more in females.

6.5 Small ant colonies produce more males

As noted above, ant colonies producing few sexuals tend to have relatively male-biased sex allocation, and ones with large production tend to have relatively female-biased sex allocation (Nonacs, 1986a; Boomsma, 1988). This pattern occurs even in queen-right monogynous colonies (Boomsma *et al.*, 1982; Herbers, 1984; Nonacs, 1986a,b), showing that worker male-production by queenless colonies is not a sufficient explanation. There is a positive correlation between quantity of sexuals produced and colony size, but it is not a strong one (Nonacs, 1986b).

Three explanations have been proposed for the association in queen-right colonies of female bias with a larger biomass of sexual production.

Queen–worker conflict. Herbers (1984) suggested that when workers are few in number they are less able to counter the sex ratio biasing actions of the queen than when they are more numerous. Because workers in monogynous colonies are selected to prefer an F/M ratio of 3 and the queen a ratio of 1, smaller colonies (which are suggested to be more queen dominated) will have more male biased sex ratios. Under this explanation, the colony sex ratio depends on the number of workers, and any correlation with the biomass of sexuals produced is a secondary one due to the correlation of worker number with sexual biomass production. This prediction is contradicted by partial correlation coefficient analyses by Nonacs (1986b) of 18 monogynous ant species: controlling for sexual biomass largely eliminates the effects of worker number on m, but controlling for worker number still leaves a strong effect of sexual biomass produced.

Resource availability. Nonacs (1986a) suggested that colonies with small production of sexuals are experiencing more stringent conditions, and are investing more in the cheaper sex (males). This argument supposes that ant females, like male mammals, benefit more individually from increased investment than do their brothers, as reflected in the size difference (ant females are typically much larger than males). In such circumstances, colonies faced with producing either solely smaller males or smaller females should produce smaller males (Charnov, 1982:38). However, the colonial sex ratios in the ant *Leptothorax longispinosus* did not respond noticably to manipulated resource availability, although such a response might have been expected (Backus and Herbers, 1992).

Local mate competition. Greater competition for mates among sons than among daughters reduces the value of having sons and leads to female biased sex allocation (Section 2.8). Although, as discussed previously (Section 2.8), it is unlikely that overall sex allocation patterns are much affected by LMC, even low levels of LMC are sufficient to cause heterogeneity between colonies. This result was found independently by Yamaguchi (1985) for aphid species in which females are able to vary the sex ratio in their brood by cytogenetic means, and Frank (1987c) for ant colonies. Specifically, LMC will cause all colonies to invest all production in males until a threshold is reached, with all further production going into females (Fig. 6.2). The ant *Technomyrmex albipes* has a colony development with wingless and winged sexuals and cycling inbreeding (Section 4.5.1). This leads to strong LMC and female-biased sex ratios among

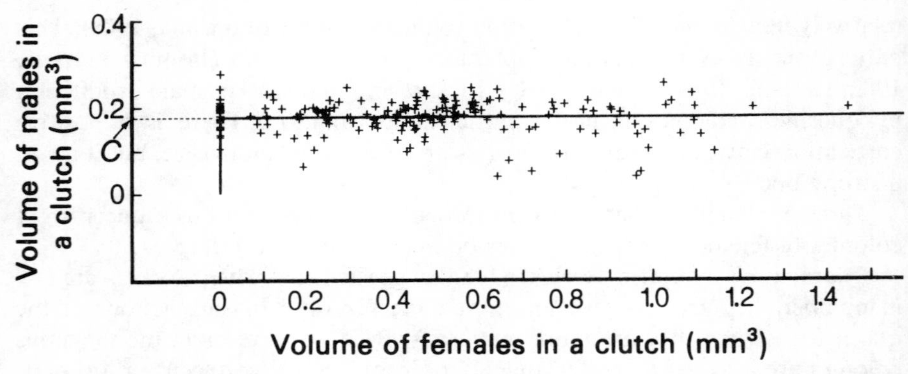

Fig. 6.2 Sex allocation in the aphid *Prociphilus oriens* compared with the predicted pattern under local resource competition theory (Yamaguchi, 1985). Females in this and other aphid species have the ability to determine the sex of individual offspring by cytogenetic means, paralleling the ability of hymenopteran females to determine offspring sex via sperm release. Males produced by a group of mother aphids mate at random with the females produced by the group, who then disperse. Yamaguchi estimated the total volume of male and of female offspring produced by each mother aphid and found that, as predicted by theory, each mother invested in males up to a threshold (*C* in the figure), with all additional investment being in females. The value of *C* is determined jointly by the total investment available and the number of mothers in the group. (Reprinted with permission from *Nature* (Yamaguchi, 1985).)

the inbreeding wingless sexuals, the female bias increasing with an increasing production of sexuals (Tsuji and Yamauchi, 1994).

A further alternative is possible (Crozier and Pamilo, 1993), stemming from the idea that the effective ratio of costs of females to males may vary with colony size. Such variation leads to different payoffs for male as against female production in colonies of different sizes, and hence leads to systematic differences between the sex ratios of colonies of different sizes.

How would effective cost differences arise? A difference in effective cost will occur even if each individual queen and male in a small colony absorbs the same amount of resource as an individual queen or male in a large colony. This variation in effective cost also depends critically on the facts that resource levels for sexual production in ant colonies vary from year to year, as shown by variation in total sexual production (e.g. Elmes, 1987a,b), and that colonies may produce sexuals even when resources are limited, again as reflected in actual sexual production (for example, Herbers (1984) mentions that in her study population of *Leptothorax longispinosus* colonies produced a median of four sexuals).

Let the actual cost of producing a single queen be c resource units and that of a male one unit ($c > 1$) and let this amount be the same for colonies of all sizes. For the purpose of initially fixing ideas, suppose that colonies specialize in producing either only queens or only males, and that, as in many species, sexuals are produced in a brood by themselves (e.g. Gösswald and Bier, 1957; Schmidt, 1974). (While this assumption simplifies the model, it is not a necessary condition for the effect to occur. In many species queen determination occurs in the larval or even egg stage (reviewed by Wheeler, 1986), and conversion of these individuals into others by cannibalism, or of workers into queens by the same method, will necessarily impose a cost whether sexuals are produced in a separate brood or not.) Let the amount of resource available for the production of sexuals be K. The amount of resource required before a queen can be produced is c, which is by definition greater than the threshold before a male can be produced. There is then a second threshold before the second individual is produced in each case, and so forth. On average, the expected leftover of resources will be $c/2$ units if a colony produces solely queens and $\frac{1}{2}$ units if solely males. Thus, if all resources are invested in females, then the mean number produced will be $(K-c/2)/c$, yielding the cost per female of $cK/(K-c/2)$. The effective ratio of costs for female to male production is then

$$c_E = \frac{c(2K-1)}{2K-c} \; . \tag{6.3}$$

As the total resource increases, c_E approaches c, but for low resource levels it is greater (Fig. 6.3). Trivially, where only fewer than c units of resource are available, the cost ratio becomes infinite because males can still be reared although queens cannot.

Clearly, ant colonies need not produce only queens or males, although even if each colony had perfect knowledge of the resources available to it for sexual production, a bias towards the production of males would still be expected in small colonies because they may have insufficient resources to produce a queen

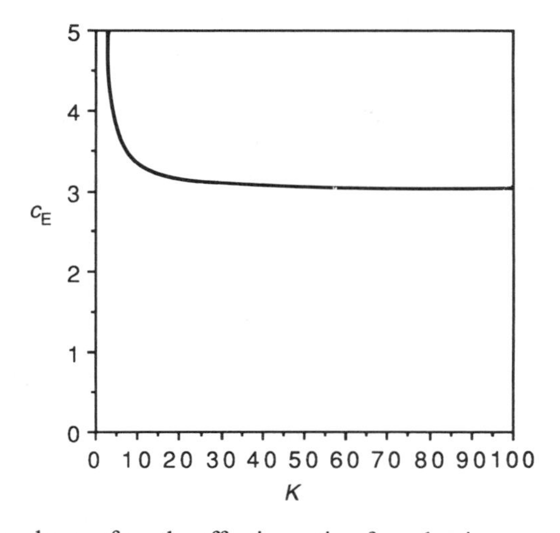

Fig. 6.3 Dependence of c_E, the effective ratio of production costs of queens to males, on the resources available, K.

but could produce one or more males. However, the inter-year variability of resources mentioned above also means that the effective cost ratio will still be greater in small than in large colonies even if both sexes can be produced. This result follows because there is uncertainty at the start of the season as to how many queen and male larvae of those initially produced can finally be carried to adulthood.

At a certain point in the season, the colony will be able to assess correctly how much resource is actually available to it, K. It then faces the linear programming problem of ensuring that

$$D''c + H'' + t(D'c + H') < K \tag{6.4}$$

where D' and H' are the numbers of abandoned sexuals, D'' and H'' are sexuals completing development ($D'' + D' = D$), and t is the proportion of resources invested in abandoned larvae which cannot be recovered.

Equation (6.4) allows a number of outcomes, which can be understood in terms of a modification of our standard approach, with selection maximizing

$$W_1 = \left\{\frac{g_{\mathrm{F1}}v_{\mathrm{F}}f}{F}\right\}\left\{\frac{c_{\mathrm{EP}}}{c_{\mathrm{E1}}}\right\} + \left\{\frac{g_{\mathrm{M1}}v_{\mathrm{M}}m}{M}\right\} \tag{6.5}$$

where c_{EP}, c_{E1} are effective costs of a queen relative to that of a male, in the entire population, and for the colony in question.

In a variable environment, the twin constraints of eqns. (6.4) and (6.5) mean that queens become effectively more expensive as K decreases, given that D and H are fixed. This effect arises because it is more expensive to abandon queen-destined than male-destined larvae.

How general is this model likely to be? Although it was convenient to assume a brood composed solely of males and females, the same principles as those used above should apply to many, perhaps most, species in which queens and workers can be produced together. When the decision on whether or not to rear a potential queen is made after the point at which caste determination has occurred, the current model is clearly directly applicable. This point varies between species: in *Formica polyctena* development as a queen or worker is fixed at about 72 hours of larval life; in *Myrmica ruginodis* this point occurs late in the last larval instar, and in *Pheidole* species development as a queen becomes irrevocable in the egg (reviewed by Wheeler, 1986). When the decision on whether to rear a potential queen or not falls before the caste determination point, the cost of reclaiming materials invested in sexuals will be much less, and liable to be more per investment unit for males (which cannot be made into workers) than for females. Even low costs can still result in the effects discussed here, as shown above.

Differentiation of this model from that of Nonacs is difficult, so that it is reasonable to regard it as a quantification of his views. We are then left with two explicit models, the cost-variation one just discussed and the local mate competition suggestion of Franks. A number of relatively weak tests are possible to choose between these models.

1. Number of males produced. This should be relatively constant if any females are produced under the local mate competition model (which posits a production threshold before any females are produced), but can be zero in larger colonies under the cost-variation model.

2. Size difference between queens and males. This should be correlated with the tendency for the population sex ratio to depart from the infinite case expectation under the cost-variation model, but no correlation is expected under the local mate competition model.

3. Detectable local mate competition. Under the local mate competition model, one would expect to find detectable local mate competition more often in populations with small-colony male bias than in populations without such bias, whereas the cost-variation model would make no prediction.

4. Association between small-colony male bias and seasonal separation of worker and sexual production. If local mate competition is the most important factor, then across species there should be no association between the timing of sexual production and a bias towards male production by small colonies. If the cost-variation effect is the important factor, however, then such an association is more likely because the cost of terminating a potential queen is greater because there are no worker larvae her substance can be fed to (although she might be fed to other sexual larvae).

An interesting result emerged from a preliminary sensitivity analysis involving just two colony types, namely that population structure (proportion of colony types) has a much greater effect on population sex allocation than the difference in cost between males and queens. Of course this result has to be regarded with caution because of the simplistic nature of the model, but this finding points up the importance of thorough population studies to solve sex allocation problems. These studies should investigate such factors as the relatedness structure within and between colonies, population size, variation in reproductive output between colonies, the numbers of queens contributing to different colonies, the numbers of times the queens have mated, the relative investment in queens, males, and workers, and the relative timing of sexual and worker production in relation to estimated sex allocation patterns.

6.6 Inter-colony communication could destabilize the sex ratio game

What happens to the predicted sex allocation patterns if colonies exchange information about their individual allocation patterns? The question is raised by those cases in which ants 'visit' between nests, as in various *Formica* species.

Herbers (1979) analysed this situation using game theory. Her model assumes that all colonies in the population have the same optimum sex investment ratio and that all are involved in information transfer: to be out of the 'coalition' is to fail to reproduce. Under these circumstances, the surprising result is that there is no predictable result. How could this be? And how general is this finding?

The approach taken by Herbers (1979) is quite different to that assumed throughout this book, and so we will use an intuitive approach to explain her result.

Consider a finite-sized group of colonies, each with its own sex allocation pattern, m_i, and each with its own share p_i of the population's sexual production ($\Sigma p_i = 1$). The overall sex allocation of the population is therefore

$$M = \Sigma m_i p_i$$

It is highly unlikely that M will correspond with the optimum investment ratio, in which case all colonies are selected to produce a sex ratio more extreme in the opposite direction from M (Section 2.1). However, if this situation is the norm, then selection has to take account of just this response, because any colony adopting the opposite allocation pattern would have a higher fitness than the rest. From this it can be seen that there is no ideal sex investment ratio that a colony could adopt, whatever the value of M.

Variability in sex ratios of the magnitude expected by this model are not generally seen, leading to the suspicion that it may not be very general. Several points could be made.

1. The breeding population is likely to be larger than any one coalition of colonies. Females and males are liable to be able to fly to mating stations over a wider area than covered by groups of interconnecting colonies. The situation then devolves to that of a small number of colonies.

2. If information is exchanged throughout the season, and if there is a cost to altering the sex ratio proportional to previous investment, then selection might be expected to favour colonies with allocation patterns close to the optimum

3. If colonies vary in their sex allocation optima, then they will not all respond in the same way to many values of M, stabilizing the situation.

6.7 The time factor

In many eusocial Hymenoptera, male and female reproductives tend to be produced at different times. If males are produced before females, the species is termed *protandrous*. If females are produced first, the population is termed *protogynous*.

Strassmann and Hughes (1986) discuss a range of factors affecting the order of production of males and females (termed the 'production schedule' by Suzuki, 1986), especially in annual vespids:

Protandry is favoured if

(1) females mate once and soon after emergence, and males may mate more than once, so that males emerging before females would be better able to find receptive females than males emerging simultaneously with females or after them (Bulmer (1983c) models this case),

(2) females have the higher mortality before mating, so that females are selected to emerge only when males are already present, thus limiting the length of time the females are liable to die, or

(3) the queen wins the queen–worker conflict by laying haploid eggs earlier (Section 5.2) and by forcing workers to rear more males than selection at their level would favour.

Protogyny is favoured if:

(1) male mortality prior to mating is greater than that for females, favouring late emergence by males to reduce the length of their vulnerable period,

(2) the number of days available for brood production each year is variable, so that it is advantageous to be able to shift development from the production of a reproductive to the production of another worker, a shift possible with females but not with males, the 'caste plasticity' hypothesis (Strassmann, 1984).

The production schedule is clearly a phylogenetically labile trait, and indeed Strassmann and Hughes (1986) found that *Polistes* wasps from southern populations (Texas) are protogynous whereas those northern US and Canadian populations (Michigan, Illinois, and Ontario) are not. Suzuki (1986) classified a range of eusocial vespids into three categories, adding that of *simultaneous* production of males and females. Suzuki also noted that roughly equal investment in males and females occurs in populations with simultaneous production schedules, whereas both protandry and protogyny tend to be associated with female-biased ratios. The data in Table 6.1 are arranged according to Suzuki's scheme.

Protandry and protogyny are widespread among social insects. Among bees, Page (1981) reported peak drone production in *Apis mellifera ligustica* prior to swarm production, and interpreted this as evidence of protandry. Bumblebees are also protandrous (Michener, 1974:325).

As discussed previously (Section 6.5), in ants there is a tendency for smaller nests to produce more males, and there is also a tendency in *Formica* species for nests to 'specialize' in producing either males or females (Pamilo and Rosengren, 1983). *Formica* species tend to be protandrous, but there is considerable internest variation in some species (Table 6.2).

Table 6.1. Production schedules of males and reproductive females in paper wasps. The investment ratios were determined by Suzuki (1986), unless stated otherwise. The investment ratio was determined from live weights, save for *Polistes exclamans* (cubed wing length) and *Parapolybia varia* (dry weight). The classification of production schedules follows that of Suzuki, unless noted otherwise.

Production schedule species	Locality	*m*	Note
Simultaneous production			
Polistes			
fuscatus	Michigan	0.50	1
jadwigae	Central Japan	0.51	2
metricus	Illinois	0.50	3
rothneyi	Central Japan	—	2
variatus	Illinois	0.49	4
Protandrous production			
Parapolybia			
varia	Taiwan	0.18	5
Polistes			
japonicus	Central Japan	0.18	2
mandarinus	Central Japan	0.35	2
snelleni	Central Japan	0.15	2
sp. cf. *nimpha*	Northern Japan	0.18	5
Protogynous production			
Parapolybia			
indica	Central Japan	0.03	2
Polistes			
chinensis	Central Japan	0.39	2
exclamans	Texas	0.34	6

1. Noonan (1978), in Suzuki (1986).
2. Suzuki (1986).
3. Recalculated by Suzuki (1986) from data of Metcalf (1980).
4. Metcalf (1980), in Suzuki (1986).
5. Yamane (1980), in Suzuki (1986).
6. Mean percentage investment for nests with their original queens present; calculated by Suzuki (1986) from the data of Strassmann (1984). Strassmann (1984) found considerable year-to-year variation in investment, as discussed in the text.

As noted by Strassmann (1984), differences between the production schedules of the two sexes are likely to lead to year-to-year variation in sex ratios. For example, in a protogynous species, any factors affecting the end of the production season will affect male production. Thus, if unfavourable years shorten that portion of the production season during which males are produced, then *M* will be less during those years. Strassmann (1984) in fact found considerable variation between years for sex ratios in a Texas population of *Polistes exclamans* (Table 6.3)

Table 6.2. Differences in emergence time in single nests of *Formica* species, given as numbers of nests in each category, from Pamilo and Rosengren (1983).

Species	Males first	Females first	No difference
Formica			
aquilonia	2	4	10
cinerea	2	—	3
exsecta	10	3	7
fusca	3	—	7
pratensis	8	—	1
pressilabris	4	—	—
rufibarbis	4	—	2
sanguinea	—	3	—
truncorum	10	—	3

Table 6.3. Parameters pertinent to sex allocation for a population of *Polistes exclamans* over 4 years, with r the population sex ratio, M the population sex allocation in males, and m the mean allocation to males per nest. Allocation was inferred from sex ratios using the estimate of investment in an individual female being 1.16 that in an individual male, arrived at by converting the difference in wing lengths to a volumetric measure. The data are from Strassmann (1984).

	1976	1977	1978	1979
Number of nests producing reproductives	19	17	25	18
Reproductives produced per nest ± SE	19 ± 5	13 ± 2	48 ± 11	30 ± 9
Total reproductives produced	363	215	1192	548
Population proportion of males (r)	0.29	0.45	0.57	0.38
Population allocation to males (M)	0.26	0.42	0.47	0.37
Mean sex allocation per nest (m) ± SE	0.19 ± 0.05	0.27 ± 0.08	0.55 ± 0.06	0.37 ± 0.08
Correlation of sex allocation with number of reproductives produced per nest	0.38*	0.64**	−0.21	0.04

* $P < 0.05$, ** $P < 0.01$.

It is not surprising that Strassmann (1984) failed to find any correlation between population sex ratio and a variety of weather and biotic factors, given that data were available for only four years. However, the two years with the lowest sex ratios had unusually low August or October temperatures, possibly reflecting reduction of the male-producing portion of the season.

Strassmann's valuable study points up the importance of long-term studies in species with different production schedules for the two sexes, unless there is negligible inter-year variation in favourability. What is expected to evolve under natural selection is a set of responses giving a long-term mean predictable by colony and population characteristics: results in any one year may depart from this mean. Herbers's (1990) massive study of sex allocation patterns over several years in two populations of the ant *Leptothorax longispinosus* underlines this conclusion.

Why are there differences between production schedules in some species, and why are these associated with sex ratio variation? Strassmann and Hughes (1986) note that males of southern US *Polistes* populations spend more time than those of northern populations as adults waiting for females to arrive at hibernacula, where mating occurs. Greater male than female mortality before mating would select for the observed protogyny in the southern populations. Strassmann and Hughes (1986) found that the southern nests are likely to have lower relatedness levels than the northern ones. Hence, selection will be stronger on workers in the northern nests to produce female-biased sex ratios.

Suzuki (1986) does not invoke relatedness structure variation, nor life-table variation, but suggests that the simultaneous production species are those in which the queen has control over both oviposition and the sex ratio. A variety of factors lead to either protandry or to protogyny, according to Suzuki. Thus, if the queen loses the ability to inhibit worker oviposition and to control the sex ratio, selection should lead her to lay male-destined eggs before any queen-destined ones and thus to force the workers to care for the queen's sons (Section 5.2). On the other hand, when there is uncertainty as to the best time to produce reproductives, protogyny will be selected for because then females can be redirected to worker development if it turns out to be too early to rear reproductives (the caste plasticity hypothesis). Suzuki (1986) suggests that in these species, too, queens have failed to control the colony's investment patterns, so leading to female-biased allocation.

Suzuki (1986) stresses the variability of polistine wasp species with regard to production schedule and sex ratio. It is also evident from this body of work that these animals will both require and repay detailed and careful demographic study of several variables if the various models on the evolution of production schedule and sex allocation are to be understood.

Termites are not male-haploid and the reproductives lack the ability to determine the sex of offspring by controlling sperm release. Yet Luykx (1986) reports that the kalotermitid *Incisitermes schwarzi* is weakly protogynous: there is an excess of females in the initial reproductive output each year relative to later in the year. The very different biologies of termites and hymenopterans enters into the explanation. The shift in the sex ratio of alates results from females developing into alates more rapidly than males. In many other termite species, including other kalotermitids, there is no difference in maturation time between males and females (Jones *et al.*, 1988).

Despite the expected production of 1:1 sex ratios in termites due to their genetic system, termites often produce extremely skewed sex ratios. In the kalotermitid *Neotermes connexus*, for example, there is a strong bias towards males. Myles and Chang (1984) found that replacement reproductives are all males, that there are seven males to one female among the rest of the population, and that among alates this ratio is still three males per female. In *Incisitermes schwarzi* Luykx (1986) also found a male bias, although this was slight. Jones *et al.* (1988) surveyed the literature and concluded that, apart from the case of *Neotermes connexus*, kalotermitids generally have 1:1 sex ratios, whereas significant differences between colonies often occur in other termite groups. Jones *et al.* (1988) interpreted these differences as an adaptation against inbreeding but, while this may yet be found to be the best explanation, we note that other explanations for skewed sex ratios have been found for hymenopteran societies. It is also possible that termite females and males can adopt different developmental strategies in order to increase their personal fitnesses. For example, *Neotermes* (Kalotermitidae) females might benefit more by being imaginal replacement reproductives than by becoming neotenics, whereas it might be better for the males to become neotenics (Roisin and Pasteels, 1991). Such differences would affect the individual reproductive values and sex allocation optima.

6.8 Workers do affect colony decisions

One of the main questions in social insect biology is to understand how the sterile workers might promote their own evolutionary interests. In the primitively social bees and wasps with behavioural caste differentiation, females can behave as workers but retain the option of reproducing either by laying haploid eggs or by mating and becoming replacement queens. In addition to direct reproduction, worker females have the potential of biasing resource allocation in their own favour. This is the reason why sex ratios form such a central part in social insect biology, including this book. Queen–worker conflict is not just another factor affecting sex ratios, it forms the core of the evolution of social behaviour.

One prediction following from the worker-control hypothesis is the female-biased population sex ratio. We already noted in Chapter 5 that there are several ways to try to test this prediction and most of them face the problem that it is difficult to eliminate confounding factors that also can move the population sex ratios from equality. One approach is to look at the variation of colony sex ratios, as this can be done whatever the population mean is. As explained in Section 6.3.2, the varying relatedness asymmetries lead to different sex ratio optima for workers in different colonies. This hypothesis was discussed briefly by Nonacs (1986a) and rigorously developed by Boomsma and Grafen (1990, 1991), This gives perhaps the strongest possibility for testing the hypothesis that workers increase their inclusive fitness by optimally biasing the colony sex ratios.

Boomsma (1991) noticed that the sex ratios of the bee *Halictus rubicunda*, collected by Yanega (1989), fitted the prediction of worker control. The bees have small eusocial colonies with a single queen and her daughters acting as workers. If the mother queen dies, one of the workers takes her place and the colony turns from eusocial to parasocial (see Table 1.1). From the point of view of workers, the relatedness asymmetries are different in these two colony types. In a eusocial colony, they raise sisters ($g = 0.75$, assuming single mating) and brothers ($g = 0.5$), but in a parasocial colony the brood consists of their nieces ($g = 0.375$) and nephews ($g = 0.75$). It would be optimal for the workers to raise mainly females in eusocial and males in parasocial colonies, and this is exactly what the sex ratios observed by Yanega showed (Boomsma, 1991). Similar trends exist also in *Halictus ligatus* (Boomsma and Eickwort, 1993) but not in the wasp *Polistes exclamans* (Strassmann, 1984; see Boomsma, 1991).

Mueller (1991) went a step further and tested experimentally whether workers make a distinction between eusocial and parasocial colonies (see also Section 5.6.2). He removed the queens from 19 eusocial colonies of the bee *Augochlorella sriata* and compared the resulting parasocial colonies with matched eusocial pairs. A comparison showed that the parasocial colonies produced significantly more males as predicted by the hypothesis of worker control. Unmatedness of the replacement queens could be excluded as an explanation.

In the above cases the workers could primarily react to the replacement of the egg-laying females without any need to measure the relatedness asymmetry directly. The result is, of course, the same. The relatedness asymmetry within colonies is significantly associated with the colony sex ratio also in the bee *Lasioglossum laevissimum* (Packer and Owen, 1994). The factor causing the variation in relatedness asymmetries in this species seems to be largely he number of foundresses establishing colonies.

The above examples from social bees were based on situations in which workers can respond to the number or quality of the reproductive females. When Boomsma and Grafen (1990, 1991) originally presented their hypothesis of split sex ratios, they used an example of multiple matings as the basis of different relatedness asymmetries. In such a case, the workers should be able to detect the number of matings from the diversity of cues (most likely odour cues) within colonies. That also represents a case where it is relatively easy to eliminate confounding factors causing sex ratio biases, as the colonies should not otherwise differ from each other in any systematic way.

The ant *Formica truncorum* has both monogynous-monodomous colonies and highly polygynous and polydomous colonial networks. Sundström (1994) studied the genetical structures of colonies in a population consisting of monodomous colonies in the southern Finnish archipelago. Electrophoretic results showed that colonies were monogynous but that the number of matings varied. The segregation of genotypes at four enzyme gene loci could be explained with a single mating in about half of the colonies. The rest of the colonies had a queen which must have mated with at least two males. As found already earlier (Pamilo and Rosengren, 1983; Rosengren *et al.*, 1986), the *F. truncorum* colonies specialize often in producing either female or male sexuals. When clustering the nests according to the queen mating frequency, Sundström was able to show that the colonial sex ratios clearly followed the intra-colonial genetic heterogeneity (Fig. 6.4). As multiple mating might have other advantages, for example it could lead to larger colonies and therefore to female-biased sex ratios, Sundström also examined total sexual production in the same colonies. Large colonies tended indeed to produce a larger fraction of females, but when plotting the colonial sex ratios against the colony

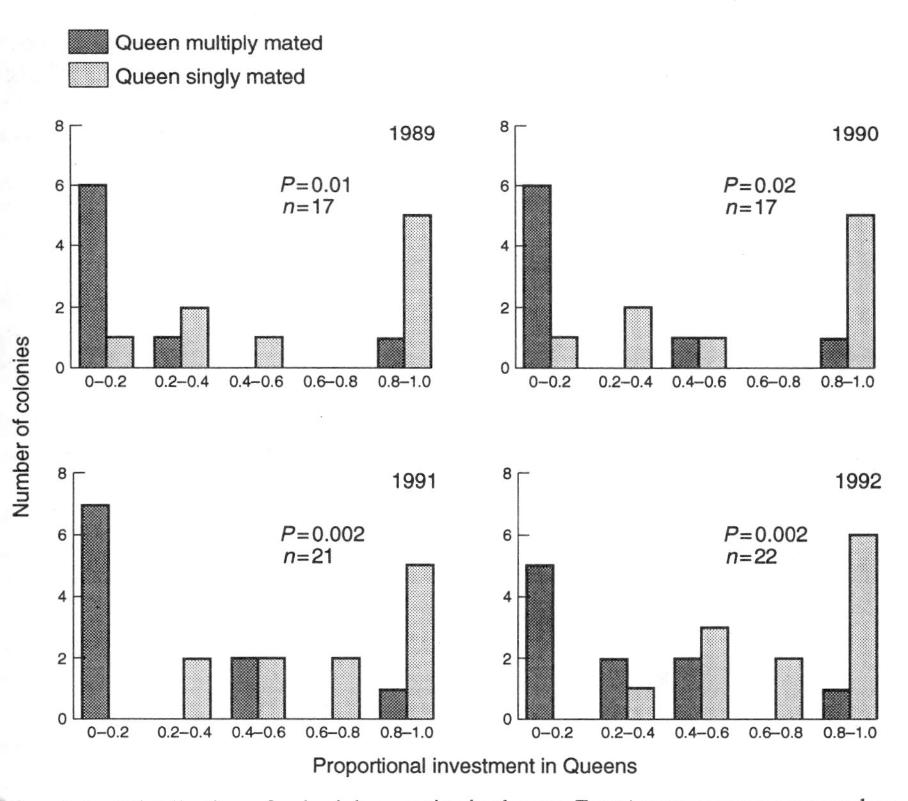

Fig. 6.4 Distribution of colonial sex ratios in the ant *Formica truncorum* expressed as relative investment (dry weight) in queens separately for colonies with singly and multiply inseminated queens. *n* is the number of colonies sampled each year, and *P* gives the significance of Mann-Whitney Rank Sum test for the difference between the two types of colonies. (Reprinted with permission from *Nature* (Sundsröm, 1994b).)

productivity, the colonies fell into two non-overlapping groups based on the queen mating frequency (Fig. 6.5).

The hypothesis of Boomsma and Grafen (1991) predicts a balancing class of colonies of the common type. Colonies of the rare type can bias the colony sex ratio to the extreme value, but those of the common type cannot do that because an extreme bias would produce a non-optimal population sex ratio. It appeared that the nests with a polyandrous queen formed the balancing class in *F. truncorum*, nicely supporting the theoretical predictions. The most parsimonious explanation is that the workers of *F. truncorum* can detect from some cues the number of times the queen has mated and use this information to bias the colony sex ratios. The sexual production in *F. truncorum* overlaps with worker production, so the queens cannot prevent workers raising females by laying only haploid eggs because that would damage the colony.

Another *Formica* ant, *F. sanguinea*, has colonies that differ from each other with respect to both the number of matings per queen and the number of queens, i.e. in both polyandry and polygyny. Pamilo and Seppä (1994) used this setting to examine the split sex ratio hypothesis. Genealogically heterogeneous colonies (polygynous and/or polyandrous) have low relatedness asymmetries and are predicted to produce males, whereas monogynous (especially if also monandrous) colonies should produce females. Pamilo and

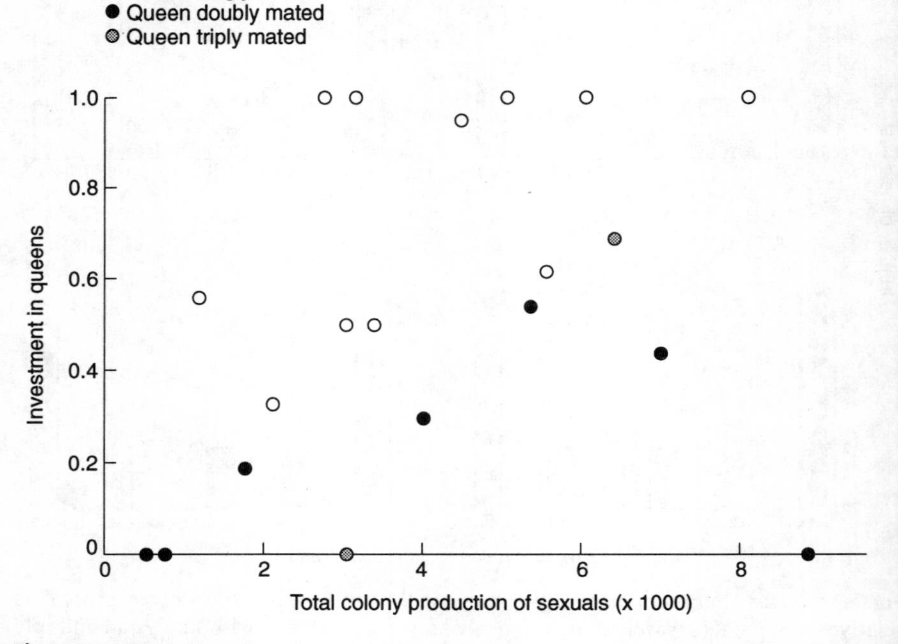

Fig. 6.5 Sex allocation in colonies of *Formica truncorum* in relation to colony productivity and queen mating frequency. (Reprinted with permission from *Nature* (Sundsröm, 1994b).)

Seppä collected sexuals from 94 nests and these could be divided into three classes based on the sex ratio: female producers (> 90% of sexuals being females, $n = 44$ colonies), male producers (> 90% of sexuals being males, $n = 24$ colonies), and mixed colonies ($n = 26$). They also estimated relatednesses among workers in these groups but there was no association between the genetic heterogeneity (measured as relatedness) and sex ratio.

Tsuji and Yamauchi (1994) predicted that the relatedness asymmetry in the ant *Technomyrmex albipes* should be highest in young and small colonies, because inbreeding within colonies of this species reduces the asymmetry when the colony grows. However, they did not find any sex ratio differences between young and old colonies. There was a correlation between the sex ratios among the wingless (inbreeding) and winged (outbreeding) sexuals produced by the same colony, leading Tsuji and Yamauchi to speculate that both sex ratios may be influenced by a common initial sex ratio with a genetic component.

Split sex ratios are also found in the third major group of eusocial Hymenoptera, the vespids. The epiponine wasps have colony cycles where young colonies have many foundresses, but the number of reproductive females declines as the colonies grow older. The relatedness asymmetry is highest in old and relatively monogynous colonies, and these colonies are expected to produce mainly females if sex allocation is optimally controlled by workers. Queller *et al.* (1993b) suggested this to be the case in epiponine wasps, although the observations were partly indirect. It is difficult to distinguish workers and reproductive females from each other, and therefore it is difficult to estimate colonial sex ratios. But the conclusions can be drawn in the same way as in the above study on *Formica sanguinea*. The genetic heterogeneity in male-producing colonies can be estimated directly from the worker genotypes. The heterogeneity within female-producing colonies has to be estimated indirectly. Assuming that nest foundresses are a random sample of nest-mate females, heterogeneity of the natal colony should equal the relatedness among nest-founding females. Using this approach, Queller *et al.* (1993b) showed that males were produced in polygynous colonies (worker relatednesses being 0.27 in *Polybia occidentalis*, 0.31 in *P. emaciata*, 0.49 in *Protopolybia exigua* and 0.30 in *Parachartergus colobopterus*), and females were produced in less polygynous or monogynous colonies (relatednesses among nest foundresses being 0.58, 0.55, 0.82 and 0.66, respectively).

The results of Queller *et al.* (1993b) lead to an interesting conclusion concerning the maintenance of the genetic integrity of colonies. If females are produced in relatively monogynous colonies, the relatedness among nest-founding females remains high. The same feature might explain why relatedness is higher among queens than among workers in some polygynous ants, such as *Myrmica scabrinodis* (Seppä, 1994b). This would prevent relatedness decreasing to low levels even if the levels of polygyny can be high in some colonies or at some stages of colonial cycles. This led Queller *et al.* (1993b) to conclude that optimal worker behaviour and conditional biasing of colonial sex ratios can help to maintain sociality by preserving the genetic integrity of colonies.

6.9 Summary

Simulation results, as well as analyses of relatively simple genetic models suggest that there will be relatively little genetic variation for the sex ratio under worker control, although there may be some under queen control. It therefore appears likely that many cases of variation will reflect conditional responses to differences between colonies in environmental circumstances, relatedness structure, demographic circumstances, or in their composition in terms of the relative numbers of the various castes present.

Somewhat unexpectedly, communication between colonies can prevent the emergence of predictable sex ratios. However, it seems unlikely that this situation occurs often, and the generally intermediate nature of observed sex ratios bears this out.

Some variation in sex ratio between colonies may be due to internal factors unconnected with reproductive competition between colonies, such as that due to differential ease of converting males and females to workers for colony maintenance. It is more likely that many interesting phenomena result from different sex allocation optima for different classes of colonies. The optimum sex allocation pattern for one type of colony modifies that of others, and is in turn affected by them. Such factors were considered earlier and we can now see that they have a general importance in the populations of eusocial Hymenoptera. The population sex ratio naturally tends to be dominated by that preferentially produced by the class of colonies with the largest output. Boomsma and Grafen (1991) note that populations may be sufficiently small for this dominant class to consist of a few large colonies.

The tendency for larger ant colonies to produce more female-biased sex allocation patterns stems from a number of causes. In many species, queenless colonies (which are dying out and tend therefore to be smaller) produce males but no females, through worker reproduction. This male-production by small colonies devalues males, leading to relatively female-biased reproduction by larger colonies: the phenomenon of reproductive compensation that we also discussed earlier. There is evidence, however, that small colonies have relatively male-biassed sex allocation in species lacking worker reproduction. Three hypotheses to explain this last point are that local mate competition is important (although perhaps not otherwise detectable) in ant populations, that the effective cost of queens is higher in small than in large colonies, leading to different sex allocation optima with colony size, and that workers are more likely to win the worker–queen conflict in larger colonies. Some relatively weak tests are available to choose between these alternatives. These tests emphasize the point that, although it would be rash to try and predict the precise sex allocation pattern in a population, detailed study of a broad range of population parameters is needed if data from different populations or species are to be useful.

Various reports point up the importance of long-term studies. Year-to-year variation in sex allocation is well demonstrated in ants, and particularly well studied in various vespine wasps. Some of these wasps are protogynous (produce queens before males), some are protandrous (produce males before queens), and some show simultaneous production of males and females. Protogynous and protandrous species appear particularly prone to variations in sex allocation due to year-to-year variation in environmental factors; for example, if the production season ends unusually early in a protandrous species, cutting short the production of females, then the result will be a relatively male-biased sex ratio. These phenomena are interconnected with the general question of why some wasps have one production schedule while others have different ones. Many more seasons of data collection are needed to provide understanding of these matters.

Studies on primitively eusocial bees, *Formica* ants, and epiponine wasps demonstrate that colonies with different relatedness structures have sex allocation patterns following those expected on the basis of worker interests, and also that these populations have split sex ratios. The results suggest that relatedness structure can be assessed by the workers of a colony, and that these adjust the sex allocation pattern accordingly.

We close by noting that sex allocation in termites is often skewed, and that this shows a phylogenetic pattern. The reasons for such skewing, and its phylogenetic distribution remain obscure; we suggest that the principles we have expounded will underly the eventual explanations, although it is doubtful that these will involve the relatedness asymmetry vital to understanding the sex allocation pattern, and indeed general biology, of eusocial Hymenoptera.

References

An asterisk (*) indicates that the authors have not seen the original reference.

Adams, E.S. 1991. Nest-mate recognition based on heritable odors in the termite *Microcerotermes arboreus*. *Proc. Natl. Acad. Sci. USA* **88**: 2031–4.

Adams, J., Rothman, E.D., Kerr, W.E. and Paulino, Z.L. 1977. Estimation of the number of sex alleles and queen matings from diploid male frequencies in a population of *Apis mellifera*. *Genetics* **86**: 583–96.

Akre, R.D., Reed, H.C. and Landolt, P.K. 1982. Nesting biology and behavior of the blackjacket *Vespula consobrina* (Hymenoptera: Vespidae). *J. Kansas Ent. Soc.* **55**: 373–405.

Alexander, R.D. 1974. The evolution of social behavior. *Annu. Rev. Ecol. Syst.* **4**: 325–84.

Alexander, R.D., Noonan, K. and Crespi, B.J. 1991. The evolution of eusociality. In *The biology of the naked mole rat* (ed. P.W. Sherman, J. Jarvis and R.D. Alexander) pp. 3–44. Princeton University Press.

Alexander, R.D. and Sherman, P.W. 1977. Local mate competition and parental investment in social insects. *Science* **196**: 494–500.

Alloway, T.M., Buschinger, A., Talbot, M., Stuart, R. and Thomas, C. 1982. Polygyny and polydomy in three north American species of the ant genus *Leptothorax* Mayr (Hymenoptera: Formicidae). *Psyche* **89**: 249–74.

Alpert, G.D. and Akre, R.D. 1973. Distribution, abundance, and behavior of the inquiline ant *Leptothorax diversipilosus*. *Ann. Entomol. Soc. Am.* **66**: 753–60.

Andersson, M. 1984. The evolution of eusociality. *Annu. Rev. Ecol. Syst.* **15**: 165–189.

Aoki, K. and Moody, M. 1981. One- and two-locus models of the origin of worker behavior in Hymenoptera. *J. Theor. Biol.* **89**: 449–474.

Aoki, S. 1982. Soldiers and altruistic dispersal in aphids. In *The biology of social insects* (ed. M.D. Breed, C.D. Michener and H.E. Evans) pp. 154–8. Westview Press, Boulder, CO.

Aoki, S. 1987. Evolution of sterile soldiers in aphids. In *Animal societies: theories and facts* (eds. Y. Itô, J.L. Brown and J. Kikkawa) pp. 53–65. Japan Science Society Press, Tokyo.

Aoki, S., Kurosu U. and Stern, D.L. 1991. Aphid soldiers discriminate between soldiers and non-soldiers, rather than between kin and non-kin, in *Ceratoglyphina bambusae*. *Anim. Behav.* **42**: 865–6.

Aron, S., Passera L. and Keller, L. 1994. Queen–worker conflict over sex ratio: a comparison of the primary and the secondary sex ratios in the Argentine ant, *Iridomyrmex humilis*. *J. Evol. Biol.* **7**: 403–18.

Autuori, M. 1950. Contribuição para o conhecimento da saúva (*Atta* spp. — Hymenoptera-Formicidae) V. Número de formas aladas e redução dos sauveiros iniciais. *Arq. Inst. Biol. São Paulo* **19**: 325–31.

Aviles, L. 1986. Sex ratio bias and possible group selection in the social spider *Anelosimus eximius. Am. Nat.* **128**: 1–12.

Aviles, L. 1993. Interdemic selection and the sex ratio: a social spider perspective. *Am. Nat.* **142**: 320–45.

Ayre, G.L. 1971. Preliminary studies on the foraging and nesting habits of *Myrmica americana* Weber (Hymenoptera: Formicidae) in eastern Canada. *Z. Angew. Entomol.* **68**: 295–9.

Backus, V.L. and Herbers, J.M. 1992. Sexual allocation ratios in forest ants: food limitation does not explain observed patterns. *Behav. Ecol. Sociobiol.* **30**: 425–9.

Bar Anon, R. and Robertson, A. 1975. Variation in sex ratio between progeny groups in dairy cattle. *Theor. Appl. Genet.* 46:63–5.

Baroni-Urbani, C. 1968. Domination et monogynie fonctionelle dans une société digynique de *Myrmecina graminicola* Latr. *Insectes Soc.* **15**: 407–12.

Bartels, P.J. 1985. Field observations of multiple matings in *Lasius alienus* Foerster (Hymenoptera: Formicidae). *Am. Midl. Nat.* **113**: 190–2.

Bartz, S.H. 1979. Evolution of eusociality in termites. *Proc. Natl. Acad. Sci., USA* **76**: 5764–8.

Bartz, S.H. and Hölldobler, B. 1982. Colony founding in *Myrmecocystus mimicus* Wheeler (Hymenoptera:Formicidae) and the evolution of foundress associations. *Behav. Ecol. Sociobiol.* **10**: 137–47

Batra, S.W.T. 1966. The life cycle and behavior of the primitively social bee *Lasioglossum zephyrum* (Halictidae). *Univ. Kansas Sci. Bull.* **46**: 359–423

Beig, D. 1972. The production of males in queenright colonies of *Trigona* (*Scaptotrigona*) *postica. J. Apic. Res.* **11**: 33–9.

Benford, F.A. 1978. Fisher's theory of the sex ratio applied to the social Hymenoptera. *J. Theor. Biol.* **72**: 701–27.

Bennett, B. 1986. *Nestmate recognition, intracolonial relatedness, and ecology of monogynous and polygynous ants.* PhD Thesis, University Microfilms International, Ann Arbor, MI.

Benois, A. 1969. Étude écologique de *Camponotus vagus* Scop. (= *pubescens* Fab.)(Hymenoptera, Formicidae) dans le région d'Antibes: nidification et architecture des nids. *Insectes Soc.* **19**: 111–29.

Berkelhamer, R.C. 1984. An electrophoretic analysis of queen number in three species of dolichoderine ants. *Insectes Soc.* **31**:132–41.

Bernstein, R.A. 1976. The adaptive value of polymorphism in an alpine ant, *Formica neorufibarbis gelida* Wheeler. *Psyche* **83**: 180–4.

Berrigan, D., Evans, J., Holway, D., Jacobs, L., Richards, M. and Seger, J. 1994. Gene flow or heterozygote advantage? *Science* **263**: 1157.

Bhatkar, A.P. and Vinson, S.B. 1987. Foraging in mono- and polydomous *Solenopsis invicta* Buren colonies. In *Chemistry and biology of social insects* (eds. Eder, J. and H. Rembold) pp. 545–6. Verlag J. Peperny, München.

Bhattacharya, G.C. 1943. Reproduction and caste determination in aggressive red ants, *Oecophylla smaragdina* Fabr. *Trans. Bose Res. Inst. Calcutta* **15**: 137–56.

Bier, K. 1952. Zur scheinbaren Thelytokie der Ameisengattung *Lasius. Naturwissenschaften* **39**: 433.

Bier, K. 1954. Über den Einfluss der Königin auf die Arbeiterinnenfertilität im Ameisenstaat. *Insectes Soc.* **3**: 177–84.

Blanchetot, A. 1991. Genetic relatedness in honeybees as established by DNA fingerprinting. *J. Hered.* **82**: 391–6.

Blanchetot, A. 1992. DNA fingerprinting analysis in the solitary bee *Megachile rotundata*—variability and nest mate genetic relationships. *Genome* **35**: 681–8.

van der Blom, J. and Verkade, A.-J. 1991. Does kin recognition in honey bees, *Apis mellifera* L., influence the workers' chances of becoming egg layers? *Anim. Behav.* **42**: 867–70.

Blows, M.W. and Schwarz, M.P. 1991. Spatial distribution of a primitively social bee: does population structure facilitate altruism? *Evolution* **45**: 680–93.

Bolton, B. 1986. Apterous females and shift of dispersal strategy in the *Monomorium salomonis*-group (Hymenoptera: Formicidae). *J. Nat. Hist.* **20**: 267–72.

Bonavita-Cougourdan, A., Clément, J.L. and Lange, C. 1987. Nestmate recognition: the role of cuticular hydrocarbons in the ant *Camponotus vagus* Scop. *J. Entomol. Sci.* **22**: 1–10.

Boomsma, J.J. 1987. The empirical analysis of sex allocation in monogynous ants. In *Chemistry and biology of social insects* (eds. J. Eder, and H. Rembold) pp 355–7. Verlag J. Peperny, Munich.

Boomsma, J.J. 1988. Empirical analysis of sex allocation in ants: from descriptive surveys to population genetics. In *Population genetics and evolution* (ed. G. de Jong) pp 42–51. Springer Verlag, Berlin.

Boomsma, J.J. 1989. Sex investment ratios in ants: has female bias been systematically overestimated? *Am. Nat.* **133**: 517–32.

Boomsma, J.J. 1991. Adaptive colony sex ratios in primitively eusocial bees. *Trends Ecol. Evol.* **6**: 92–5.

Boomsma, J.J. and Eickwort, G.C. 1993. Colony structure, provisioning and sex allocation in the sweat bee *Halictus ligatus* (Hymenoptera: Halictidae). *Biol. J. Linn. Soc.* **48**: 355–77.

Boomsma, J.J. and Grafen, A. 1990. Intraspecific variation in ant sex ratios and the Trivers-Hare hypothesis. *Evolution* **44**: 1026–34.

Boomsma, J.J. and Grafen, A. 1991. Colony-level sex-ratio selection in the eusocial Hymenoptera. *J. Evol. Biol.* **4**: 383–407.

Boomsma, J.J. and Isaaks, J.A. 1985. Energy investment and respiration in queens and males of *Lasius niger* (Hymenoptera: Formicidae). *Behav. Ecol. Sociobiol.* **18**: 19–27.

Boomsma, J.J., van der Lee, G.A. and van der Have, T.M. 1982. On the production ecology of *Lasius niger* (Hymenoptera: Formicidae) in successive coastal dune valleys. *J. Anim. Ecol.* **51**: 975–91.

Boomsma, J.J., Brouwer A.H. and van Loon, A.J. 1990. A new polygynous *Lasius* species (Hymenoptera; Formicidae) from central Europe. II. Allozymatic confirmation of species status and social structure. *Insectes Soc.* **37**: 363–75.

Boomsma, J.J., Wright, P.J. and Brouwer, A.H. 1993. Social structure in the ant *Lasius flavus*: multi-queen nests or multi-nest mounds? *Ecol. Entomol.* **18**: 47–53.

Bourke, A.F.G. 1988. Worker reproduction in the higher eusocial Hymenoptera. *Q. Rev. Biol.* **63**: 291–311.

Bourke, A.F.G. 1989. Comparative analysis of sex-investment ratios in slave-making ants. *Evolution* **43**: 913–18.

Bourke, A.F.G. and Franks, N.R. 1991. Alternative adaptations, sympatric speciation and the evolution of parasitic, inquiline ants. *Biol. J. Linn.* Soc. **43**: 157–78.

Bourke, A.F.G. and Heinze, J. 1994. The ecology of communal breeding: the case of multiple-queen leptothoracine ants. *Philos. Trans. R. Soc. London Ser B* **345**: 359–72.

Bourke, A.F.G., van der Have, T.M. and Franks, N.R. 1988. Sex ratio determination and worker reproduction in the slave-making ant *Harpagoxenus sublaevis*. *Behav. Ecol. Sociobiol.* **23**: 233–45.

Breed, M.D. 1976. The evolution of social behavior in primitively social bees: a multivariate analysis. *Evolution* **30**: 234–40.

Brian, M.V. 1953. Oviposition by workers of the ant *Myrmica*. *Physiol. Comp. Oecol.* **3**: 25–36.

Brian, M.V. 1972. Population turnover in wild colonies of the ant, *Myrmica*. *Ekol. Pol.* **20**: 43–53.

Brian, M.V. 1979a. Caste differentiation and division of labor. In *Social insects* (ed. H.R. Hermann) pp 121–222. Academic Press, New York.

Brian, M.V. 1979b. Habitat differences in sexual production by two coexistent ants. *J. Anim. Ecol.* **48**: 387–405.

Brian, M.V. 1983. *Social insects*. Chapman and Hall, London.

Brian, M.V. and Brian, A.D. 1951. Insolation and ant population in the west of Scotland. *Trans. R. Entomol. Soc. London* **102**: 303–30.

Brian, M.V. and Elmes, G.W. 1974. Production by the ant *Tetramorium caespitum* in a southern English heath. *J. Anim. Ecol.* **43**: 889–903.

Brian, M.V., Clarke, R.T. and Jones, R.M. 1981. A numerical model of an ant society. *J. Anim. Ecol.* **50**: 387–405.

Brockmann, H.J. and Grafen, A. 1989. Male conflict and male behaviour in a solitary wasp, *Trypoxylon (Trypargilum) politum* (Hymenoptera: Sphecidae). *Anim. Behav.* **37**: 232–55.

Brockmann, H.J. and Grafen, A. 1992. Sex ratios and life-history patterns of a solitary wasp, *Trypoxylon (Trypargilum) politum* (Hymenoptera: Sphecidae). *Behav. Ecol. Sociobiol.* **30**: 7–27.

Brown, J.L. 1987. *Helping and communal breeding in birds*. Princeton University Press.

Bruniquel, S. 1972. La ponte de la fourmi *Aphaenogaster subterranea* (Latr.): oeufs reproducteurs–oeufs alimentaires. *C. R. Acad. Sci. Paris* D275: 397–9.

Buckle, G.R. and Greenberg, L. 1981. Nestmate recognition in sweat bees (*Lasioglossum zephyrum*): does an individual recognize its own odour or only odours of its nestmates? *Anim. Behav.* **29**: 802–9.

Bull, J.J. 1983. *Evolution of sex determining mechanisms*. Benjamin Cummings, Menlo Park, CA.

Bulmer, M.G. 1981. Worker–queen conflict in annual social Hymenoptera. *J. Theor. Biol.* **83**: 239–52.

Bulmer, M.G. 1983a. Sex ratio evolution in social Hymenoptera under worker control with behavioral dominance. *Am. Nat.* **121**: 899–902.

Bulmer, M.G. 1983b. Sex ratio theory in social insects with swarming. *J. Theor. Biol.* **100**: 329–39.

Bulmer, M.G. 1983c. The significance of protandry in social Hymenoptera. *Am. Nat.* **121**: 540–51.

Bulmer, M.G. 1986. Sex ratio theory in geographically structured populations. *Heredity* **56**: 69–73.

Bulmer, M.G. and Taylor, P.D. 1980. Sex ratio under the haystack model. *J. Theor. Biol.* **86**: 83–9.

Bulmer, M.G. and Taylor, P.D. 1981. Worker–queen conflict and sex ratio theory in social Hymenoptera. *Heredity* **47**: 197–207.

Burda, H. and Kawalika, M. 1993. Evolution of eusociality in the Bathyergidae. The case of the giant mole rats (*Cryptomys mechowi*). *Naturwissenschaften* **80**: 235–7.

Burke, T. 1989. DNA finger printing and other methods for the study of mating success. *Trends Ecol. Evol.* **4**: 139–44.

Buschinger, A. 1968. Mono- und Polygynie bei Arten der Gattung *Leptothorax* Mayr (Hymenoptera Formicidae). *Insectes Soc.* **15**: 217–26.

Buschinger, A. 1970. Zur Frage der Monogynie oder Polygynie bei *Myrmecina graminicola* (Latr.)(Hym., Form.). *Insectes Soc.* **17**: 177–82.

Buschinger, A. 1974. Monogynie und Polygynie in Insektensozietäten. In *Sozialpolymorphismus bei Insekten* (ed. G.H. Schmidt) pp. 862–896. Wissenschaftliche Verlagsgesellschaft, Stuttgart.

Buschinger, A. 1975. Eine genetische Komponente im Polymorphismus der dulotischen Ameise *Harpagoxenus sublaevis. Naturwissenschaften* **62**: 239.

Buschinger, A. 1978. Genetisch bedingte Enstehung gefügelter Weibchen bei der sklavenhaltenden Ameise *Harpagoxenus sublaevis* (Nyl.) (Hym., Form.). *Insectes Soc.* **25**: 163–72.

Buschinger, A. 1979. Functional monogyny in the American guest ant *Formicoxenus hirticornis* (Emery)(=*Leptothorax hirticornis*), (Hym., Form.). *Insectes Soc.* **1**: 61–8.

Buschinger, A. 1986. Evolution of social parasitism in ants. *Trends Ecol. Evol.* **1**: 155–60.

Buschinger, A. 1989. Evolution, speciation, and inbreeding in the parasitic ant genus *Epimyrma* (Hym., Formicidae). *J. Evol. Biol.* **2**: 265–83.

Buschinger, A. 1990. Sympatric speciation and radiative evolution of socially parasitic ants—Heretic hypotheses and their factual background. *Z. Zool. Syst. Evolutionsforsch.* **28**: 241–60.

Buschinger, A. and Alloway, T.M. 1977. Population structure and polymorphism in the slave-making ant *Harpagoxenus americanus* (Emery) (Hymenoptera: Formicidae). *Psyche* **83**: 233–42

Buschinger, A. and Alloway, T.M. 1978. Caste polymorphism in *Harpagoxenus canadensis* (Hym., Formicidae). *Insectes Soc.* **25**: 339–50.

Buschinger, A. and Fischer, K. 1991. Hybridization of chromosome-polymorphic populations of the inquiline ant, *Doronomyrmex kutteri* (Hym., Formicidae). *Insectes Soc.* **38**: 95–103.

Buschinger, A. and Heinze, J. 1992. Polymorphism of female reproductives in ants. In *Biology and evolution of social insects*. (ed. Billen J.) pp. 11–23. Leuven University Press.

Buschinger, A., Francoeur A. and Fischer, K. 1980. Functional monogyny, sexual behavior, and karyotype of the guest ant, *Leptothorax provancheri* Emery (Hymenoptera, Formicidae). *Psyche* **87**: 1–12.

Buschinger, A., Cagniant, H., Ehrhardt, E., and Heinze, J. 1988. *Chalepoxenus brunneus*, a workerless 'degenerate slave-maker' ant (Hymenoptera, Formicidae). *Psyche* **95**: 253–63

Buskirk, R.E. 1981. Sociality in the Arachnida. In *Social insects* (ed. H.R. Hermann) Vol. II pp. 282–393. Academic Press, New York.

Cagniant, H. 1979. La parthenogenese thelytoque et arrhenotoque chez la fourmi *Cataglyphis cursor* Fonsc. (Hym. Form.). Cycle biologique en elevage des colonies avec reine et des colonies sans reines. *Insectes Soc.* **26**: 51–60.

Cagniant, H. 1982. La parthenogenese thelytoque et arrhenotoque chez la fourmi *Cataglyphis cursor* Fonscolombe (Hymenoptera, Formicidae) étude des oeufs pondus par les reines et les ouvrieres: morphologie, devenir, influence sur le determinisme de la caste reine. *Insectes Soc.* **29**: 175–88.

Cagniant, H. 1988. Étude expérimentale du role des ouvrières dans le développement des sexués ailés chez la fourmi *Cataglyphis cursor* (Fonsc.) (Hyménoptères, Formicidae). *Insectes Soc.* **35**: 271–92.

Calderone, N.W. and Page, R.E., Jr. 1988. Genotypic variability in age polyethism and task specialization in the honey bee, *Apis mellifera* (Hymenoptera: Apidae). *Behav. Ecol. Sociobiol.* **22**: 17–25.

Calderone, N.W. and Page, R.E., Jr. 1991. Evolutionary genetics of division of labor in colonies of the honey bee (*Apis mellifera*). *Am. Nat.* **138**: 69–92.

Calderone, N.W. and Page, R.E., Jr. 1992. Effects of interactions among genotypically diverse nestmates on task specialization by foraging honey bees (*Apis mellifera*). *Behav. Ecol. Sociobiol.* **30**: 219–26.

Cameron, S.A. 1993. Multiple origins of advanced eusociality in bees inferred from mitochondrial DNA sequences. *Proc. Natl. Acad. Sci., USA* **90**: 8687–91.

Cannings, C. and Cruz Orive, L.M. 1975. On the adjustment of the sex ratio and the gregarious behaviour of animal populations. *J. Theor. Biol.* **55**: 115–36.

Carlin, N.F. and Hölldobler, B. 1983. Nestmate and kin recognition in interspecific mixed colonies of ants. *Science* **222**: 1027–9.

Carlin, N.F. and Hölldobler, B. 1986. The kin recognition system of carpenter ants. I. Hierarchical cues in small colonies. *Behav. Ecol. Sociobiol.* **19**: 123–34.

Carlin, N.F. and Hölldobler, B. 1987. The kin recognition system of carpenter ants. II. Larger colonies. *Behav. Ecol. Sociobiol.* **20**: 209–17.

Carlin, N.F., Hölldobler, B. and Gladstein, D.S. 1987. The kin recognition system of carpenter ants (*Camponotus* spp.) III. Within-colony discrimination. *Behav. Ecol. Sociobiol.* **20**: 219–27.

Carpenter, J.M., Strassmann, J.E., Turillazzi, S., Hughes, C.R., Solis, C.R. and Cervo, R. 1993. Phylogenetic relationships among paper wasp social parasites and their hosts (Hymenoptera: Vespidae; Polistinae). *Cladistics* **9**: 129–46.

Cavalli-Sforza, L.L. and Bodmer, W.F. 1971. *The genetics of human populations.* Freeman, San Francisco, CA.

Charlesworth, B. 1978. Some models of the evolution of altruistic behaviour between siblings. *J. Theor. Biol.* **72**: 297–319.

Charlesworth, B. and Toro, M.A. 1982. Female-biased sex ratios. *Nature* **298**:494.

Charnov, E.L. 1978. Sex-ratio evolution in eusocial Hymenoptera. *Am. Nat.* **112**: 317–26.

Charnov, E.L. 1979. Simultaneous hermaphroditism and sexual selection. *Proc. Natl. Acad. Sci. USA* **76**: 2480–4.

Charnov, E.L. 1981. Kin selection and helpers at the nest: effects of paternity and biparental care. *Anim. Behav.* **29**: 631–2.

Charnov, E.L. 1982. *The theory of sex allocation.* Princeton University Press.

Chaud-Netto, J. 1975. Sex determination in bees. II. Additivity of maleness genes in *Apis mellifera. Genetics* **79**: 213–17.

Choe, J.C. 1988. Worker reproduction and social evolution in ants (Hymenoptera: Formicidae). In *Advances in myrmecology* (ed. J.C. Trager), pp. 163–187. E.J. Britt, Leiden.

Choudhary, M., Strassmann, J.E., Solís, C.R. and Queller, D.C. 1993. Microsatellite variation in a social insect. *Biochem. Genet.* **31**: 87–96.

Clark, A.B. 1978. Sex ratio and local resource competition in a prosimian primate. *Science* **201**: 163–5.

Clark, A.M., Bertrend, H. A. and Smith, R.E. 1963. Lifespan differences between haploid and diploid males of *Habrobracon serinopae* after exposure as adults to X-rays. *Am. Nat.* **97**:203–8.

Cline, T.W. 1993. The *Drosophila* sex determination signal: how do flies count to two? *Trends Genet.* **9**: 385–90.

Clutton-Brock, T.H. and Iason, G.R. 1986. Sex ratio variation in mammals. *Q. Rev. Biol.* **61**: 339–74.

Cole, A.C. 1939. The life history of a fungus-growing ant of the Mississippi Gulf Coast. *Lloydia* **2**: 153–60.

Cole, B.J. 1981. Dominance hierarchies in *Leptothorax* ants. *Science* **212**: 83–4.

Cole, B.J. 1983. Multiple mating and the evolution of social behavior in the Hymenoptera. *Behav. Ecol. Sociobiol.* **12**: 191–201.

Cole, B.J. 1986. The social behavior of *Leptothorax allardycei* (Hymenoptera: Formicidae): time budgets and the evolution of worker reproduction. *Behav. Ecol. Sociobiol.* **18**: 165–73.

Colombel, P. 1972. Recherches sur la biologie et l'ethologie D'*Odontomachus haematodes* L. (Hym. Formicoidea, Poneridae) Biologie des ouvrieres. *Insectes Soc.* **19**: 171–94.

Colwell, R.K. 1981. Group selection is implicated in the evolution of female-biassed sex ratios. *Nature* **290**: 401–4.

Contel, E.P.B. and Kerr, W.E. 1976. Origin of males in *Melipona subnitida* estimated from data of an isozymic polymorphic system. *Genetica* **46**: 271–7.

Cook, J.M. 1993a. Empirical tests of sex determination on *Goniozus nephantidis* (Hymenoptera: Bethylidae). *Heredity* **71**: 130–7.

Cook, J.M. 1993b. Sex determination in the Hymenoptera: a review of models and evidence. *Heredity* **71**: 421–35.

Cornuet, J.M. 1980. Rapid estimation of the number of sex alleles in panmictic honeybee populations. *J. Apic. Res.* **19**: 3–5.

Corso, C.R. and Serzedello, A. 1981. A study of multiple mating habit in *Atta laevigata* based on the DNA content. *Comp. Biochem. Physiol. B* **69**: 901–2.

Cowan, D.P. 1979. Sibling matings in a hunting wasp: adaptive inbreeding? *Science* **205**: 1403–5.

Cowan, D.P. 1981. Parental investment in two solitary wasps *Ancistrocerus adiabatus* and *Euodynerus foraminatus* (Eumenidae: Hymenoptera). *Behav. Ecol. Sociobiol.* **9**: 95–102.

Craig, R. 1979. Parental manipulation, kin selection, and the evolution of altruism. *Evolution* **33**: 319–34.

Craig, R. 1980. Sex investment ratios in social Hymenoptera. *Am. Nat.* **116**: 311–23.

Craig, R. 1982. Evolution of male workers in the Hymenoptera. *J. Theor. Biol.* **94**: 95–105.

Craig, R. 1983. Subfertility and the evolution of eusociality by kin selection. *J. Theor. Biol.* **100**: 379–97.

Craig, R. and Crozier, R.H. 1978. No evidence for role of heterozygosity in ant caste determination. *Isozyme Bull.* **11**: 66.

Craig, R. and Crozier, R.H. 1979. Relatedness in the polygynous ant *Myrmecia pilosula*. *Evolution* **33**: 335–41.

Crawley, W.C. 1912. Parthenogenesis in worker ants, with special reference to two colonies of *Lasius niger*. *Trans. Entomol. Soc. London* **1911**(4): 657–63.

Crespi, B.J. 1992a. Eusociality in Australian gall thrips. *Nature* **359**: 724–6.

Crespi, B.J. 1992b. Cannibalism and trophic eggs in subsocial and eusocial insects. In *Cannibalism: ecology and evolution among diverse taxa* (ed. M.A. Elgar, and B.J. Crespi) pp. 176–213. Oxford University Press.

Crosland, M.W.J. 1990. The influence of the queen, colony size and worker ovarian development on nestmate recognition in the ant *Rhytidoponera confusa*. *Anim. Behav.* **39**: 413–25.

Crow, J.F. and Kimura, M. 1970. *An introduction to population genetics theory*. Harper & Row, New York.

Crozier, R.H. 1969. *Genetic and phylogenetic studies on ants*. PhD Thesis, Cornell University, Ithaca, NY.

Crozier, R.H. 1970a. Coefficients of relationship and the identity of genes by descent in the Hymenoptera. *Am. Nat.* **104**: 216–17.

Crozier, R.H. 1970b. Karyotypes of twenty-one ant species (Hymenoptera: Formicidae), with reviews of the known ant karyotypes. *Can. J. Genet. Cytol.* **12**: 109–28.

Crozier, R.H. 1971. Heterozygosity and sex determination in haplo-diploidy. *Am. Nat.* **105**: 399–412.

Crozier, R.H. 1974. Allozyme analysis of reproductive strategy in the ant *Aphaenogaster rudis. Isozyme Bull.* **7**: 18.

Crozier, R.H. 1975. *Animal cytogenetics 3 Insecta 7 Hymenoptera.* Gebrüder Bornträger, Berlin.

Crozier, R.H. 1976. Why male-haploid and sex-linked genetic systems seem to have unusually sex-limited mutational genetic loads. *Evolution* **30**: 623–4.

Crozier, R.H. 1977a. Evolutionary genetics of the Hymenoptera. *Annu. Rev. Entomol.* **22**: 263–88.

Crozier, R.H. 1977b. Genetic differentiation between populations of the ant *Aphaenogaster 'rudis'* in the southeastern United States. *Genetica* **47**: 17–36.

Crozier, R.H. 1979. Genetics of sociality. In *Social insects*, (ed. H.R. Hermann) *Vol. I*, pp. 223–86. Academic Press, New York.

Crozier, R.H. 1980. Genetical structure of social insect populations. In *Evolution of social behavior: hypotheses and empirical tests* (ed. H. Markl) pp. 129–46. Verlag Chemie GmbH, Weinheim.

Crozier, R.H. 1982. On insects and insects: twists and turns in our understanding of the evolution of eusociality. In *The biology of social insects* (ed. M. Breed, C.D. Michener and H.E. Evans) pp 4–9. Westview Press, Boulder.

Crozier, R.H. 1985. Adaptive consequences of male-haploidy. In *Spider mites. Their biology, natural enemies and control* (ed. W. Helle and M.W. Sabelis), Vol. 1A, Chapter 1.3.4, pp. 201–22. Elsevier, Amsterdam.

Crozier, R.H. 1986. Genetic clonal recognition abilities in marine invertebrates must be maintained by selection for something else. *Evolution* **40**: 1100–1.

Crozier, R.H. 1987a. Selection, adaptation and evolution. *Proc. R. Soc. New South Wales* **120**: 21–37.

Crozier, R.H. 1987b. Genetic aspects of kin recognition: concepts, models, and synthesis. In *Kin recognition in animals* (ed. D.J.C. Fletcher and C.D. Michener), pp 55–73. Wiley, Chichester.

Crozier, R.H. 1988. Kin recognition using innate labels: a central role for piggy-backing? In *The mechanisms ecology, and evolution of historecognition in marine invertebrates* (ed. J. Clegg), pp 143–56. Plenum, New York.

Crozier, R.H. 1992. All about (eusocial) wasps. *Evolution* **46**: 1979–81.

Crozier, R.H. 1993. Molecular methods for insect phylogenetics. In *Molecular approaches to fundamental and applied entomology* (ed. J. Oakeshott and M.J. Whitten), pp 164–221. Springer-Verlag, New York.

Crozier, R.H. 1994. The second sociality. *Science* **265**: 1255–7.

Crozier, R.H. and Brückner, D. 1981. Sperm clumping and the population genetics of Hymenoptera. *Am. Nat.* **117**: 561–3.

Crozier, R.H. and Consul, P.C. 1976. Conditions for genetic polymorphism in social Hymenoptera under selection at the colony level. *Theor. Popul Biol.* **10**: 1–9.

Crozier, R.H. and Dix, M.W. 1979. Analysis of two genetic models for the innate components of colony odor in social Hymenoptera. *Behav. Ecol. Sociobiol.* **4**: 217–24.

Crozier, R.H. and Luykx, P. 1985. The evolution of termite eusociality is unlikely to have been based on a male-haploid analogy. *Am. Nat.* **126**: 867–869.

Crozier, R.H. and Page, R.E. 1985. On being the right size: male contributions and multiple mating in social Hymenoptera. *Behav. Ecol. Sociobiol.* **18**: 105–15.

Crozier, R.H. and Pamilo, P. 1980. Asymmetry in relatedness: Who is related to whom? *Nature* **283**: 604.

Crozier, R.H. and Pamilo, P. 1986. Relatedness within and between colonies of a queenless ant species of the genus *Rhytidoponera* (Hymenoptera: Formicidae). *Entomol. General.* **11**: 113–17.

Crozier, R.H. and Pamilo, P. 1993. Sex allocation in social insects: problems in prediction and estimation. In *Evolution and diversity of sex ratio in haplodiploid insects and mites* (ed. D.L. Wrensch and M.A. Ebbert), pp. 369–383. Chapman and Hall, New York.

Crozier, R.H., Pamilo, P. and Crozier, Y.C. 1984. Relatedness and microgeographic genetic variation in *Rhytidoponera mayri*, an Australian arid-zone ant. *Behav. Ecol. Sociobiol.* **15**: 143–50.

Crozier, R.H., Smith, B.H. and Crozier, Y.C. 1987. Relatedness and population structure of the primitively eusocial bee *Lasioglossum zephyrum* (Hymenoptera: Halictidae) in Kansas. *Evolution* **41**: 902–10.

Cruz, Y.P. 1981. A sterile defender morph in a polyembryonic hymenopterous parasite. *Nature* **294**: 446–7

Cumber, R.A. 1949. The biology of humble-bees, with special reference to the production of the worker caste. *Trans. R. Entomol. Soc. London* **100**: 1–45.

Curtsinger, J.W. 1991. X-chromosome segregation distortion in *Drosophila. Am. Nat.* **137**: 344–8.

Dallai, R. 1972. Fine structure of the spermathecal gland of *Apis mellifera. Redia* **53**: 413–25.

Dallai, R. 1975. Fine structure of the spermatheca of *Apis mellifera. J. Inst. Physiol.* **21**: 89–109.

Danforth, B.N. 1990. Provisioning behavior and the estimation of investment ratios in a solitary bee, *Calliopis (Hypomacroptera) persimilis* (Cockerell) (Hymenoptera: Andrenidae). *Behav. Ecol. Sociobiol.* **27**: 159–68.

Danks, H.V. 1983. Differences between generations in the sex ratio of aculeate Hymenoptera. *Evolution* **37**: 414–16.

*Darling, D.C. 1978. *Reproductive ratio of investment in the ants (Family Formicidae).* Master's Thesis, Department of Biology, University of Utah, Salt Lake City UT.

Darlington, J.P.E.C. 1985. Multiple primary reproductives in the termite *Macrotermes michaelsoni* (Sjöstedt). In *Caste differentiation in the social insects* (ed. J.A.L. Watson, B.M. Okot-Kotber and C. Noirot), pp 187–200. Pergamon, Oxford.

Darlington, J.P.E.C. 1988. Multiple reproductives in nests of *Macrotermes herus* (Isoptera: Termitidae). *Sociobiol.* **14**: 347–51.

Dartigues, L., and Passera, L. 1979. La ponte des ouvrieres chez la fourmi *Camponotus aethiops* Latreille (Hym. Formicidae). *Ann. Soc. Ent. Fr.* (N. S.) **15**: 109–16.

Darwin C. 1979. *The illustrated origin of species*, selections from the 6th edn. (ed. R.E. Leakey). Oxford University Press.

Davidson, D.W. 1982. Sexual selection in harvester ants (Hymenoptera: Formicidae: *Pogonomyrmex*). *Behav. Ecol. Sociobiol.* **10**: 245–50.

Davis, S.K., Strassmann, J.E., Hughes, C., Pletscher, L.S. and Templeton, A.R. 1990. Population structure and kinship in *Polistes* (Hymenoptera, Vespidae): an analysis using ribosomal DNA and protein electrophoresis. *Evolution* **44**: 1242–53.

Dawkins, R. 1982. *The extended phenotype*. Freeman, Oxford.

Dawkins, R. and Krebs, J.R. 1979. Arms races between and within species. *Proc. R. Soc. London, ser. B* **205**: 489–511.

Dejean, A. and Passera, L. 1974. Ponte des ouvrieres et inhibition royale chez la fourmi *Temnothorax recedens* (Nyl.) (Formicidae, Myrmicinae). *Insectes Soc.* **21**: 343–56.

Delage, B. 1968. Recherches sur les fourmis moissonneuses du bassin aquitain: écologie et biologie. *Bull. Biol. Fr. Belg.* **102**: 315–67

Delage-Darchen, B. 1974. Ecologie et biologie de *Crematogaster impressa* Emery, fourmi savanicole d'Afrique. *Insectes Soc.* **21**: 13–34

DeSalle R., Gatesy, J., Wheeler, W. and Grimaldi, D. 1992. DNA sequences from a fossil termite in Oligo-Miocene amber and their phylogenetic implications. *Science* **257**: 1933–6.

Douwes, P., Sivusaari, L., Niklasson, M. and Stille, B. 1987. Relatedness among queens in polygynous nests of the ant *Leptothorax acervorum*. *Genetica* **75**: 23–9.

Douwes, P., Jessen, K. and Buschinger, A. 1988. *Epimyrma adlerzia* sp. n. (Hymenoptera: Formicidae) from Greece: morphology and life history. *Entomol. Scand.* **19**:239–49.

Drescher, W. and Rothenbuhler, W.C. 1964. Sex determination in the honey bee. *J. Hered.* **55**: 91–6.

Droual, R. 1982. Sex ratios in the ant *Pheidole desertorum*. *Am. Zool.* **22**: 971.

Duchateau, M.J. and Velthuis, H.H.W. 1988. Development and reproductive strategies in *Bombus terrestris* colonies. *Behaviour* **107**: 186–207.

Edwards, J.P. 1991. Caste regulation in the pharaoh's ant *Monomorium pharaonis*: Recognition and cannibalism of sexual brood by workers. *Physiol. Entomol.* **16**: 263–71.

Ehrhardt, H.-J. 1962. Ablage übergroßer Eier durch Arbeiterinnen von *Formica polyctena* Föster (*Ins., Hym.*) in Gegenwart von Königinnen. *Naturwissenschaften* **22**: 524–5.

Eickwort, G.C. 1981. Presocial insects. In *Social insects* (ed. H.R. Hermann), Vol. II, pp. 199–280. Academic Press, New York.

Eickwort, G.C. and Eickwort, K.R. 1969. Aspects of the biology of Costa Rican halictine bees, I. *Agapostemon nasutus*. *J. Kansas Entomol. Soc.* **42**: 421–52.

Elmes, G.W. 1973. Observations on the density of queens in natural colonies of *Myrmica rubra* L. (Hymenoptera: Formicidae). *J. Anim. Ecol.* **42**: 761–71

Elmes, G.W. 1974. Colony populations of *Myrmica sulcinodis* Nyl. (Hym. Formicidae). *Oecologia* **15**: 337–43

Elmes, G.W. 1987a. Temporal variation in colony populations of the ant *Myrmica sulcinodis* I. Changes in queen number, worker number and spring production. *J. Anim. Ecol.* **56**: 559–71

Elmes, G.W. 1987b. Temporal variation in colony populations of the ant *Myrmica sulcinodis*. II. Sexual production and sex ratios. *J. Anim. Ecol.* **56**: 573–83.

Elmes, G.W. and Wardlaw, J.C. 1982. A population study of the ants *Myrmica sabuleti* and *Myrmica scabrinodis*, living at two sites in the south of England. I. A comparison of colony populations. *J. Anim. Ecol.* **51**: 651–64.

Emery, C. 1909. Über den Ursprung der dulotischen, parasitischen und myrmekophilen Ameisen. *Biol. Zentralbl.* **29**: 352–62.

Emlen, S.T., Emlen, J.M. and Levin, S.A. 1986. Sex-ratio selection in species with helpers-at-the-nest. *Am. Nat.* **127**: 1–8.

Estoup, A., Solignac, M., Harry, M. and Cornuet, J.-M. 1993. Characterization of $(GT)_n$ and $(CT)_n$ microsatellites in two insect species: *Apis mellifera* and *Bombus terrestris*. *Nucleic Acids Res.* **21**: 1427–31.

Estoup, A., Solignac, M. and Cornuet, J.-M. 1994. Precise assessment of the number of patrilines and of genetic relatedness in honeybee colonies. *Proc. R. Soc. London B* **258**: 1–7.

Estoup, A., Scholl, A., Pouvreau, A. and Solignac, M. 1995. Monoandry and polyandry in bumble bees (Hymenoptera; Bombinae) as evidenced by highly variable microsatellites. *Mol. Ecol.* **4**: 89–93.

Evans, H.E. 1977. Extrinsic versus intrinsic factors in the evolution of insect eusociality. *BioScience* **27**: 613–17.

Evans, J.D. 1993. Parentage analyses in ant colonies using simple sequence repeat loci. *Mol. Ecol.* **2**: 393–7.

Fielde, A. 1905. Observations on the progeny of virgin ants. *Biol. Bull.* **9**: 355–60.

Fischer, K. 1987. *Karyotypuntersuchungen an selbständigen und sozialparatischen Ameisen der Tribus Leptothoracini* (Hymenoptera, Formicidae) *im Hinblick auf ihre Verwandtschaftsbeziehungen.* PhD Thesis, TH Darmstadt.

Fisher, R.A. 1930. *The genetical theory of natural selection.* Oxford University Press.

Fisher, R.M. 1987. Queen–worker conflict and social parasitism in bumble bees (Hymenoptera: Apidae). *Anim. Behav.* **35**: 1026–36.

Fisher, R.M. and Pomeroy, N. 1990. Sex discrimination and infanticide by queens of the bumble bee *Bombus terrestris* (Hymenoptera: Apidae). *Anim. Behav.* **39**: 801–2.

Fletcher, D.J.C. and Blum, M.S. 1983. Regulation of queen number by workers in colonies of social insects. *Science* **219**: 312–14.

Fletcher, D.J.C. and Ross, K.G. 1985. Regulation of reproduction in eusocial Hymenoptera. *Annu. Rev. Entomol.* **30**: 319–43.

Fondrk, M.K., Page, R.E., Jr. and Hunt, G.J. 1993. Paternity analysis of worker honeybees using random amplified polymorphic DNA. *Naturwiss.* **80**: 226–31.

Fontana, F. 1991. Multiple reciprocal translocations and their role in the evolution of eusociality in termites. *Ethol. Ecol. Evol.* **1**: 15–19.

Forel, A. 1928. *The social world of the ants compared with that of a man.* G.P. Putnam, London.

Forsyth, A. 1980. Worker control of queen density in hymenopteran societies. *Am. Nat.* **116**: 895–8.

Forsyth, A. 1981. Sex ratio and parental investment in an ant population. *Evolution* **35**: 1252–3.

Fortelius, W., Pamilo, P., Rosengren, R. and Sundström, L. 1987. Male size dimorphism and alternative reproductive tactics in *Formica exsecta* ants (Hymenoptera, Formicidae). *Ann. Zool. Fenn.* **24**: 45–54.

Foster, R.L. 1992. Nestmate recognition as an inbreeding avoidance mechanism in bumble bees (Hymenoptera: Apidae). *J. Kansas Entomol. Soc.* **65**: 238–43.

Fowler, H.G. 1982. Male induction and function of workers' excitability during swarming in leaf-cutting ants (*Atta* and *Acromyrmex*) (Hymenoptera, Formicidae). *Int. J. Invert. Repr.* **4**: 333–5.

Fowler, H.G. and Roberts, R.B. 1982. Seasonal occurrence of founding queens and the sex ratio of *Camponotus pennsylvanicus* (Hymenoptera: Formicidae) in New Jersey. *N. Y. Entomol. Soc.* **90**: 247–51

Frank, S.A. 1987a. Individual and population sex allocation patterns. *Theor. Popul. Biol.* **31**: 47–74.

Frank, S.A. 1987b. Demography and sex ratio in social spiders. *Evolution* **41**: 1267–81.

Frank, S.A. 1987c. Variable sex ratio among colonies of ants. *Behav. Ecol. Sociobiol.* **20**: 195–201.

Frank, S.A. and Crespi, B.J. 1989. Synergism between sib-rearing and sex ratio in Hymenoptera. *Behav. Ecol. Sociobiol.* **24**: 155–62.

Franks, N.R. 1985. Reproduction, foraging efficiency and worker polymorphism in army ants. In *Experimental behavioral ecology and sociobiology* (ed. B. Hölldobler and M. Lindauer), pp. 91–107. Gustav Fischer, Stuttgart.

Franks, N.R. and Hölldobler, B. 1987. Sexual competition during colony reproduction in army ants. *Biol. J. Linn. Soc.* **30**: 229–43.

Franks, N.R., Ireland, B. and Bourke, A.F. 1990. Conflicts, social economics and life history strategies in ants. *Behav. Ecol. Sociobiol.* **27**: 175–81

Freeland, J. 1958. Biological and social patterns in the Australian bulldog ants of the genus *Myrmecia*. *Aust. J. Zool.* **6**: 1–18.

Freeman, B.E. 1981. Parental investment and its ecological consequences in the solitary wasp *Sceliphron assimile* (Dahlbom) (Sphecidae). *Behav. Ecol. Sociobiol.* **9**: 261–8.

Fresneau, D. 1984. Développement ovarien et statut social chez une fourmi primitive *Neoponera obscuricornis* Emery (Hym. Formicidae, Ponerinae). *Insectes Soc.* **31**: 387–402.

Frohlich, D.R. and Tepedino, V.J. 1986. Sex ratio, parental investment, and interparent variability in nesting success in a solitary bee. *Evolution* **40**: 142–51.

Frumhoff, P.C. and Baker, J. 1988. A genetic component to division of labour within honey bee colonies. *Nature* **333**: 358–61.

Frumhoff, P.C. and Ward, P.S. 1992. Individual-level selection, colony-level selection, and the association between polygyny and worker monomorphism in ants. *Am. Nat.* **139**: 559–90.

Fye, R.C. 1965. The biology of the Vespidae, Pompilidae, and Sphecidae (Hymenoptera) from trap nests in northwestern Ontario. *Can. Entomol.* **97**: 716–44.

Gadagkar, R. 1990. Evolution of eusociality: the advantage of assured fitness returns. *Philos. Trans. R. Soc. London, Ser. B.* **329**: 17–25.

Gadagkar, R. 1991a. Demographic predisposition to the evolution of eusociality: A hierarchy of models. *Proc. Natl. Acad. Sci. USA* **88**: 10993–7.

Gadagkar, R. 1991b. On testing the role of genetic asymmetries created by haplodiploidy in the evolution of eusociality in the Hymenoptera. *J. Genet.* **70**: 1–31.

Gamboa, G.J. 1978. Intraspecific defense: advantage of social cooperation among paper wasp foundresses. *Science* **199**: 1463–5.

Garofalo, C.A. 1973. Occurrence of diploid drones in a Neotropical bumblebee. *Experientia* **29**: 726–7.

Gerber, H.S. and Klostermeyer, E.C. 1970. Sex control by bees: a voluntary act of fertilization during oviposition. *Science* **162**: 82–4.

Gertsch, P., Pamilo, P. and Varvio, S-L. 1995. Microsatellites reveal high genetic diversity within colonies of *Camponotus* ants. *Mol. Ecol.* **4**: 257–60.

Getz, W.M. 1982. An analysis of learned kin recognition in Hymenoptera. *J. Theor. Biol.* **99**: 585–97.

Gibo, D.L. 1974. A laboratory study on the selective advantage of foundress association in *Polistes fuscatus* (Hymenoptera: Vespidae). *Can. Entomel.* **106**: 101–6.

Gibo, D.L. 1978. The selective advantage of foundress associations in *Polistes fuscatus* (Hymenoptera: Vespidae): a field study of the effects of predation on productivity. *Can. Entomol.* **110**: 519–40.

Godfray, H.C.J. and Grafen, A. 1988. Unmatedness and the evolution of eusociality. *Am. Nat.* **131**: 303–5.

Gösswald, K. and Bier, K. 1957. Untersuchungen zur Kastendetermination in der Gattung *Formica*. 5. Der Einfluss der Temperatur auf die Eiablage und Geschlechtsbestimmung. *Insectes Soc.* **4**: 335–48.

Götsch, W. and Käthmer, B. 1937. Die Koloniegründung der Formicinen und ihre experimentelle Beeinflussung. *Z. Morphol. Ökol. Tiere* **33**: 201–60.

Grafen, A. 1985. A geometric view of relatedness. *Oxford Surv. Evol. Biol.* **2**: 28–89.

Grafen, A. 1986. Split sex ratios and the evolutionary origins of eusociality. *J. Theor. Biol.* **122**: 95–121.

Grafen, A. 1990. Do animals really recognize kin? *Anim. Behav.* **39**:42–54.

Grbic, M., Ode, P.J. and Strand, M.R. 1992. Sibling rivalry and brood sex ratios in polyembryonic wasps. *Nature* **360**: 254–6.

Green, R.F., Gordh, G. and Hawkins, B.A. 1982. Precise sex ratios in highly inbred parasitic wasps. *Am. Nat.* **120**: 653–65.

Greenberg, L. 1979. Genetic component of bee odor in kin recognition. *Science* **206**: 1095–7.

Greene, A., Akre, R.D. and Landolt, P. 1976. The aerial yellowjacket, *Dolichovespula arenaria* (Fab.): nesting biology, reproductive production, and behavior (Hymenoptera: Vespidae). *Melanderia* **26**: 1–34.

Greenslade, P.J.M. and Halliday, R.B. 1983. Colony dispersion and relationships of meat ants *Iridomyrmex purpureus* and allies in an arid locality in South Australia. *Insectes Soc.* **30**: 82–99.

Grosberg, R.K. and Quinn, J.F. 1988. The evolution of allorecognition specificity: a theoretical analysis. In *The mechanisms, ecology, and evolution of historecognition in marine invertebrates* (ed. J. Clegg), pp. 157–67. Plenum, New York.

Grosberg, R.K. and Quinn, J.F. 1989. The evolution of selective aggression based on allorecognition specificity. *Evolution* **43**: 504–15.

Hagen, R.H., Smith, D.R. and Rissing, S.W. 1988. Genetic relatedness among cofoundresses of two desert ants, *Veromessor pergandei* and *Acromyrmex versicolor* (Hymenoptera: Formicidae). *Psyche* **95**: 191–201.

Halliday, R.B. 1983. Social organization of meat ants *Iridomyrmex purpureus* analysed by gel electrophoresis of enzymes. *Insectes Soc.* **30**: 45–56.

Halverson, D.D., Wheeler, J. and Wheeler, G.C. 1976. Natural history of the sandhill ant, *Formica bradleyi. J. Kansas Entomol. Soc.* **49**: 280–303.

Hamaguchi, K., Itô, Y. and Takenaka, O. 1993. GT dinucleotide repeat polymorphisms in a polygynous ant, *Leptothorax spinosior* and their use for measurement of relatedness. *Naturwissenschaften* **80**: 179–81.

Hamilton, W.D. 1963. The evolution of altruistic behavior. *Am. Nat.* **97**: 354–6.

Hamilton, W.D. 1964a. The genetical evolution of social behaviour, I. *J. Theor. Biol.* **7**: 1–16.

Hamilton, W.D. 1964b. The genetical evolution of social behaviour, II. *J. Theor. Biol.* **7**: 17–32.

Hamilton, W.D. 1967. Extraordinary sex ratios. *Science* **156**: 477–88.

Hamilton, W.D. 1972. Altruism and related phenomena, mainly in social insects. *Annu. Rev. Ecol. Syst.* **3**: 193–232.

Hamilton, W.D. 1975. Gamblers since life began: barnacles, aphids, elms. *Q. Rev. Biol.* **50**: 175–80.

Hamilton, W.D. 1987. Kinship, recognition, disease, and intelligence: constraints of social evolution. In *Animal societies: theories and facts* (ed. Y. Itô, J.L. Brown and J. Kikkawa), pp. 81–102. Japan Science Society Press, Tokyo.

Hasegawa, E. 1992. Annual life cycle and timing of male-egg production in the ant *Colobopsis nipponicus* (Wheeler). *Insects Soc.* **39**: 439–46.

Hasegawa, E. 1994. Sex allocation in the ant *Colobopsis nipponicus* (Wheeler). I. Population sex ratio. *Evolution* **48**: 1121–9.

Haskins, C.P. and Enzmann, E.V. 1945. On the occurrence of impaternate females in the Formicidae. *J. N. Y. Entomol. Soc.* **53**: 263–77.

Haskins, C.P. and Haskins, E.F. 1950. Notes on the biology and social behavior of the archaic ponerine ants of the genera *Myrmecia* and *Promyrmecia. Ann. Entomol. Soc. Am.* **43**: 461–91.

Haskins, C.P. and Haskins, E.F. 1979. Worker compatibilities within and between populations of *Rhytidoponera metallica*. *Psyche* **86**: 299–312.

van der Have, T.M., Boomsma, J.J and Menken, S.B.J. 1988. Sex investment ratios and relatedness in the monogynous ant *Lasius niger* (L.). *Evolution* **42**: 160–72.

Headley, A.E. 1943. Population studies of two species of ants, *Leptothorax longispinosus* Roger and *Leptothorax curvispinosus* Mayr. *Ann. Entonol. Soc. Am.* **36**: 743–53.

Headley, A.E. 1949. A population study of the ant *Aphaenogaster fulva* ssp. *aquia* Buckley (Hymenoptera, Formicidae). *Ann. Entomol. Soc. Am.* **42**: 265–72.

Hedderwick, M.P., El Agoze, M., Garaud, P. and Periquet, G. 1985. Mise en evidence de males heterozygotes chez l'hymenoptere *Diadromus pulchellus* (Ichneumonidae). *Genet. Sel. Evol.* **17**: 303–10.

Hedrick, P.W. 1985. *Genetics of populations.* Jones and Bartlett, Portola Valley, CA.

Hefetz, A., Bergström, G. and Tengö, J. 1986. Species, individual and kin specific blends in Dufour's Gland secretions of halictine bees. Chemical evidence. *J. Chem. Ecol.* **12**: 197–208.

Heinze, J. and Buschinger, A. 1989. Queen polymorphism in *Leptothorax* sp. A: its genetic and ecological background (Hymenoptera: Formicidae). *Insectes Soc.* **36**: 139–55.

Heinze, J., Hölldobler, B. and Cover, S.P. 1992. Queen polymorphism in the North American harvester ant, *Ephebomyrmex imberbiculus*. *Insectes Soc.* **39**: 267–73.

Heinze, J., Gadau, J., Hölldobler, B., Nanda, I., Schmid, M. and Scheller, K. 1994a. Genetic variability in the ant *Camponotus floridanus* detected by multilocus DNA fingerprinting. *Naturwissenschaften* **81**: 34–6.

Heinze, J., Hölldobler, B. and Peeters, C. 1994b. Conflict and cooperation in ant societies. *Naturwissenschaften* **81**: 489–97.

Helms, K.R. 1994. Sexual size dimorphism and sex ratios in bees and wasps. *Am. Nat.* **143**: 418–34.

Helms, K.R. and Rissing, S.W. 1990. Single sex alate production by colonies of *Pheidole desertorum* and *Pheidole xerophila tucsonica* (Hymenoptera: Formicidae). *Psyche* **97**: 213–16.

Herbers, J.M. 1979. The evolution of sex ratio strategies in hymenopteran societies. *Am. Nat.* **114**: 818–34.

Herbers, J.M. 1984. Queen–worker conflict and eusocial evolution in a polygynous ant species. *Evolution* **38**: 631–43.

Herbers, J.M. 1986. Ecological genetics of queen number in *Leptothorax longispinosus* (Hymenoptera: Formicidae). *Entomol. General.* **11**: 119–23.

Herbers, J.M. 1989. Community structure in north temperate ants: temporal and spatial variation. *Oecologia* **81**: 201–11.

Herbers, J.M. 1990. Reproductive investment and allocation ratios for the ant *Leptothorax longispinosus*: sorting out the variation. *Am. Nat.* **136**: 178–208.

Herbers, J.M. 1991. The population biology of *Tapinoma minutum* (Hymenoptera: Formicidae) in Australia. *Insectes Soc.* **38**: 195–204.

Herbers, J.M. 1993. Ecological determinants of queen number in ants. In *Queen number and sociality in insects* (ed. L. Keller), pp. 262–93. Oxford University Press.

Herbers, J.M. and Grieco, S. 1994. Population structure of *Leptothorax ambiguus*, a facultatively polygynous and polydomous species. *J. Evol. Biol.* **7**: 581–98.

Higashi, M., Yamamura, N., Abe, T. and Burns, T.P. 1991 Why don't all termite species have a sterile worker caste? *Proc. R. Soc. London Ser.* B **246**: 25–9.

Higashi, S. 1976. Nest proliferation by budding and nest growth pattern in *Formica* (*Formica*) *yessensis* in Ishikari Shore. *J. Fac. Sci. Hokkaido Univ. Ser. VI, Zool.* **20**: 359–89.

*Hirashima, Y. 1958. Comparative studies of the cocoon-spinning habits of *Osmia excavata* Alfken and *Osmia pedicornis* Cockerell. *Sci. Bull. Fac. Agr., Kyushu Univ.* **16**: 481–97.

Hobbs, G.A. 1965. Ecology of species of *Bombus* Latr. (Hymenoptera: Apidae) in southern Alberta. II. Subgenus *Bombias* Robt. *Can. Entomol.* **97**: 120–8.

Hobbs, G.A. 1967. Ecology of species of *Bombus* (Hymenoptera: Apidae) in southern Alberta. VI. Subgenus *Pyrobombus*. *Can. Entomol.* **99**: 1271–92.

Hocking, B. 1970. Insect associations with the swollen thorn acacias. *Trans. R. Entomol. Soc. London.* **122**: 211–55.

Hogendoorn, K. and Velthuis, H.H.W. 1988. Influence of multiple mating on kin recognition by worker honeybees. *Naturwissenschaften.* **75**: 412–13.

Hölldobler, B. 1962. Zur Frage der Oligogynie bei *Camponotus ligniperda* Latr. und *Camponotus herculeanus* L. (Hymenoptera: Formicidae). *Z. Angew. Entomol.* **49**: 337–52.

Hölldobler, B. 1976. Behavioral ecology of mating in *Pogonomyrmex* harvesting ants. *Behav. Ecol. Sociobiol.* **1**: 405–23.

Hölldobler, B. 1984. Evolution of insect communication. In *Insect communication* (ed. T. Lewis), pp. 349–77. Academic Press, London.

Hölldobler, B. 1995. The chemistry of social regulation: multicomponent signals in ant societies. *Proc. Natl. Acad. Sci. USA* **92**: 19–22.

Hölldobler, B. and Bartz, S. 1985. Sociobiology of reproduction in ants. In *Experimental behavioral ecology* (ed. B. Hölldobler and M. Lindauer), pp. 237–257. Gustav Fischer, Stuttgart.

Hölldobler, B. and Carlin, N.F. 1985. Colony founding, queen dominance and oligogyny in the Australian meat ant *Iridomyrmex purpureus*. *Behav. Ecol. Sociobiol.* **18**: 45–58.

Hölldobler, B. and Carlin, N.F. 1989. Colony founding, queen control and worker reproduction in the ant *Aphaenogaster* (=*Novomessor*) *cockerelli* (Hymenoptera: Formicidae). *Psyche* **96**: 131–51.

Hölldobler, B. and Lumsden, C.J. 1980. Territorial strategies in ants. *Science* **210**: 732–9.

Hölldobler, B. and Maschwitz, U. 1965. Der Hochzeitsschwarm der Rossameise *Camponotus herculeanus* L. (Hym. Formicidae). *Z. Vgl. Physiol.* **50**: 551–68.

Hölldobler, B. and Wilson, E.O. 1977. The number of queens: an important trait in ant evolution. *Naturwissenchaften* **64**: 8–15.

Hölldobler, B. and Wilson, E.O. 1983. Queen control in colonies of weaver ants (Hymenoptera: Formicidae). *Ann. Entomol. Soc. Am.* **76**: 235–8.

Hölldobler, B. and Wilson, E.O. 1990. *The ants.* Harvard University Press, Cambridge, MA.

van Honk, C. and Hogeweg, P. 1981. The ontogeny of the social structure in a captive *Bombus terrestris* colony. *Behav. Ecol. Sociobiol.* **9**: 111–19.

van Honk, C.G.J., Röseler, P.-F., Velthuis, H.H.W. and Hoogeveen, J.C. 1981. Factors influencing the egg-laying of workers in a captive *Bombus terrestris* colony. *Behav. Ecol. Sociobiol.* **9**: 9–14.

Hoshiba, H., Okada, I. and Kusanagi, A. 1981. The diploid drone of *Apis cerana japonica* and its chromosomes. *J. Apic. Res.* **20**: 143–7.

Hughes, C.R. and Queller, D.C. 1993. Detection of highly polymorphic microsatellite loci in a species with little allozyme polymorphism. *Mol. Ecol.* **2**: 131–7.

Hull, D.L. 1980. Individuality and selection. *Annu. Rev. Ecol. Syst.* **11**: 311–32.

Hung, A.C.F. 1973. Reproductive biology in dulotic ants: preliminary report (Hymenoptera: Formicidae). *Entomol. News,* **84**: 253–9.

Hung, A.C.F and Vinson, S.B. 1976. Biochemical evidence for queen monogamy and sterile male diploidy in the fire ant *Solenopsis invicta*. *Isozyme Bull.* **9**: 55.

Hung, A.C.F., Imai, H.T. and Kubota, M. 1972. The chromosomes of nine ant species from Taiwan, Republic of China. *Ann. Entomol. Soc. Am.* **65**: 1023–5.

Hunt, G.J. and Page, R.E., Jr. 1992. Patterns of inheritance with RAPD molecular markers reveal novel types of polymorphism in the honey bee. *Theor. Appl. Genet.* **85**: 15–20.

Hurst, L.D. 1991. The incidences and evolution of cytoplasmic male killers. *Proc. R. Soc. London Ser. B* **24**: 91–9.

Ichinose, K. 1986. Occurrence of polydomy in a monogynous ant, *Paratrechina flavipes* (Hymenoptera, Formicidae). *Kontyû* **54**: 208–17.

Ichinose, K. 1991. Seasonal variation in nestmate recognition in *Paratrechina flavipes* (Smith) worker ants (Hymenoptera: Formicidae). *Anim. Behav.* **41**: 1–6.

Imai, H.T. 1966. The chromosome observation techniques of ants and the chromosomes of Formicinae and Myrmicinae. *Acta Hymenopterol.* **2**: 119–31.

Imai, H.T., Crozier, R.H. and Taylor, R.W. 1977. Karyotype evolution in Australian ants. *Chromosoma* **59**: 341–93.

Ishay, I., Ikan, R. and Bergmann, E.D. 1965. The presence of pheromones in the oriental hornet, *Vespa orientalis* F. *J. Insect Physiol.* **11**: 1307–9.

Ito, M. and Imamura, S. 1974. Observations on the nuptial flight and internidal relationship in a polydomous ant, *Formica* (*Formica*) *yessensis* Forel. *J. Fac. Sci. Hokkaido Univ. Ser. VI, Zool.* **19**: 681–94.

Itô, Y. 1986. Social behaviour of *Ropalidia fasciata* (Hymenoptera: Vespidae) females on satellite nests and on a nest with multiple combs. *J. Ethol.* **4**: 73–80.

Itô, Y. 1987a. Social behaviour of the Australian paper wasp, *Ropalidia revolutionalis* (de Saussure)(Hymenoptera: Vespidae). *J. Ethol.* **5**: 115–24.

Itô, Y. 1987b. Role of pleometrosis in the evolution of eusociality in wasps. In *Animal societies: theories and facts* (ed. Y. Itô, J.L. Brown and J. Kikkawa), pp. 17–24. Japan Scientific Society, Tokyo.

Itô, Y. 1989. The evolutionary biology of sterile soldiers in aphids. *Trends Ecol. Evol.* **4**: 69–73

Itow, T., Kobayashi, K., Kubota, M., Ogata. K., Imai, H.T. and Crozier, R.H. 1984. The reproductive cycle of the queenless ant *Pristomyrmex pungens*. *Insectes Soc.* **31**: 87–102.

Iwasa, Y. 1981. Role of sex ratio in the evolution of eusociality in haplodiploid social insects. *J. Theor. Biol.* **93**: 125–42.

Jacquard, A. 1975. Inbreeding: one word, several meanings. *Theor. Popul. Biol.* **7**: 338–63.

Janzen, D.H. 1973. Evolution of polygynous obligate acacia-ants in Western Mexico. *J. Anim. Ecol.* **42**: 727–50.

Janzen, D.H. 1975. *Pseudomyrmex nigropilosa*: a parasite of a mutualism. *Science* **188**: 936–7.

Jayakar, S.D. and Spurway, H. 1966. Sex ratios of some mason wasps (Vespoidea and Sphecoidea). *Nature* **212**: 306–7.

Jeanne, R.L. 1972. Social biology of the neotropical wasp, *Miscochyttarus drewsenii. Bull. Mus. Comp. Zool. Harv. Univ.* **144**: 63–150.

Jeanne, R.L. 1980. Evolution of social behavior in the Vespidae. *Annu. Rev. Entomol.* **25**: 371–96.

Jessen, K. 1987. Biosystematic revision of the parasitic ant genus *Epimyrma*. In *Chemistry and biology of social insects* (ed. J. Eder and H. Rembold), pp.41–42. J. Peperny, Verlag, Munich.

Jones, S.J., La Page, J.P. and Howard, R.W. 1988. Isopteran sex ratios: phylogenetic trends. *Sociobiology* **14**: 89–156.

Jonkman, J.C.M. 1980. The external and internal structure and growth of nests of the leaf-cutting ant *Atta vollenweideri* Forel, 1893 (Hym.: Formicidae). Part II. *Z. Angew. Entomol.* **89**: 217–46.

Jouvenaz, D.P., Wojcik, D.P. and Vander Meer, R.K. 1989. First observation of polygyny in fire ants, *Solenopsis* spp., in South America. *Psyche* **96**: 161–5.

Kannowski, P.B. 1963. The flight activities of formicine ants. *Symp. Genet. Biol. Ital.* **12**: 74–102.

Kannowski, P.B. and Johnson, R.L. 1969. Male patrolling behaviour and sex attraction in ants of the genus *Formica. Anim. Behav.* **17**: 425–9.

Kannowski, P.B. and Kannowski, P.M. 1957. The mating activities of the ant *Myrmica americana* Weber. *Ohio J. Sci.* **57**: 371–4.

Kapil, R.P. and Kumar, S. 1969. Biology of *Ceratina binghami* Ckll. *J. Res.* (Ludhiana Univ.) **6**: 359–71.

Karlin, S. and Lessard, S. 1986. *Theoretical studies on sex ratio evolution.* Princeton University Press.

Katayama, E. 1971. Observations on the brood development in *Bombus ignitus* (Hymenoptera, Apidae) I. Egg-laying habits of queens and workers. *Kontyû* **39**: 189–203.

Katayama, E. 1988. Workerlike new queens in a colony of *Bombus ardens* (Hymenoptera, Apidae). *Kontyû* **56**: 879–91.

Kaufmann, B., Boomsma, J.J., Passera, L. and Petersen, K.N. 1992. Relatedness and inbreeding in a French population of the unicolonial ant *Iridomyrmex humilis* (Mayr). *Insectes Soc.* **39**: 195–200.

Keller, L. (ed.) 1993. *Queen number and sociality in insects.* Oxford University Press.

Keller, L. and Nonacs, P. 1993. The role of queen pheromones in social insects: queen control or queen signal? *Anim. Behav.* **45**: 787–94.

Keller, L. and Passera, L. 1989. Size and fat content of gynes in relation to the mode of colony founding in ants (Hymenoptera; Formicidae). *Oecologia* **80**: 236–40.

Keller, L. and Passera, L. 1993. Incest avoidance, fluctuating asymmetry, and the consequences of inbreeding in *Iridomyrmex humilis*, an ant with multiple queen colonies. *Behav. Ecol. Sociobiol.* **33**: 191–9.

Keller, L. and Ross, K.G. 1993a. Phenotypic basis of reproductive success in a social insect: genetic and social determinants. *Science* **260**: 1107–10.

Keller, L. and Ross, K.G. 1993b. Phenotypic plasticity and 'cultural transmission' of alternative social organizations in the fire ant *Solenopsis invicta. Behav. Ecol. Sociobiol.* **33**: 121–9.

Keller, L., Passera, L. and Suzzoni, J.-P. 1989. Queen execution in the Argentine ant, *Iridomyrmex humilis. Physiol. Entomol.* **14**: 157–63.

Kent, D.S. and Simpson, J.A. 1992. Eusociality in the beetle *Austroplatypus incompertus* (Coleoptera: Curculionidae). *Naturwissenschaften* **79**: 86–7

Kerr, W.E. 1950. Genetic determination of caste in the genus *Melipona. Genetics* **35**: 143–52.

Kerr, W.E. 1969. Some aspects of the evolution of social bees (Apidae). *Evol. Biol.* **3**: 119–75.

Kerr, W.E. 1974. Advances in cytology and genetics of bees. *Annu. Rev. Entomol.* **19**: 253–68.

Kerr, W.E. 1975. Evolution of the population structure in bees. *Genetics* **79**: 73–84.

Kerr, W.E. 1976. Population genetic studies in Hymenoptera. 2. Sex limited genes. *Evolution* **30**: 94–9.

Kerr, W.E. 1986. Mutation in bees. 3. Application in bee populations of a mutation rate of $m = 1.6 \times 10^{-6}$. *Rev. Brasil. Genet.* **9**: 1–10.

Kerr, W.E. 1987. Genetic parameters of populations of social bees. In *Chemistry and biology of social insects* (ed. J. Eder and H. Rembold), p. 362. J. Peperny Verlag, Munich.

Kerr, W.E. 1990. Why are workers in social Hymenoptera not males? *Rev. Brasil. Genet.* **13**: 133–6.

Keyes, L. N., Cline, T. W. and Schedl, P. 1992. The primary sex determination signal of *Drosophila* acts at the level of transcription. *Cell* **68**: 933–43.

Kim, C.H. and Murakami, Y. 1980. Ecological studies on *Formica yessensis* Forel, with special reference to its effectiveness as a biological control agent of the pine caterpillar moth in Korea II. Bionomics of *Formica yessensis* Forel (Hymenoptera: Formicidae) in Korea. *J. Fac. Agr. Kyushu Univ.* **25**: 119–33.

Kinomura, K. and Yamauchi, K. 1987. Fighting and mating behaviours of dimorphic males in *Cardiocondyla wroughtoni. J. Ethol.* **5**: 75–81.

Knerer, G. and Plateaux-Quénu, C. 1970. The life cycle and social level of *Evylaeus nigripes* (Hymenoptera: Halictinae), a Mediterranean halictine bee. *Can. Entomol.* **102**: 185–96.

Krombein, K.V. 1967. *Trap-nesting wasps and bees: life histories, nests, and associates.* Smithsonian Press, Washington, DC.

Kukuk, P.F. 1989. Evolutionary genetics of a primitively eusocial halictine bee, *Dialictus zephyrus.* In *The genetics of social evolution* (ed. M.D. Breed and R.E. Page, Jr.), pp. 183–202. Westview Press, Boulder, CO.

Kukuk, P.F. 1992. Cannibalism in social bees. In *Cannibalism: ecology and evolution among diverse taxa* (ed. M.A. Elgar and B.J. Crespi), pp. 213–237. Oxford University Press.

Kukuk, P.F. and Decelles, P.C. 1986. Behavioral evidence for population structure in *Lasioglossum (Dialictus) zephyrum* female dispersion patterns. *Behav. Ecol. Sociobiol.* **19**: 233–9.

Kukuk, P.F. and May, B. 1990. Diploid males in a primitively eusocial bee, *Lasioglossum (Dialictus) zephyrum* (Hymenotera: Halictidae). *Evolution* **44**: 1522–8.

Kukuk, P.F. and May, B. 1991. Colony dynamics in a primitively eusocial halictine bee *Lasioglossum (Dialictus) zephyrum* (Hymenoptera: Halictidae). *Insectes Soc.* **38**: 171–89.

Kukuk, P.F. and Sage, G.K. 1994. Reproductivity and relatedness in a communal halictine bee *Lasioglossum (Chilalictus) hemichalceum. Insectes Soc.* **41**: 443–55.

Kukuk, P.F., Eickwort, G.C., Raveret-Richter, M., Alexander, B., Gibson, R., Morse, R.A. and Ratnieks, F. 1989. Importance of the sting in the evolution of sociality in the Hymenoptera. *Ann. Entomol. Soc. Am.* **82**: 1–5.

Kusnezov, N. 1962. El vuelo nupcial de las hormigas. *Acta Zool. Lilloana, Tucamán* **18**: 385–442

Lacy, R.C. and Sherman, P.W. 1983. Kin recognition by phenotype matching. *Am. Nat.* **121**: 489–512.

Laidlaw, H.H. and Page, R.E. 1984. Polyandry in honey bees (*Apis mellifera* L.): sperm utilization and intracolony genetic relationships. *Genetics* **108**: 985–97.

Laidlaw, H.H. Jr. and Page, R.E. Jr. 1986. Mating designs. In *Bee genetics and breeding* (ed. T.E. Rinderer), pp. 323–44. Academic Press, New York.

Ledoux, A. 1950. Recherche sur la biologie de la fourmi fileuse (*Oecophylla longinoda* Latr.). *Ann. Sci. Nat. Zool. Ser.* 11 **12**: 313–461.

Ledoux, A. 1954. Recherches sur le cycle chromosomique de la fourmi fileuse *Oecophylla longinoda* Latr. *Insectes Soc.* **1**: 149–75.

Ledoux, M.A. and Dargagnon, D. 1973. La formation des castes chez la fourmi *Aphaenogaster senilis. C. R. Acad. Sci. Paris* **276**: 551–3.

Lee, P.E. and Wilkes, A. 1965. Polymorphic spermatozoa in the hymenopterous wasp *Dahlbominus*. *Science* **147**: 1445–6.

*Le Masne, G. 1953a. Observations sur les relations entre le couvain et les adultes chez les fourmis. *Ann. Sci. Nat.* (11) **15**: 1–56.

*Le Masne, G. 1953b. Observations sur la biologie de la fourmi *Ponera eduardi* Forel. La descendance der ouvrieres fécondées par les mâles ergatoïdes. *C. R. Acad. Sci. Paris* **236**: 1096–8.

Lenoir, A. and Cagniant, H. 1986. Role of worker thelytoky in colonies of the ant *Cataglyphis cursor* (Hymenoptera:Formicidae). *Entomol. General.* **11**: 153–7.

Lenoir, A., Querard, L. and Berton, F. 1987. Colony founding and role of parthenogenesis in *Cataglyphis cursor* ants (Hymenoptera—Formicidae). In *Chemistry and biology of social insects* (ed. J. Eder and H. Rembold) p. 260. J. Peperny Verlag, Munich.

Lenoir, A., Querard, L., Pondicq, N. and Berton, F. 1988. Reproduction and dispersal in the ant *Cataglyphis cursor* (Hymenoptera: Formicidae). *Psyche* **95**: 21–44.

Lepage, M.G. and Darlington, J.P.E.C. 1984. Observations on the ant *Carebara vidua* F. Smith preying on termites in Kenya. *J. Nat. Hist.* **18**: 293–302.

Lester, L.J. and Selander, R.K. 1981. Genetic relatedness and the social organization of *Polistes* colonies. *Am. Nat.* **117**: 147–66.

Leutert, W. 1963. Systematics of ants. *Nature* **200**: 496–7.

Leutert, W. 1965. Phenotypic variability in worker ants of *Lasius flavus* de Geer and their progeny (Hym., Formicidae). *J. Entomol. Soc. Aust. (NSW)* **2**: 1–3.

Lewis, T. 1975. Colony size, density and distribution of the leafcutting ant, *Acromyrmex octospinosus* (Reich) in cultivated fields. *Trans. R. Entomol. Soc. London.* **127**: 51–64.

Lewontin, R.C. 1970. The units of selection. *Annu. Rev. Ecol. Syst.* **1**: 1–18.

Lin, N. and Michener, C.D. 1972. Evolution of eusociality in insects. *Q. Rev. Biol.* **47**: 131–59.

Lipski, N., Heinze, J. and Hölldobler, B. 1992. Social organization of three European *Leptothorax* species (Hymenoptera, Formicidae). In *Biology and evolution of social insects* (ed. J. Billen), pp. 287–290. Leuven University Press.

Litte, M. 1979. *Mischocyttarus flavitarsis* in Arizona: social and nesting biology of a polistine wasp. *Z. Tierpsychol.* **50**: 282–312.

Lloyd, J.E. 1981. Sexual selection: individuality, identification and recognition in a bumblebee and other insects. *Fla. Entomol.* **64**: 89–118.

Loiselle, R., Francouer, A. and Buschinger, A. 1990. Variations and taxonomic significance of the chromosome numbers in the Nearctic species of the genus *Leptothorax* (s.s).(Formicidae: Hymenoptera). *Caryologia* **43**: 321–34.

Longair, R.W. 1981. Sex ratio variations in xylophilous aculeate Hymenoptera. *Evolution* **35**(3): 597–600.

Lubbock, J. 1885. *Ants, bees, and wasps.* Kegan Paul, Trench and Co., London.

Luck, R.F., Stouthamer, R. and Nunney, L. 1993. Sex determination and sex ratio patterns in parasitic Hymenoptera. In *Evolution and diversity of sex ratio in haplodiploid insects and mites* (ed. D.L. Wrensch and M.A. Ebbert), pp. 442–76. Chapman and Hall, New York.

Lumsden, C.J. 1982. The social regulation of physical caste: the superorganism revived. *J. Theor. Biol.* **95**: 749–81.

Luykx, P. 1985. Genetic relations among castes in lower termites. In *Caste differentiation in social insects* (ed. J.A.L. Watson, B.M. Okot-Kotber and C. Noirot), pp. 17–25. Pergamon, New York.

Luykx, P. 1986. Termite colony dynamics as revealed by the sex- and caste-ratios of whole colonies of *Incisitermes schwarzi* Banks (Isoptera: Kalotermitidae). *Insectes Soc.* **33**: 221–48.

Mabelis, A. A. 1979. Wood ant wars: the relationship between aggression and predation in the red wood ant (*Formica polyctena* Först.). *Netherlands J. Zool.* **29**: 451–620.

MacArthur, R.H. 1965. Ecological consequences of natural selection. In *Theoretical and mathematical biology* (ed. T.H. Waterman and H. Morowitz), pp 388–397. Blaisdell, New York.

McCorquodale, D.B. 1988. Relatedness among nestmates in a primitively social wasp, *Cerceris antipodes* (Hymenoptera: Sphecidae). *Behav. Ecol. Sociobiol.* **23**: 401–6.

MacDonald, J.F. and Matthews, R.W. 1976. Nest structure and colony composition of *Vespula vidua* and *V. consobrina* (Hymenoptera: Vespidae). *Ann. Entomol. Soc. Am.* **69**: 471–5.

MacDonald, J.F. and Matthews, R.W. 1981. Nesting biology of the eastern yellowjacket, *Vespula maculifrons* (Hymenoptera: Vespidae). *J. Kansas Entomol. Soc.* **54**: 433–57.

MacDonald, J.F., Akre, R.D. and Hill, W.B. 1974. Comparative biology and behavior of *Vespula atropilosa* and *V. pennsylvanica* (Hymenoptera: Vespidae). *Melanderia* **18**: 1–93.

Macewicz, S. 1979. Some consequences of Fisher's sex ratio principle for social Hymenoptera that reproduce by colony fission. *Am. Nat.* **113**: 363–71.

Machado, M.F.P.S., Contel, E.P.B. and Kerr, W.E. 1984. Proportion of males sons-of-the-queen and sons-of-workers in *Plebeia droryana* (Hymenoptera, Apidae) estimated from data of an MDH isozymic polymorphic system. *Genetica* **65**: 193–8.

MacKay, W.P. 1981. A comparison of the nest phenologies of three species of *Pogonomyrmex harvester* ants (Hymenoptera: Formicidae). *Psyche* **88**: 25–74.

MacKay, W.P. 1985. A comparison of the energy budgets of three species of *Pogonomyrmex* harvester ants (Hymenoptera: Formicidae). *Oecologia* **66**: 484–94.

Macy, R.M. and Whiting, P.W. 1969. Tetraploid females in *Mormoniella Genetics* **61**: 619–30.

Malyshev, S.I. 1968. *Genesis of the Hymenoptera and the phases of their evolution.* Methuen, London.

Mamsch, E. and Bier, K. 1966. Das Verhalten von Ameisenarbeiterinnen gegenüber der Königin nach vorangegangener Weisellosigkeit. *Insectes Soc.* **13**: 277–84.

Markin, G.P. 1970. The seasonal life cycle of the Argentine ant, *Iridomyrmex humilis* (Hymenoptera: Formicidae), in Southern California. *Ann. Entomol. Soc. Am.* **63**: 1238–42.

Markin, G.P. and Dillier, J.H. 1971. The seasonal life cycle of the imported fire ant, *Solenopsis saevissima richteri* on the Gulf Coast of Mississippi. *Ann. Entomol. Soc. Am.* **64**: 562–5.

Marlin, J.C. 1971. The mating, nesting and ant enemies of *Polyergus lucidus* Mayr (Hymenoptera: Formicidae). *Am. Midl. Nat.* **86**: 181–9.

Martin, P. 1963. Die Steuerung der Volksteilung beim Schwarmen der Bienen. Zugleich ein Beitrag zum Problem der Wanderschwarme. *Insectes Soc.* **10**: 13–42.

Matessi, C. and Eshel, I. 1992. Sex ratio in the social Hymenoptera: a population-genetics study of long-term evolution. *Am. Nat.* **139**: 276–312.

May, R.M. and Seger, J. 1985. Sex ratios in wasps and aphids. *Nature* **318**: 408–9.

Maynard Smith, J. 1964. Group selection and kin selection. *Nature* **201**: 1145–7.

Maynard Smith, J. 1978. *The evolution of sex.* Cambridge University Press.

Maynard Smith, J. 1980. A new theory of sexual investment. *Behav. Ecol. Sociobiol.* **7**: 247–51.

Maynard Smith, J. 1982. *Evolution and the theory of games.* Cambridge University Press.

Metcalf, R.A. 1980. Sex ratios, parent–offspring conflict, and local competition for mates in the social wasps *Polistes metricus and Polistes variatus*. *Am. Nat.* **116**: 642–54.

Metcalf, R.A. and Whitt, G.S. 1977a. Intra-nest relatedness in the social wasp *Polistes metricus*. A genetic analysis. *Behav. Ecol. Sociobiol.* **2**: 339–51.

Metcalf, R.A. and Whitt, G.S. 1977b. Relative inclusive fitness in the social wasp *Polistes metricus*. *Behav. Ecol. Sociobiol.* **2**: 353–60.

Michener, C.D. 1969. Comparative social behavior of bees. *Annu Rev. Entomol.* **14**: 299–342.

Michener, C.D. 1971. Biologies of African allodapine bees. *Bull. Am. Mus. Nat. Hist.* **145**: 221–301.

Michener, C.D. 1974. *The social behavior of the bees.* Harvard University Press, Cambridge, MA.

Michener, C.D. 1985. From solitary to eusocial: need there be a series of intervening species?. In *Experimental behavioral ecology*, (ed. B. Hölldobler and M. Lindauer), pp. 293–305. Sinauer Associates., Sunderland.

Michener, C.D. 1990. Reproduction and castes in social halictine bees. In *Social insects. An evolutionary approach to castes and reproduction* (ed. W. Engels), pp. 77–121. Springer-Verlag, Berlin.

Michener, C.D. and Brothers, D.J. 1974. Were workers of eusocial Hymenoptera initially altruistic or oppressed? *Proc. Natl. Acad. Sci. USA* **71**: 671–4.

Michener, C.D. and Kerfoot, W.B. 1967. Nests and social behavior of three species of *Pseudaugochloropsis. J. Kansas Entomol. Soc.* **30**: 214–32.

Michener, C.D. and Lange, R.B. 1958a. Observations on the behavior of Brasilian halictid bees, III. *Univ. Kansas Sci. Bull.* **39**: 473–505.

Michener, C.D. and Lange, R.B. 1958b. Observations on the behavior of Brasilian halictid bees, V. *Chloralictus. Insectes Soc.* **5**: 379–407.

Michener, C.D. and Lange, R.B. 1959. Observations on the behavior of Brasilian halictid bees IV. *Augochloropsis*, with notes on extralimital forms. *Am. Mus. Novitates.* **1924**: 1–41.

Michener, C.D. and Smith, B.H. 1987. Kin recognition in primitively eusocial insects. In *Kin recognition in animals* (ed. D.J.C. Fletcher and C.D. Michener), pp. 209–42. Wiley, Chichester.

Michener, C.D. and Wille, A. 1961. The bionomics of a primitively social bee, *Lasioglossum inconspicuum. Univ. Kansas Sci. Bull.* **42**: 1123–202.

Michener, C.D., Kerfoot, W.B. and Ramirez, W. 1966. Nests of *Neocorynura* in Costa Rica. *J. Kansas Entomol. Soc.* **39**: 245–58.

Michod, R.E. 1982. The theory of kin selection. *Annu. Rev. Ecol. Syst.* **13**: 23–55.

Michod, R.E. and Abugov, R. 1980. Adaptive topography in family-structured models of kin selection. *Science* **210**: 667–9.

Michod, R.E. and Hamilton, W.D. 1980. Coefficients of relatedness in sociobiology. *Nature* **288**: 694–7.

Mintzer, A. 1982a. Copulatory behavior and mate selection in the harvester ant, *Pogonomyrmex californicus* (Hymenoptera: Formicidae). *Ann. Entomol. Soc.* **75**: 323–6.

Mintzer, A. 1982b. Nestmate recognition and incompatibility between colonies of the acacia-ant *Pseudomyrmex ferruginea. Behav. Ecol. Sociobiol.* **10**: 165–8.

Mintzer, A. 1987. Primary polygyny in the ant *Atta texana*: number and weight of females and colony foundation success in the laboratory. *Insectes Soc.* **34**: 108–17.

Mintzer, A. and Vinson, S.B. 1985. Kinship and incompatibility between colonies of the acacia ant *Pseudomyrmex ferruginea. Behav. Ecol. Sociobiol.* **17**: 75–8.

Miyano, S. 1986. Colony development, worker behavior and male production in orphaned colonies of a Japanese paper wasp, *Polistes chinensis antennalis* Pérez (Hymenoptera: Vespidae). *Res. Popul. Ecol.* **28**: 347–61.

Miyano, S. 1991. Worker reproduction and related behavior in orphan colonies of a Japanese paper wasp, *Polistes jadwigae* (Hymenoptera, Vespidae). *J. Ethol.* **9**: 135–46.

Mizutani, A. 1981. On the two forms of the ant *Myrmica ruginodis* Nylander (Hymenoptera, Formicidae) from Sapporo and its vicinity, Japan. *Jpn. J. Ecol.* **31**: 131–7.

Mock, D.W. and Forbes, L.S. 1992. Parent–offspring conflict: a case of arrested development. *Trends Ecol. Evol.* **7**: 409–13.

Montagner, H. 1966. Sur l'origine des males dans les societes de guepes du genre *Vespa*. *C. R. Acad. Sci. Paris* **263**: 785–7.

Moritz, R.F.A. 1985. The effects of multiple mating on the worker–queen conflict in *Apis mellifera* L. *Behav. Ecol. Sociobiol.* **16**: 375–7.

Moritz, R.F.A and Hillesheim, E. 1985. Inheritance of dominance in honeybees (*Apis mellifera capensis* Esch.). *Behav. Ecol. Sociobiol.* **17**: 87–9.

Moritz, R.F.A. and Southwick, E.E. 1992. *Bees as superorganisms. An evolutionary reality*. Springer-Verlag, Berlin.

Moritz, R.F.A., Meusel, M.S. and Haberl, M. 1991. Oligonucleotide DNA fingerprinting discriminates super- and half-sisters in honeybee colonies (*Apis mellifera* L.). *Naturwissenschaften* **78**: 422–4.

Morrill, W.L. 1974. Production and flight of alate red imported fire ants. *Environ. Entomol.* **3**: 265–71.

Moser, J.C. 1967. Mating activities of *Atta texana* (Hymenoptera, Formicidae). *Insectes Soc.* **14**: 295–312.

Mueller, U.G. 1991. Haplodiploidy and the evolution of facultative sex ratios in a primitively eusocial bee. *Science* **254**: 442–4.

Mueller, U.G., Eickwort, G.C. and Aguadro, C.F. 1994. DNA fingerprinting analysis of parent-offspring conflict in a bee. *Proc. Natl. Acad. Sci. USA* **91**: 5143–7.

Muralidharan, K., Shaila, M.S. and Gadagkar, R. 1986. Evidence for multiple mating in the primitively eusocial wasp *Ropalidia marginata* (Lep.)(Hymenoptera: Vespidae). *Technical Report No. 17*, Indian Institute of Science, Centre for Ecological Studies, Bangalore.

Myles, T.G. 1988. Resource inheritance in social evolution from termites to man. In *The ecology of social behavior* (ed. C.N. Slobodchikoff), pp. 379–423. Academic Press, New York.

Myles, T.G. and Chang, F. 1984. The caste system and caste mechanisms of *Neotermes connexus* (Isoptera: Kalotermitidae). *Sociobiol.* **9**: 163–321.

Myles, T.G. and Nutting, W.L. 1988. Termite eusocial evolution: a re-examination of Bartz's hypothesis and assumptions. *Q. Rev. Biol.* **63**: 1–23.

Naito, T. and Suzuki, H. 1991. Sex determination in the sawfly, *Athalia rosae ruficornis* (Hymenoptera): occurrence of triploid males. *J. Hered.* **82**: 101–4.

Nalepa, C.A. and Jones, S.C. 1991. Evolution of monogamy in termites. *Biol. Rev.* **66**: 83–97.

Noirot, C. 1990. Sexual castes and reproductive strategies in termites. In *Social insects. An evolutionary approach to castes and reproduction* (ed. W. Engels), pp. 5–35. Springer-Verlag, Berlin.

Noirot, C. and Pasteels, J.M. 1987. Ontogenetic development and evolution of the worker caste in termites. *Experientia* **43**: 851–60.

Nonacs, P. 1986a. Ant reproductive strategies and sex allocation theory. *Q. Rev. Biol.* **61**: 1–21.

Nonacs, P. 1986b. Sex-ratio determination within colonies of ants. *Evolution* **40**: 199–204.

Nonacs, P. 1988. Queen number in colonies of social Hymenoptera as a kin-selected adaptation. *Evolution* **42**: 566–80.

Nonacs, P. 1989. Competition and kin discrimination in colony founding by social Hymenoptera. *Evol. Ecol.* **3**: 221–35.

Nonacs, P. 1990. Size and kinship affect success of co-founding *Lasius pallitarsis* queens. *Psyche* **97**: 217–28.

Nonacs, P. 1991. Alloparental care and eusocial evolution: the limits of Queller's head-start advantage. *Oikos* **61**: 122–5.

Nonacs, P. 1993. Male parentage and sexual deception in the social Hymenoptera. In *Evolution and diversity of sex ratio in insects and mites* (ed. D.L. Wrensch and M.A. Ebbert), pp. 384–401. Chapman & Hall, New York.

Nonacs, P. and Carlin, N.F. 1990. When can ants discriminate the sex of brood? A new aspect of queen–worker conflict. *Proc. Natl. Acad. Sci. USA* **87**: 9670–3.

Noonan, K.C. 1986. Recognition of queen larvae by worker honeybees (*Apis mellifera*). *Ethology* **73**: 295–306.

Noonan, K.M. 1978. Sex ratio of parental investment in colonies of the social wasp *Polistes fuscatus*. *Science* **199**: 1354–6.

Noonan, K.M. 1981. Individual strategies of inclusive-fitness-maximizing in *Polistes fuscatus* foundresses. In *Natural selection and social behavior* (ed. R.D. Alexander and D.W. Tinkle), pp. 18–44. Chiron Press, New York.

Nur, U. 1972. Diploid arrhenotoky and automictic thelytoky in soft scale insects (Lecaniidae: Coccoidea: Homoptera). *Chromosoma* **39**: 381–401.

Nur, U., Werren, J.H., Eickbush, D.G., Burke, W.D. and Eickbush, T.H. 1988. A 'selfish' B chromosome that enhances its transmission by eliminating the paternal genome. *Science* **240**: 512–14.

Nutting, W.L. 1969. Flight and colony foundation. In *Biology of termites* (ed. K. Krishna and F.M. Weesner), Vol I, pp. 233–282. Academic Press, New York.

Oldroyd, B.P., Rinderer, T.E., Harbo, J.R. and Buco, S.M. 1992. Effects of intracolonial genetic diversity on honey bee (Hymenoptera: Apidae) colony performance. *Ann. Entomol. Soc. Am.* **85**: 335–43.

Oldroyd, B.P., Smolenski, A.J., Cornuet, J-M. and Crozier, R.H. 1994a. Anarchy in the beehive. *Nature* **371**: 749.

Oldroyd, B.P., Sylvester, H.A., Wongsiri, S. and Rinderer, T.E. 1994b. Task specialization in a wild bee, *Apis florea* (Hymenoptera: Apidae), revealed by RFLP banding. *Behav. Ecol. Sociobiol.* **34**: 25–30.

O'Neill, K.M. 1994. The male mating strategy of the ant. *Formica subpolita* Mayr (Hymenoptera: Formicidae): swarming, mating, and predation risk. *Psyche* **101**: 93–108.

Orlove, M.J. and Wood, C.L. 1978. Coefficients of relationship and coefficients of relatedness in kin selection: a covariance form for the rho formula. *J. Theor. Biol.* **73**: 679–86.

Orzack, S.H. 1986. Sex-ratio control in a parasitic wasp, *Nasonia vitripennis*. II. Experimental analysis of an optimal sex-ratio model. *Evolution* **40**: 341–56.

Orzack, S.H. 1993. Sex ratio evolution in parasitic wasps. In *Evolution and diversity of sex ratio in haplodiploid insects and mites* (ed. D.L. Wrensch and M.A. Ebbert), pp. 477–511. Chapman and Hall, New York.

Orzack, S.H. and Parker, E.D. 1986. Sex-ratio control in a parasitic wasp, *Nasonia vitripennis*. I. Genetic variation in facultative sex-ratio adjustment. *Evolution* **40**: 331–40.

Oster, G.F., Eshel, I. and Cohen, D. 1977. Worker–queen conflict and the evolution of social castes. *Theor. Popul. Biol.* **12**: 49–85.

Oster, G.F. and Wilson, E.O. 1978. *Caste and ecology in the social insects*. Princeton University Press.

Otto, D. 1960. Zur Erscheinung der Arbeiterinnenfertilität und Parthenogenese bei der Kahlrueckigen Roten Waldameise (*Formica polyctena* Foerst.), (Hym.). *Dtsch. Entomol. Z. N.F.* **7**: 1–9

Owen, R.E. 1985. The opportunity for polymorphism and genetic variation in social Hymenoptera with worker-produced males. *Heredity* **54**: 25–36.

Owen, R.E. 1986. Colony-level selection in the social insects: single locus additive and nonadditive models. *Theor. Popul. Biol.* **29**: 198–234.

Owen, R.E. and Plowright, R.C. 1980. Abdominal pile color dimorphism in the bumble bee, *Bombus melanopygus*. *J. Hered.* **71**: 241–7.

Owen, R.E. and Plowright, R.C. 1982. Worker–queen conflict and male parentage in bumble bees. *Behav. Ecol. Sociobiol.* **11**: 91–9.

Owen, R.E., Rodd, F.H. and Plowright, R.C. 1980. Sex ratios in bumble bee colonies: complications due to orphaning? *Behav. Ecol. Sociobiol.* **7**: 287–91.

Packer, L. 1986a. The social organization of *Halictus ligatus* (Hymenoptera; Halictidae) in southern Ontario. *Can. J. Zool.* **64**: 2317–24.

Packer, L. 1986b. Multiple-foundress associations in a temperate population of *Halictus ligatus* (Hymenoptera; Halictidae). *Can. J. Zool.* **64**: 2325–32.

Packer, L. 1987. An unexpectedly high female bias to the sex ratio in a temperate population of *Halictus ligatus*—a primitively social sweat bee. In *Chemistry and biology of social insects* (ed. J. Eder and H. Rembold), pp. 353–4. J. Peperny, Verlag, Munich.

Packer, L. 1991. The evolution of social behavior and nest architecture in sweat bees of the subgenus *Evylaeus* (Hymenoptera: Halictidae): a phylogenetic approach. *Behav. Ecol. Sociobiol.* **29**: 153–60.

Packer, L. and Knerer, G. 1985. Social evolution and its correlates in bees of the subgenus *Evylaeus* (Hymenoptera; Halictidae). *Behav. Ecol. Sociobiol.* **17**: 143–9.

Packer, L. and Knerer, G. 1986. The biology of a subtropical population of *Halictus ligatus* Say (Hymenoptera: Halictidae). *Behav. Ecol. Sociobiol.* **18**: 363–75.

Packer, L and Owen, R.E. 1990. Allozyme variation, linkage disequilibrium and diploid male production in a primitive social bee *Augochorella striata*. *Heredity* **65**: 241–8.

Packer, L. and Owen, R.E. 1994. Relatedness and sex ratio in a primitively eusocial halictine bee. *Behav. Ecol. Sociobiol.* **34**: 1–10.

Page, R.E. 1980. The evolution of multiple mating behavior by honey bee queens (*Apis mellifera* L.). *Genetics* **96**: 263–73.

Page, R.E. Jr. 1981. Protandrous reproduction in honey bees. *Environ. Entomol.* **10**: 359–62.

Page, R.E. 1986. Sperm utilization in social insects. *Annu. Rev. Entomol.* **31**: 297–320.

Page, R.E. and Erickson, E.H. 1988. Reproduction by worker honey bees (*Apis mellifera* L.). *Behav. Ecol. Sociobiol.* **23**: 117–26.

Page, R.E. and Metcalf, R.A. 1984. A population investment sex ratio for the honey bee (*Apis mellifera* L.). *Am. Nat.* **124**: 680–702.

Page, R.E., Jr., Robinson, G.E. and Fondrk, M.K. 1989a. Genetic specialists, kin recognition and nepotism in honeybee colonies. *Nature* **338**: 576–9.

Page, R.E., Jr., Post, D.C. and Metcalf, R.A. 1989b. Satellite nests, early males, and plasticity of reproductive behavior in a paper wasp. *Am. Nat.* **134**: 731–48.

Page, R.E., Jr., Fondrk, M.K. and Robinson, G.E. 1993. Selectable components of sex allocation in colonies of the honeybee (*Apis mellifera* L.). *Behav. Ecol.* **4**: 239–45.

Pamilo, P. 1981. Genetic organization of *Formica sanguinea* populations. *Behav. Ecol. Sociobiol.* **9**: 45–50.

Pamilo, P. 1982a. Genetic population structure in polygynous *Formica* ants. *Heredity* **48**: 95–106.

Pamilo, P. 1982b. Genetic evolution of sex ratios in eusocial Hymenoptera: allele frequency simulations. *Am. Nat.* **119**: 638–56.

Pamilo, P. 1982c. Multiple mating in *Formica ants, Hereditas* **97**: 37–45.

Pamilo, P. 1983. Genetic differentiation within subdivided populations of *Formica* ants. *Evolution* **37**: 1010–22.

Pamilo, P. 1984a. Genotypic correlation and regression in social groups: multiple alleles, multiple loci and subdivided populations. *Genetics* **107**: 307–20.

Pamilo, P. 1984b. Genetic relatedness and evolution of insect sociality. *Behav. Ecol. Sociobiol.* **15**: 241–8.

Pamilo, P. 1987. Sex ratios and the evolution of eusociality in the Hymenoptera. *J. Genet.* **66**: 111–22.

Pamilo, P. 1989. Estimating relatedness in social groups. *TREE* **4**: 353–5.

Pamilo, P. 1990a. Comparisons of relatedness estimators. *Evolution* **44**: 1378–82.

Pamilo, P. 1990b. Sex allocation and queen–worker conflict in polygynous ants. *Behav. Ecol. Sociobiol.* **27**: 31–6.

Pamilo, P. 1991a. Evolution of colony characteristics in social insects. I. Sex allocation. *Am. Nat.* **137**: 83–107.

Pamilo, P. 1991b. Evolution of colony characteristics in social insects. II. Number of reproductive individuals. *Am. Nat.* **138**: 412–33.

Pamilo, P. 1991c. Evolution of the sterile caste. *J. Theor. Biol.* **149**: 75–95.

Pamilo, P. 1991d. Life span of queens in the ant *Formica exsecta. Insectes Soc.* **38**: 111–19.

Pamilo, P. 1993. Polyandry and allele frequency differences between the sexes in the ant *Formica aquilonia. Heredity* **70**: 472–80.

Pamilo, P. and Crozier, R.H. 1982. Measuring genetic relatedness in natural populations: methodology. *Theor. Popul. Biol.* **21**: 171–93.

Pamilo, P. and Rosengren, R. 1983. Sex ratio strategies in *Formica* ants. *Oikos* **40**: 24–35.

Pamilo, P. and Rosengren, R. 1984. Evolution of nesting strategies of ants: genetic evidence from different population types of *Formica* ants. *Biol. J. Linn. Soc.* **21**: 331–48.

Pamilo, P. and Seppä, P. 1994. Reproductive competition and conflicts in colonies of the ant *Formica sanguinea. Anim. Behav.* **48**: 1201–6.

Pamilo, P. and Varvio-Aho, S.-H. 1979. Genetic structure of nests in the ant *Formica sanguinea. Behav. Evol. Sociobiol.* **6**: 91–8.

Pamilo, P., Pekkarinen, A. and Varvio-Aho, S. 1981. Phylogenetic relationships and the origin of social parasitism in Vespidae and in *Bombus* and *Psithyrus* as revealed by enzyme genes. In *Biosystematics of social insects* (ed. P.E. Howse and J-L. Clement), pp. 37–48. Academic Press, London and New York.

Pamilo, P., Crozier, R.H. and Fraser, J. 1985. Inter-nest interactions, nest autonomy, and reproductive specialization in an Australian arid-zone ant, *Rhytidoponera* sp. 12. *Psyche* **92**: 217–36.

Pamilo, P., Chautems, D. and Cherix, D. 1992. Genetic differentiaiton of disjunct populations of the ants *Formica aquilonia* and *Formica lugubris* in Europe. *Insectes Soc*. **39**: 15–29.

Pamilo, P., Sundström, L., Fortelius, W. and Rosengren, R. 1994. Diploid males and colony level selection in *Formica ants. Ethol. Ecol. Evol*. **6**: 221–35.

Pardi, L. 1943. Ricerche sui Polistini. 6. Sulla sproporzione numerica dei sessi nei nidi dei Polistini. *Atti Soc. Toscana. Sci. Nat*. **52**(2): 1–9.

Parker, E.D. and Orzack, S.H. 1985. Genetic variation for the sex ratio in *Nasonia vitripennis. Genetics* **110**: 93–105.

Parker, F.D. 1988. Nesting biology of two North American species of *Cheliostoma* (Hymenoptera: Megachilidae). *Pan-Pacific Entomol*. **64**: 1–7.

Passera, L. 1966. La ponte des ouvriéres de la fourmi *Plagiolepis pygmaea* Latr. (Hym., Formicidae): oeufs reproducteurs et oeufs alimentaires. *C. R. Acad. Sci. Paris* **D263**: 1095–8.

Passera, L. 1984. *L'Organisation sociale des fourmis*. Privat, Toulouse.

Peacock, A.D. 1951. Studies in Pharaoh's Ant, Monomorium pharaonis (L.) (5) Pupal and adult sex ratios. *Entomol. Mon. Mag*. **87**: 185–91.

Peacock, A.D., Smith, I.C., Hall, D.W. and Baxter, A.T. 1954. Studies in Pharaoh's Ant, *Monomorium pharaonis* (L.). (8) Male production by parthenogenesis. *Entomol. Mon. Mag*. **90**: 154–8.

Pearson, B. 1982. Relatedness of normal queens (macrogynes) in nests of the polygynous ant *Myrmica rubra* Latreille. *Evolution* **36**: 107–12.

Pearson, B. 1983a. Intra-colonial relatedness amongst workers in a population of nests of the polygynous ant, *Myrmica rubra* Latreille. *Behav. Ecol. Sociobiol*. **12**: 1–4.

Pearson, B. 1983b. Hybridisation between the ant species *Lasius niger* and *Lasius alienus*: the genetic evidence. *Insectes Soc*. **30**: 402–11.

Pearson, B. 1987. The sex ratio of heathland populations of the ants *Lasius alienus*, *Lasius niger* and their hybrids. *Insectes Soc*. **34**: 194–203.

Peeters, C.P. 1987a. The reproductive division of labour in the queenless ponerine ant *Rhytidoponera* sp. 12. *Psyche* **34**: 75–86.

Peeters, C.P. 1987b. The diversity of reproductive systems in ponerine ants. In *Chemistry and biology of social insects* (ed. J. Eder and H. Rembold), pp. 253–4. J. Peperny Verlag, Munich.

Peeters, C.P. 1991. The occurrence of sexual reproduction among ant workers. *Biol. J. Linn. Soc*. **44**: 141–52.

Peeters, C.P. and Crewe, R.M. 1984. Insemination controls the reproductive division of labour in a ponerine ant. *Naturwissenschaften* **71**: 50–1.

Peeters, C. and Crewe, R.M. 1985. Worker reproduction in the ponerine ant *Opthalmopone berthoudi*: an alternative form of eusocial organization. *Behav. Ecol. Sociobiol*. **18**: 29–37.

Peeters, C. and Crewe, R. 1988. Worker laying in the absence of an ergatoid queen in the Ponerine ant genus *Plectroctena. S. Afr. J. Zool*. **23**: 78–80.

Peeters, C. and Crozier, R.H. 1988. Caste and reproduction in ants: not all mated egg-layers are 'queens'. *Psyche* **95**: 283–8.

Peeters, C. and Higashi, S. 1989. Reproductive dominance controlled by mutilation in the queenless ant *Diacamma australe. Naturwissenschaften* **76**: 177–80.

Petersen, M. and Buschinger, A. 1971. Das Begattungsverhalten der Pharaoameise, *Monomorium pharaonis* (L.). *Z. angew. Entomol*. **68**: 168–75.

Pfennig, D.W. and Reeve, H.K. 1993. Nepotism in a solitary wasp as revealed by DNA fingerprinting. *Evolution* **47**: 700–4

Pickering, J. 1980. *Sex ratio, social behavior and ecology in* Polistes (*Hymenoptera, Vespidae*), Pachysomoides (*Hymenoptera, Ichneumonidae*) *and* Plasmodium (*Protozoa, Haemosporidia*). Ph.D. Thesis, Harvard University.

Pickles, W. 1940. Fluctuations in the populations, weights and biomasses of ants at Thornhill, Yorkshire, from 1935 to 1939. *Trans. R. Entomol. Soc. London.* **90**: 467–85.

Plateaux, L. 1981. The *pallens* morph of the ant *Leptothorax nylanderi*: description, formal genetics, and study of populations. In *Biosystematics of social insects* (ed. P.E. Howse and J.-L. Clement), pp. 63–74. Academic Press, New York.

Plateaux-Quénu, C. 1959. Un nouveau type de société d'insectes: *Halictus marginatus* Brullé (Hymenoptera, Apidae). *Année Biol.* **35**: 325–444.

Plowright, R.C. and Pallett, M.J. 1979. Worker–male conflict and inbreeding in bumble bees (Hymenoptera: Apidae). *Can. Entomol.* **111**: 289–94.

Poirié, M., Périquet, G. and Beukeboom, L. 1992. The hymenopteran way of determining sex. *Sem. Dev. Biol.* **3**: 357–61.

Pollock, G.B. and Rissing, S.W. 1985. Mating season and colony foundation of the seed-harvester ant, *Veromessor pergandei. Psyche* **92**: 125–34.

Pomeroy, N. 1979. Brood bionomics of *Bombus ruderatus* in New Zealand (Hymenoptera: Apidae). *Can. Entomol.* **111**: 865–74.

Porter, S.D. and Jorgensen, C.D. 1988. Longevity of harvester ant colonies in southern Idaho. *J. Range Manag.* **41**: 104–7.

Porter, S.D., van Eimeren, B. and Gilbert, L.E. 1988. Invasion of fire ants (Hymenoptera: Formicidae): microgeography and competitive replacement. *Ann. Entomol. Soc. Am.* **81**: 777–81.

Pratte, M., Gervet, J. and Semenoff, S. 1984. L'évolution de la production de descendance dans le guêpier de poliste (*Polistes gallicus* L.). *Insectes Soc.* **31**: 34–50.

Pricer, J.L. 1908. The life history of the carpenter ant. *Biol. Bull.* **14**: 177–217.

Provost, E. 1991. Nonnestmate kin recognition in the ant *Leptothorax lichtensteini*: evidence that genetic factors regulate colony recognition. *Behav. Genet.* **21**: 151–67.

Queller, D.C. 1989. The evolution of eusociality: reproductive head starts of workers. *Proc. Natl. Acad. Sci. USA* **86**: 3224–6.

Queller, D.C. 1992. A general model for kin selection. *Evolution* **46**: 376–80.

Queller, D.C. 1993. Worker control of sex ratios and selection for extreme multiple matings by queens. *Am. Nat.* **142**: 346–51.

Queller, D.C. and Goodnight, K.F. 1989. Estimating relatedness using genetic markers. *Evolution* **43**: 258–75.

Queller, D.C. and Strassmann, J.E. 1988. Reproductive success and group nesting in the paper wasp, *Polistes annularis*. In *Reproductive success: studies of individual variation in contrasting breeding systems* (ed. T. Clutton-Brock), pp. 76–96. Chicago University Press.

Queller, D.C. and Strassmann, J.E. 1989. Measuring inclusive fitness in social wasps. In *The genetics of social evolution* (ed. M.D. Breed and R.E., Page, Jr.), pp. 103–22. Westview Press, Boulder, CO.

Queller, D.A., Strassmann, J.E. and Hughes, C.R. 1988. Genetic relatedness in colonies of tropical wasps with multiple queens. *Science* **242**: 1155–7.

Queller, D.C., Hughes, C.R. and Strassmann, J.E. 1990. Wasps fail to make distinctions. *Nature* **344**: 388.

Queller, D.A., Strassmann, J.E. and Hughes, C.R. 1992. Genetic relatedness and population structure in primitively eusocial wasps in the genus *Mischocyttarus. J. Hymenopt. Res.* **1**: 115–45.

Queller, D.C., Negron-Sotomayor, J.A., Strassmann, J.E. and Hughes, C.R. 1993a. Queen number and genetic relatedness in a neotropical wasp, *Polybia occidentalis*. *Behav. Ecol.* **4**: 7–13.

Queller, D.C., Strassmann, J.E., Solis, C.R., Hughes, C.R. and Moralez DeLoach, D. 1993b. A selfish strategy of social insect workers that promotes social cohesion. *Nature* **365**: 639–41.

Raignier, A. 1972. Sur l'origine des nouvelles societes des fourmis voyageuse Africaines (Hymenopteres Formicidae, Dorylinae). *Insectes Soc.* **19**: 153–70.

Ratnieks, F.L.W. 1988. Reproductive harmony via mutual policing by workers in eusocial Hymenoptera. *Am. Nat.* **132**: 217–36.

Ratnieks, F.L.W. 1990. The evolution of polyandry by queens in social Hymenoptera: the significance of the timing of removal of diploid males. *Behav. Ecol. Sociobiol.* **26**: 343–8.

Ratnieks, F.L.W. and Reeve, H.K. 1991. The evolution of queen-rearing nepotism in social Hymenoptera: effects of discrimination costs in swarming species. *J. Evol. Biol.* **4**: 93–115.

Ratnieks, F.L.W. and Reeve, H.K. 1992. Conflict in single-queen hymenopteran societies: the structure of conflict and processes that reduce conflict in advanced eusocial societies. *J. Theor. Biol.* **158**: 33–65.

Ratnieks, F.L.W. and Visscher, P.K. 1989. Worker policing in the honeybee. *Nature* **342**: 796–7.

Reeve, H.K. 1993. Haplodiploidy, eusociality and absence of male parental and alloparental care in Hymenoptera: a unifying genetic hypothesis distinct from kin selection theory. *Philos. Trans. R. Soc. London Ser. B* **342**: 335–52.

Reeve, H.K. and Nonacs, P. 1992. Social contracts in wasp societies. *Nature* **359**: 823–5.

Reeve, H.K., Westneat, D.F., Noon, W.A., Sherman, P.W. and Aquadro, C.F. 1990. DNA 'fingerprinting' reveals high levels of inbreeding in colonies of the eusocial naked mole-rat. *Proc. Natl. Acad. Sci. USA* **87**: 2496–500.

Reeve, H.K., Westneat, D.F. and Queller, D.C. 1992. Estimating average within-group relatedness from DNA fingerprints. *Mol. Ecol.* **1**: 223–32.

Reilly, L.M. 1987. Measurements of inbreeding and average relatedness in a termite population. *Am. Nat.* **130**: 339–49.

Richards, M.H., Packer, L. and Seger, J. 1995. Unexpected patterns of parentage and relatedness in a primitively eusocial bee. *Nature* **373**: 239–41

Rissing, S.W. and Pollock, G.B. 1987. Queen aggression, pleometrotic advantage and brood raiding in the ant *Veromessor pergandei* (Hymenoptera: Formicidae). *Anim. Behav.* **35**: 975–81.

Rissing, S.W. and Pollock, G.B. 1988 Pleometrosis and polygyny in ants. In *Interindividual behavioral variability in social insects* (ed. R.L. Jeanne), pp. 223–55. Westview Press, Boulder, CO.

Rissing, S.W., Pollock, G.B., Higgins, M.R., Hagen, R.R. and Smith, D.R. 1989. Foraging specialization without relatedness or dominance among co-founding ant queens. *Nature* **338**: 420–2.

Robinson, G.E. and Page, R.E., Jr. 1988. Genetic determination of guarding and undertaking in honey-bee colonies. *Nature* **333**: 356–8.

Robinson, G.E., Page, R.E., Jr. and Arensen, N. 1994. Genotypic differences in brood rearing in honey bee colonies: context-specific? *Behav. Ecol. Sociobiol.* **34**: 125–37.

Roisin, Y. 1987. Polygyny in *Nasutitermes* species: field data and theoretical approaches. *Experientia* (suppl). **54**: 379–404.

Roisin, Y. 1994. Intragroup conflicts and the evolution of sterile castes in termites. *Am. Nat.* **143**: 751–65.

Roisin, Y. and Pasteels, J.M. 1985. Imaginal polymorphism and polygyny in the Neo-guinean termite *Nasutitermes princeps* (Desneux). *Insectes Soc.* **32**: 140–57.

Roisin, Y. and Pasteels, J.M. 1986. Reproductive mechanisms in termites: polycalism and polygyny in *Nasutitermes polygynous* and *N. costalis*. *Insectes Soc.* **33**: 149–67.

Roisin, Y. and Pasteels, J.M. 1991. Sex ratio asymmetry between the sexes in the production of replacement reproductives in the termite, *Neotermes papua* (Desneux). *Ethol. Ecol. Evol.* **3**: 327–35.

Rojas-Rousse, D. and Palevody, C. 1981. Structure et fonctionnement de la spermatheque chez l'endoparasite solitaire *Diadromus pulchellus* Wesmael (Hymenoptera: Ichneumonidae). *Int. J. Insect Morphol. Embryol.* **10**: 309–20.

Roonwal, M.L. and Rathore, N.S. 1975. Swarming, egg-laying and hatching in the Indian desert harvester termite, *Acanthotermes macrocephalus* (Hodotermitidae). *Ann. Arid Zone* **14**: 329–38.

Röseler, P.-F. 1973. Die Anzahl der Spermen im Receptaculum seminis von Hummelköniginnen (*Hym., Apoidea, Bombinae*). *Apidologie* **4**: 267–74.

Rosengaus, R.B. and Traniello, J.F.A. 1993. Disease risk as a cost of outbreeding in the termite *Zootermopsis angusticollis*. *Proc. Natl. Acad. Sci. USA* **90**: 6641–5.

Rosengren, R. and Pamilo, P. 1983. The evolution of polygyny and polydomy in mound-building *Formica* ants. *Acta Entomol. Fenn.* **42**: 65–77.

Rosengren, R. and Pamilo, P. 1986. Sex ratio strategy as related to queen number, dispersal behaviour and habitat quality in *Formica* ants (Hymenoptera: Formicidae). Entomol. General. 11: 139–51.

Rosengren, R., Cherix, D. and Pamilo, P. 1985. Insular ecology of the red wood ant *Formica truncorum* Fabr. I. Polydomous nesting, population size and foraging. *Mitt. Schweiz. Entomol. Ges.* **58**: 147–75.

Rosengren, R., Cherix, D. and Pamilo, P. 1986. Insular ecology of the red wood ant *Formica truncorum* Fabr. II. Distribution, reproductive strategy and competition. *Mitt. Schweiz. Entomol. Ges.* **59**: 63–94.

Ross, K.G. 1985. Aspects of worker reproduction in four social wasp species (Insecta: Hymenoptera: Vespidae). *J. Zool.* **205**: 411–24.

Ross, K.G. 1986. Kin selection and the problem of sperm utilization in social insects. *Nature* **323**: 798–9.

Ross, K.G. 1988. Differential reproduction in multiple-queen colonies of the fire ant, *Solenopsis invicta* (Hymenoptera: Formicidae). *Behav. Ecol. Sociobiol.* **23**: 341–55.

Ross, K.G. 1992. Strong selection on a gene that influences reproductive competition in a social insect. *Nature* **355**: 347–9.

Ross, K.G. 1993. The breeding system of the fire ant *Solenopsis invicta*: effects of colony genetic structure. *Am. Nat.* **141**: 554–76.

Ross, K.G. and Carpenter, J.M. 1991. Phylogenetic analysis and the evolution of queen number in eusocial Hymenoptera. *J. Evol. Biol.* **4**: 117–30.

Ross, K.G. and Fletcher, D.J.C. 1985a. Genetic origin of male diploidy in the fire ant, *Solenopsis invicta* (Hymenoptera: Formicidae), and its evolutionary significance. *Evolution* **39**: 888–903.

Ross, K.G. and Fletcher, D.J.C. 1985b. Comparative study of genetic and social structure in two forms of the fire ant *Solenopsis invicta* (Hymenoptera: Formicidae). *Behav. Ecol. Sociobiol.* **17**: 349–56.

Ross, K.G. and Fletcher, D.J.C. 1986. Diploid male production—a significant colony mortality factor in the fire ant *Solenopsis invicta* (Hymenoptera: Formicidae). *Behav. Ecol. Sociobiol.* **19**: 283–91.

Ross, K.G. and Matthews, R.W. 1989a. New evidence for eusociality in the sphecid wasp *Microstigmus comes*. *Anim. Behav.* **38**: 613–19.

Ross, K.G. and Matthews, R.W. 1989b. Population genetic structure and social evolution in the sphecid wasp *Microstigmus comes* (Hymenoptera). *Am. Nat.* **134**: 574–98.

Ross, K.G. and Trager, J.C. 1990. Systematics and population genetics of fire ants (*Solenopsis saevissima* complex) from Argentina. *Evolution* **44**: 2113–34.

Ross, K.G. and Visscher, P.K. 1983. Reproductive plasticity in yellowjacket wasps: a polygynous, perennial colony of *Vespula maculifrons*. *Psyche* **90**: 179–91.

Ross, K.G., Vargo, E.L. and Fletcher, D.J.C. 1987. Comparative biochemical genetics of three fire ant species in North America, with special reference to the two social forms of *Solenopsis invicta* (Hymenoptera: Formicidae). *Evolution* **41**: 979–90.

Ross, K.G., Vargo, E.L. and Fletcher, D.J.C. 1988. Colony genetic structure and queen mating frequency in fire ants of the subgenus *Solenopsis* (Hymenoptera: Formicidae). *Biol. J. Linn. Soc.* **34**: 105–17.

Ross, K.G., Vargo, E.L., Keller, L. and Trager, J.C. 1993. Effect of a founder event on variation in the genetic sex-determining system of the fire ant *Solenopsis invicta*. *Genetics* **135**: 843–54.

Roughgarden, J. 1979. *Theory of population genetics and evolutionary ecology: an introduction*. Macmillan, New York.

Rowell, D.M. 1987. Complex sex-linked translocation heterozygosity: its genetics and biological significance. *Trends Ecol. Evol.* **2**: 242–6.

Ruttner, F. 1988. *Biogeography and taxonomy of honeybees*. Springer-Verlag, Berlin.

Ryan, R. and Gamboa, G.J. 1986. Nestmate recognition between males and gynes of the social wasp *Polistes fuscatus* (Hymenoptera: Vespidae). *Ann. Entomol. Soc. Am.* **79**: 572–5.

Saito, Y. 1994. Is sterility by deleterious recessives an origin of inequalities in the evolution of eusociality? *J. Theor. Biol.* **166**: 113–5.

Sakagami, S.F. 1976. Specific differences in the bionomic characters of bumblebees. A comparative review. *J. Fac. Sci. Hokkaido Univ. Ser. VI, Zool.* **20**: 390–447.

Sakata, K. and Itô, Y. 1991. Life history characteristics and behaviour of the bamboo aphid, *Pseudoregma bambucicola* (Hemiptera: Pemphigidae), having sterile soldiers. *Insectes Soc.* **38**: 317–26.

Scherba, G. 1961. Nest structure and reproduction in the mound-building ant *Formica opaciventris* Emery in Wyoming. *J. N. Y. Entomol. Soc.* **69**: 71–87.

Schmidt, G.H. 1974. Steuerung der Kastenbildung und Geschlechtsregulation im Waldameisenstaat. In *Sozialpolymorphismus bei Insekten* (ed. G.H. Schmidt), pp. 404–512. Wissenschaftliche Verlagsgesellschaft, Stuttgart.

Schneirla, T.C. 1971. *Army ants*. Freeman, San Francisco, CA.

Schwarz, M.P. 1986. Persistent multi-female nests in an Australian allodapine bee, *Exoneura bicolor* (Hymenoptera, Anthophoridae). *Insectes Soc.* **33**: 258–77.

Schwarz, M.P. 1987. Intra-colony relatedness and sociality in the allodapine bee *Exoneura bicolor*. *Behav. Ecol. Sociobiol.* **21**: 387–92.

Schwarz, M.P. 1988a. Local resource enhancement and sex ratios in a primitively social bee. *Nature* **331**: 346–8.

Schwarz, M.P. 1988b. Intra-specific mutualism and kin-association of cofoundresses in allodapine bees (Hymenoptera: Anthroporidae). *Monit. Zool. Ital. (N.S.)* **22**: 245–54.

Seeley, T.D. 1985. *Honeybee ecology: a study of adaptation in social life*. Princeton University Press.

Seeley, T.D. 1989. The honey bee colony as a superorganism. *Am. Sci.* **77**: 546–53.

Seger, J. 1983. Partial bivoltinism may cause alternating sex-ratio biases that favour eusociality. *Nature* **301**: 59–62.

Seppä, P. 1992. Genetic relatedness of worker nestmates in *Myrmica ruginodis* (Hymenoptera: Formicidae) populations. *Behav. Ecol. Sociobiol.* **30**: 253–60.

Seppä, P. 1994a. Sociogenetic organization of *Myrmica ruginodis* and *Myrmica lobicornis* (Hymenoptera: Formicidae) colonies and populations: number, relatedness and longevity of reproducing individuals. *J. Evol. Biol.* **7**: 71–95.

Seppä, P. 1994b. *Social and genetic structure* of *Myrmica populations*. PhD Thesis University of Helsinki.

Seppä, P. and Pamilo, P. 1995. Gene flow and population viscosity in *Myrmica* ants *Heredity* **74**: 200–9.

Sewell, J.J. and Watson, J.A.L. 1981. Developmental pathways in Australian species of *Kalotermes* Hagen (Isoptera). *Sociobiology* **6**: 243–324.

Shattuck, S.O. 1992. Review of the dolichoderine ant genus *Iridomyrmex* Mayr with descriptions of three new genera (Hymenoptera: Formicidae). *J. Aust. Entomol. Soc.* **31**: 13–18.

Shaw, R.F. and Mohler, J.D. 1953. The selective advantage of the sex ratio. *Am. Nat.* **87**: 337–42.

Sherman, P.W. 1988. The levels of analysis. *Anim. Behav.* **36**: 616–19.

Sherman, P.W., Seeley, T.D. and Reeve, H.K. 1988. Parasites, pathogens, and polyandry in social Hymenoptera. *Am. Nat.* **131**: 602–10.

Sherman, P.W, Jarvis, J. and Alexander, R.D. (eds.) 1991. *The biology of the naked mole rat.* Princeton Univeristy Press.

Shoemaker, D.D., Ross, K.G. and Arnold, M.L. 1994. Development of RAPD markers in two introduced fire ants, *Solenopsis invicta* and *S. richteri*, and their application to the study of a hybrid zone. *Mol. Ecol.* **3**: 531–9.

Shykoff, J.A. and Schmid-Hempel, P. 1991a. Parasites and the advantage of genetic variability within social insect colonies. *Proc. R. Soc. London Ser. B* **243**: 55–8.

Shykoff, J.A. and Schmid-Hempel, P. 1991b. Genetic relatedness and eusociality: parasite-mediated selection on the genetic composition of groups. *Behav. Ecol. Sociobiol.* **28**: 371–6.

Shykoff, J. A. and Schmid-Hempel, P. 1991c. Parasites delay worker reproduction in bumblebees: consequences for eusociality. *Behav. Ecol.* **2**: 242–8.

Smeeton, L. 1981. The source of males in *Myrmica rubra* L. (Hym. Formicidae). *Insectes Soc.* **28**: 263–78.

Smith, B.H. 1983. Recognition of female kin by male bees through olfactory signals. *Proc. Natl. Acad. Sci. USA* **80**: 4551–3.

Smith, R.H. and Shaw, M.R. 1980. Haplodiploid sex ratios and the mutation rate. *Nature* **287**: 728–9.

Smith, S.G. and Wallace, D.R. 1971. Allelic sex determination in a lower hymenopteron *Neodiprion nigroscutum* Midd. *Can. J. Genet. Cytol.* **13**: 617–21.

Snell, G.D. 1935. The determination of sex in *Habrobracon*. *Proc. Natl. Acad. Sci. USA* **21**: 446–53.

Snelling, R.R. 1981. Systematics of social Hymenoptera. In *Social insects* (ed. H.R. Hermann), Vol II, pp. 369–453. Academic Press, New York.

Snyder, L.E. 1992. The genetics of social behavior in a polygynous ant. *Naturwissenschaften* **79**: 525–7.

Snyder, L.E. and Herbers, J.M. 1991. Polydomy and sexual allocation ratios in the ant *Myrmica punctiventris*. *Behav. Ecol. Sociobiol.* **28**: 409–15.

Sober, E. 1985. *The nature of selection.* MIT Press, Cambridge, MA.

Sokal, R.R. and Rohlf, F.J. 1981. *Biometry*, 2nd edn. Freeman, San Francisco, CA.

Sommer, K. and Hölldobler, B. 1992. Coexistence and dominance among queens and mated workers in the ant *Pachycondyla tridentata*. *Naturwissenschaften* **79**: 470–2

Soulié, J. 1960. Des considerations ecologiques peuvent-elles apporter une contribution a la connaissance du cycle biologique des colonies de *Cremastogaster* (Hymenoptera-Formicoidea)? *Insectes Soc.* **7**: 283–95.

Speicher, B.R. and Speicher, K.G. 1940. The occurrence of diploid males in *Habrobracon brevicornis. Am. Nat.* **74**: 379–82.

Spradbery, J.P. 1973. *Wasps*. Sidgwick & Jackson, London.

Spradbery, J.P. 1986. Polygyny in the Vespinae with special reference to the hornet *Vespa affinis picea* Buysson (Hymenoptera: Vespidae) in New Guinea. *Monit. Zool. Ital. (N.S.)* **20**: 101–18.

Starr, C.K. 1979. Origin and evolution of insect sociality. In *Social insects* (ed. H.R. Hermann), *Vol. I*, pp. 35–79. Academic Press, New York.

Starr, C.K. 1984. Sperm competition, kinship, and sociality in the aculeate Hymenoptera. In *Sperm competition and the evolution of animal mating systems* (ed. R.L. Smith), pp. 427–64. Academic Press, Orlando, FL.

Starr, C.K. 1985. Enabling mechanisms in the origin of sociality in the Hymenoptera—the sting's the thing. *Ann. Entomol. Soc. Am.* **78**: 836–40.

*Steiner, W.W.M and Teig, D.A. 1989. *Microplitis croceipes*: genetic characterization and developing insecticide resistance biotypes. *Southwest. Entomol.* **12**: 81–7.

Stille, B and Dävring, L. 1980. Meiosis and reproductive strategies in the parthenogenetic gall wasp *Diplolepis rosae*. *Hereditas* **92**: 353–62.

Stille, M. and Stille, B. 1992. Intra- and inter-nest variation in mitochondrial DNA in the polygynous ant *Leptothorax acervorum* (Hymenoptera; Formicidae). *Insectes Soc.* **39**: 335–40.

Stille, M. and Stille, B. 1993. Intrapopulation nestclusters of maternal mtDNA lineages in the polygynous ant *Leptothorax acervorum* (Hymenoptera: Formicidae). *Insect Mol. Biol.* **1**: 117–21.

Stille, M., Stille, B. and Douwes, P. 1991. Polygyny, relatedness and nest founding in the polygynous myrmicine ant *Leptothorax acervorum* (Hymenoptera; Formicidae). *Behav. Ecol. Sociobiol.* **28**: 91–6.

Stockhammer, K.A. 1966. Nesting habits and life cycle of a sweat bee, *Augochlora pura*. *J. Kansas Entomol. Soc.* **39**: 157–92.

Stouthamer, R., Luck, R.F. and Hamilton, W.D. 1990. Antibiotics cause parthenogenetic *Trichogramma* (Hymenoptera: Trichogrammatidae) to revert to sex. *Proc. Natl. Acad. Sci. USA* **78**: 2424–7.

Stouthamer, R., Luck, R.F and Werren, J.H. 1992. Genetics of sex determination and the improvement of biological control using parasitoids. *Environ. Entomol.* **21**: 427–35.

Strassmann, J.E. 1981. Evolutionary implications of early male and satellite nest production in *Polistes exclamans* colony cycles. *Behav. Ecol. Sociobiol.* **8**: 55–64.

Strassmann, J.E. 1984. Female-biased sex ratios in social insects lacking morphological castes. *Evolution* **38**: 256–66.

Strassmann, J.E. 1989. Altruism and relatedness at colony foundation in social insects. *TREE* **4**: 371–4.

Strassmann, J.E. 1993. Weak queen or social contract? *Nature* **363**: 502–3.

Strassmann, J.E. and Hughes, C. 1986. Latitudinal variation in protandry and protogyny in polistine wasps. *Monit. Zool. Ital. (N.S.)* **20**: 87–100.

Strassmann, J.E. and Queller, D.C. 1989. Ecological determinants of social evolution. In *The genetics of social evolution* (ed. M.D. Breed and R.E., Page, Jr.), pp. 81–101. Westview Press, Boulder, CO.

Strassmann, J.E., Hughes, C.R., Queller, D.C., Turillazzi, S., Cervo, R., Davis, S.K. and Goodnight, K.F. 1989. Genetic relatedness in primitively eusocial wasps. *Nature* **342**: 268–9.

Strassmann, J.E., Queller, D.C., Solis, C.R. and Hughes, C.R. 1991. Relatedness and queen number in the Neotropical wasp, *Parachartergus colobopterus*. *Anim. Behav.* **42**: 461–70.

Strassmann, J.E., Gastreich, K.R., Queller, D.C. and Hughes, C.R. 1992. Demographic and genetic evidence for cyclical changes in queen number in a Neotropical wasp, *Polybia emaciata*. *Am. Nat.* **140**: 363–72.

Stuart, R.J. 1987. Transient nestmate recognition cues contribute to a multicolonial population structure in the ant, *Leptothorax curvispinosus*. *Behav. Ecol. Sociobiol.* **21**: 229–35.

Stuart, R.J. 1988. Collective cues as a basis for nestmate recognition in polygynous leptothoracine ants. *Proc. Natl. Acad. Sci. USA* **85**: 4572–5.

Stuart, R.J. 1992. Nestmate recognition and the ontogeny of acceptability in the ant, *Leptothorax curvispinosus*. *Behav. Ecol. Sociobiol.* **30**: 403–8.

Stuart, R.J. and Page, R.E. 1991. Genetic component to division of labor among workers of a leptothoracine ant. *Naturwissenschaften* **78**: 375–7.

Stubblefield, J.W. and Charnov, E.L. 1986. Some conceptual issues in the origin of eusociality. *Heredity* **57**: 181–7.

Sturtevant, A.H. 1938. Essays on evolution. II. On the effects of selection on social insects. *Q. Rev. Biol.* **13**: 74–6.

Sundström, L. 1989. Genetic relatedness and population structure in *Formica truncorum* Fabr (Hymenoptera, Formicidae). *Actes Colloq. Insectes Soc.* **5**: 93–100.

Sundström, L. 1993. Genetic population structure and sociogenetic organization in *Formica truncorum*. *Behav. Ecol. Sociobiol.* **33**: 345–54.

Sundström, L. 1994. Sex ratio bias, relatedness asymmetry and queen mating frequency in ants. *Nature* **367**: 266–8.

Suzuki, T. 1985. Mating and laying of female-producing eggs by orphaned workers of a paper wasp, *Polistes snelleni* (Hymenoptera: Vespidae). *Ann. Entomol. Soc. Am.* **78**: 736–9.

Suzuki, T. 1986. Production schedules of males and reproductive females, investment sex ratios, and worker–queen conflict in paper wasps. *Am. Nat.* **128**: 366–78.

Syren, R.M. and Luykx, P. 1977. Permanent segmental interchange complex in the termite *Incisitermes schwarzi*. *Nature* **266**: 167–8.

Talbot, M. 1943. Population studies of the ant, *Prenolepis imparis* Say. *Ecology* **24**: 31–44.

Talbot, M. 1945a. Population studies of the ant *Myrmica schencki* ssp. *emeryana* Forel. *Ann. Entomol. Soc. Am.* **38**: 365–72.

Talbot, M. 1945b. A comparison of flights of four species of ants. *Am. Midl. Nat.* **34**: 504–10.

Talbot, M. 1948. A comparison of two ants of the genus *Formica*. *Ecology* **29**: 316–25.

Talbot, M. 1951. Populations and hibernating conditions of the ant *Aphaenogaster* (*Attomyrma*) *rudis* Emery. *Ann. Entomol. Soc. Am.* **44**: 302–7.

Talbot, M. 1954. Populations of the ant *Aphaenogaster* (*Attomyrma*) *treatae* Forel on abandoned fields on the Edwin S. George Reserve. *Contrib. Lab. Verteb. Biol. Univ. Mich.* **69**: 1–9.

Talbot, M. 1971. Flights of the ant *Formica dakotensis* Emery. *Psyche* **78**: 169–79.

Talbot, M. 1975. Habitats and populations of the ant *Stenamma diecki* Emery in southern Michigan. *Great Lakes Entomol.* **8**: 241–4.

Tanner, J.E. 1892. *Oecodoma cephalotes*. Second paper. *Trinidad Field Naturalists' Club* **1**: 123–7.

Taylor, P.D. 1981. Sex ratio compensation in ant populations. *Evolution* **35**: 1250–1.

Taylor, P.D. 1988. Inclusive fitness models with two sexes. *Theoret. Popul Biol.* **34**: 145–68.

Taylor, P.D. 1989. Evolutionary stability in one-parameter models under weak selection. *Theor. Popul. Biol.* **36**: 125–43.

Taylor, P.D., and Sauer, A. 1980. The selective advantage of sex-ratio homeostasis. *Am. Nat.* **116**: 305–10.

Taylor, R.W. 1978. *Nothomyrmecia macrops*: a living-fossil ant rediscovered. *Science* **201**: 979–85.

Tepedino, V.J. and Torchio, P.F. 1982. Temporal variability in the sex ratio of a non-social bee, *Osmia lignaria propinqua*: extrinsic determination or the tracking of an optimum? *Oikos* **38**: 177–82.

Terron, G. 1972. La ponte des ouvriéres fécondées chez une fourmi Camerounaise du genre *Technomyrmex* Mayr: mise en évidence d'une descendance ouvriére. *C. R. Acad. Sci. Paris* **D274**: 1516–17.

Thorne, B.L. 1982. Polygyny in termites: multiple primary queens in colonies of *Nasutitermes corniger* (Motschulsky)(Isoptera: Termitidae). *Insectes Soc.* **29**: 102–17.

Thorne, B.L. 1984. Polygyny in the Neotropical termite *Nasutitermes corniger*: life history consequences of queen mutualism. *Behav. Ecol. Sociobiol.* **14**: 117–36.

Thorne, B.L. 1985. Termite polygyny: the ecological dynamics of queen mutualism. In *Experimental behavioral ecology* (ed. B. Hölldobler and M. Lindauer), pp. 325–41. G. Fischer Verlag, Stuttgart.

Thorne, B.L. 1991. Ancestral transfer of symbionts between cockroaches and termites: an alternative hypothesis. *Proc. R. Soc. London Ser. B* **246**: 191–5.

Torchio, P.F. and Youssef, N.N. 1968. The biology of *Anthophora* (*Micranthophora*) *flexipes* and its cleptoparasite, *Zacosmia maculata*, including a description of the immature stages of the parasite. *J. Kansas Entomol. Soc.* **41**: 289–302.

Toro, M.A. and Charlesworth, B. 1982. An attempt to detect genetic variation in sex ratio in *Drosophila melanogaster*. *Heredity* **49**: 199–209.

Torossian, C. 1967a. Recherches sur la biologie et l'éthologie de *Dolichoderus quadripunctatus* (L.) (Hym. Formicoidea Dolichoderidae). I. Etude des populations dans leur milieu naturel. *Insectes Soc.* **14**: 105–22.

Torossian, C. 1967b. Recherches sur la biologie et l'éthologie de *Dolichoderus quadripunctatus* (L.) (Hym. Formicoidea Dolichoderidae). IV. Etude des possibilités évolutives des colonies avec reine, et des femelles isolées désailées. *Insectes Soc.* **14**: 259–80.

Torossian, C. 1968a. Recherches sur la biologie et l'éthologie de *Dolichoderus quadripunctatus* (L.) (Hym. Formicoidea Dolichoderidae). VII. Etude des mécanismes permettant l'inhibition de la ponte des ouvrières en présence de leur reine: rôle des pheromones. *Insectes Soc.* **15**: 105–44.

Torossian, C. 1968b. Recherches sur la biologie et l'éthologie de *Dolichoderus quadripunctatus* (L.) (Hym. Formicoidea Dolichoderidae). VIII. Mode de reproduction et cycle biologique des colonies. *Insectes Soc.* **15**: 375–88.

Trivers, R.L. 1974. Parent–offspring conflict. *Am. Zool.* **14**: 249–64.

Trivers, R.L. and Hare, H. 1976. Haplodiploidy and the evolution of the social insects. *Science* **191**: 249–63.

Tschinkel, W.R. 1987. Fire ant queen longevity and age: estimation by sperm depletion. *Ann. Entomol. Soc. Am.* **80**: 263–6.

Tschinkel, W.R. 1992. Brood raiding and the population dynamics of founding and incipient colonies of the fire ant, *Solenopsis invicta*. *Ecol. Entomol.* **17**: 179–88.

Tschinkel, W.R. and Howard, D.F. 1978. Queen replacement in orphaned colonies of the fire ant, *Solenopsis invicta*. *Behav. Ecol. Sociobiol.* **3**: 297–310.

Tschinkel, W.R. and Howard, D.F. 1983. Colony founding by pleometrosis in the fire ant, *Solenopsis invicta*. *Behav. Ecol. Sociobiol.* **12**: 103–13.

Tsuchida, K. 1994. Genetic relatedness and the breeding structure of the Japanese paper wasp, *Polistes jadwigae*. *Ethol. Ecol. Evol.* **6**: 237–42.

Tsuji, K. and Yamauchi, K. 1994. Colony level sex allocation in a polygynous and polydomous ant. *Behav. Ecol. Sociobiol.* **34**: 157–67.

Unruh, T.R., Gordh, G. and Gonsalez, D. 1984. Electrophoretic studies on parasitic Hymenoptera and implications for biological control. *Proc. XVII Int. Congress of Entomology, Hamburg*, p. 705.

Uyenoyama, M.K. 1984. Inbreeding and the evolution of altruism under kin selection: effects on relatedness and group structure. *Evolution* **38**: 778–95.

Uyenoyama, M. and Feldman, M.W. 1980. Theories of kin and group selection: a population genetics perspective. *Theor. Popul. Biol.* **17**: 380–414.

Uyenoyama, M. and Feldman, M.W. 1981. On relatedness and adaptive topography in kin selection. *Theor. Popul. Biol.* **19**: 87–123.

Vanderplank, F.L. 1960. The bionomics and ecology of the red tree ant, *Oecophylla* sp., and its relationship to the coconut bug *Pseudotheraptus wayi* Brown (Coreidae). *J. Anim. Ecol.* **29**: 15–33.

Vawter, L. 1991. *Evolution of blattoid insects and of the small subunit ribosomal RNA gene*. PhD Dissertation, University of Michigan, University Microfilms, Ann Arbor, MI.

Velthuis, H.H.W. 1985. The honeybee queen and the social organization of her colony. In *Experimental behavioral ecology* (ed. B. Hölldobler and M. Lindauer), pp. 343–57. G. Fischer Verlag, Stuttgart.

Verma, S., and Ruttner, F. 1983. Cytological analysis of the thelytokous parthenogenesis in the cape honeybee (*Apis mellifera capensis* Escholtz). *Apidologie* **14**: 41–57.

Verner, J. 1965. Selection for sex ratio. *Am. Nat.* **99**: 419–21.

Visscher, P.K. 1986. Kinship discrimination in queen rearing by honey bees (*Apis mellifera*). *Behav. Ecol. Sociobiol.* **18**: 453–60.

Vollrath, F. 1986. Eusociality and extraordinary sex ratios in the spider *Anelosimus eximius* (Araneae: Theridiidae). *Behav. Ecol. Sociobiol*. **18**: 283–7

Wade, M.J. 1980. An experimental study of kin selection. *Evolution* **34**: 844–55.

Wade, M.J. 1985. Soft selection, hard selection, kin selection, and group selection. *Am. Nat.* **125**:61–73.

Waldman, B., Frumhoff, P.C. and Sherman, P.W. 1988. Problems of kin recognition. *Trends Ecol. Evol.* **3**: 8–13.

Walloff, N. 1957. The effect of the number of queens of the ant *Lasius flavus* (Fab.)(Hym., Formicidae) on their survival and on the rate of development of the first brood. *Insectes Soc.* **4**: 391–408.

Ward, P.S. 1978. *Genetic variation, colony structure, and social behaviour in the* Rhytidoponera impressa *group, a species complex of ponerine ants*. PhD Thesis, University of Sydney.

Ward, P.S. 1981. Ecology and life history of the *Rhytidoponera impressa* group (Hymenoptera: Formicidae). II. Colony origin, seasonal cycles, and reproduction. *Psyche* **88**: 109–26.

Ward, P.S. 1983a. Genetic relatedness and colony organization in a species complex of ponerine ants. I. Phenotypic and genotypic composition of colonies. *Behav. Ecol. Sociobiol.* **12**: 285–99.

Ward, P.S. 1983b. Genetic relatedness and colony organization in a species complex of ponerine ants. II. Patterns of sex ratio investment. *Behav. Ecol. Sociobiol.* **12**: 301–7.

Ward, P.S. and Taylor, R.W. 1981. Allozyme variation, colony structure and genetic relatedness in the primitive ant *Nothomyrmecia macrops* Clark (Hymenoptera: Formicidae). *J. Aust. Entomol. Soc.* **20**: 177–83.

Watmough, R.H. 1983. Mortality, sex ratio and fecundity in natural populations of large carpenter bees (*Xylocopa* spp.). *J. Anim. Ecol.* **52**: 111–25.

Watson, J.A.L. and Sewell, J.J. 1981. The origin and evolution of caste systems in termites. *Sociobiology* **6**: 101–18.

Watson, J.A.L. and Sewell, J.J. 1985. Caste development in *Mastotermes* and *Kalotermes*: which is primitive? In *Caste differentiation in social insects* (ed. J.A.L. Watson, B.M. Okot-Kotber and C. Noirot), pp. 27–40. Pergamon, Oxford.

Way, M.J. 1954. Studies of the life history and ecology of the ant *Oecophylla longinoda* (Latreille). *Bull. Entomol. Res.* **45**: 93–112.

*Webb, M.C. 1961. *The biology of the bumblebees of a limited area in eastern Nebraska.* PhD Dissertation, University of Nebraska (data taken from Trivers and Hare 1976).

Weber, N.A. 1972. *Gardening ants—the attines. Mem. Am. Phil. Soc.*, Philadelphia, PA.

Weiss, K. 1962. Untersuchungen Über die Drohnenseugung im Bienenvolk. *Arch. Bienenkd.* **39**: 1–7.

Welsh, J. and McClelland, M. 1990 Fingerprinting genomes using PCR with arbitrary primers. *Nucleic Acids Res.* **18**: 7213–18.

Werren, J.H. 1983. Sex ratio evolution under local mate competition in a parasitic wasp. *Evolution* **37**: 116–24.

Werren, J.H. 1987. Labile sex ratios in wasps and bees. *Bioscience* **37**: 498–506.

Werren, J.H. 1991. The paternal-sex-ratio chromosome of Nasonia. *Am. Nat.* **137**: 392–402.

Werren, J.H., Skinner, S.W. and Huger, A.M. 1986. Male-killing bacteria in a parasitic wasp. *Science* **231**: 990–2.

Werren, J. H., Nur, U. and Wu, C-I. 1988. Selfish genetic elements. *Trends. Ecol. Evol.* **3**: 297–302.

Wesson, L.G. 1939. Contributions to the natural history of *Harpagoxenus americanus* Emery (Hymenoptera: Formcidae). *Trans. Am. Entomol. Soc.* **65**: 97–122.

West, M.J. 1967. Foundress associations in polistine wasps: dominance hierarchies and the evolution of social behavior. *Science* **157**: 1584–5.

West-Eberhard, M.J. 1975. The evolution of social behavior by kin selection. *Q. Rev. Biol.* **50**: 1–33.

West-Eberhard, M.J. 1978. Polygyny and the evolution of social behavior in wasps. *J. Kansas Entomol. Soc.* **51**: 832–56.

West-Eberhard, M.J. 1981. Intragroup selection and social behavior. In *Natural selection and social behavior* (ed. R.D Alexander and D.W. Tinkle), pp. 3–17. Chiron, New York.

West-Eberhard, M.J. 1990. The genetic and social structure of polygynous social wasp colonies (Vespidae: Polistinae). In *Social insects and the environment* (ed. G.K. Veeresh, B. Mallik and C.A. Viraktamath) pp. 254–5. Oxford & IBH Publishing, New Delhi.

Wheeler, D.E. 1986. Devlopmental and physiological determinants of caste in social Hymenoptera: evolutionary implications. *Am. Nat.* **128**: 13–34.

Wheeler, W.M. 1903. The origin of female and worker ants from the eggs of parthenogenetic workers. *Science* **18**: 830–3.

Wheeler, W.M. 1911. The ant-colony as an organism. *J. Morphol.* **22**: 307–25.

Whiting, A.R. 1946. Motherless males from irradiated eggs. *Science* **103**: 219–20.

Whiting, A.R. 1961. Genetics of *Habrobracon. Adv. Genet.* **10**: 295–348.

Whiting, A.R. 1967. The biology of the parasitic wasp *Mormoniella vitripennis* (*Nasonia brevicornis*) (Walker). *Q. Rev. Biol.* **42**: 333–406.

Whiting, P.W. 1939. Sex determination and reproductive economy in *Habrobracon. Genetics* **24**: 110–11.

Whiting, P.W. 1943. Multiple alleles in complementary sex determination of *Habrobracon. Genetics* **28**: 365–82.

Whiting, P.W. 1960. Polyploidy in *Mormoniella. Genetics* **45**: 949–70.

Wilkes, A. 1966. Sperm utilization following multiple insemination in the wasp *Dahlbominus fuscipennis. Can. J. Genet. Cytol.* **8**: 451–61.

Wilkes, A. and Lee, P.E. 1965. The ultrastructure of dimorphic spermatozoa in the hymenopteron *Dahlbominus fuscipennis* (Zett.)(Eulophidae). *Can. J. Genet. Cytol.* **8**: 451–61.

Williams, G.C. 1979. The question of adaptive sex ratio in outcrossed vertebrates. *Proc. R. Soc. London Ser. B* **205**: 567–80.

Williams, J. G. K., Kubelik, A. R., Livak, K. J., Rafalski, J. A. and Tingey, S.V. 1990. DNA polymorphisms amplified by arbitrary primers are useful as genetic markers. *Nucleic Acids Res.* **18**: 6531–5

Wilson, D.S. and Colwell, R.K. 1981. The evolution of sex ratio in structured demes. *Evolution* **35**: 882–97.

Wilson, E.O. 1971. *The insect societies.* Harvard University Press, Cambridge, MA.

Wilson, E.O. 1974a. Aversive behavior and competition within colonies of the ant *Leptothorax curvispinosus. Ann. Entomol. Soc. Am.* **67**: 777–80.

Wilson, E.O. 1974b. The population consequences of polygyny in the ant *Leptothorax curvispinosus. Ann. Entomol. Soc. Am.* **67**: 781–6.

Wilson, E.O. 1976. A social ethogram of the Neotropical arboreal ant *Zacryptocerus varians* (Fr. Smith). *Anim. Behav.* **24**: 354–63.

Wilson, E.O. 1990. *Success and dominance in ecosystems: the case of the social insects.* Ecology Institute, Oldendorf/Luhe.

Winter, U. and Buschinger, A. 1983. The reproductive biology of a slavemaker ant, *Epimyrma ravouxi*, and a degenerate slavemaker, *E. kraussi* (Hymenoptera: Formicidae). *Entomol. General.* **9**: 1–15.

Winter, U. and Buschinger, A. 1986. Genetically mediated queen polymorphism and caste determination in the slave-making ant, *Harpagoxenus sublaevis* (Hymenoptera Formicidae). *Entomol. General.* **11**: 125–37.

Woyciechowski, M. 1990a. Mating behaviour in the ant *Myrmica rubra* (Hymenoptera Formicidae). *Acta Zool. Cracov.* **33**: 565–74.

Woyciechowski, M. 1990b. Do honeybee, *Apis mellifera* L., workers favour sibling eggs and larvae in queen rearing? *Anim. Behav.* **39**: 1220–2.

Woyciechowski, M. and Łomnicki, A. 1987. Multiple mating of queens and the sterility of workers among eusocial Hymenoptera. *J. Theoret. Biol.* **128**: 317–27.

Woyke, J. 1969. A method of rearing diploid drones in a honeybee colony. *J. Apic. Res.* **8**: 65–74.

Woyke, J. 1978. Comparative biometrical investigation on diploid drones of the honeybee. III. The abdomen, and weight. *J. Apic. Res.* **17**: 206–17.

Wright, S. 1943. Isolation by distance. *Genetics* **28**: 114–38.

Yamaguchi, Y. 1985. Sex ratios of an aphid subject to local mate competition with variable maternal condition. *Nature* **318**: 460–2.

Yamamura, N. 1993. Different evolutionary conditions for worker and soldier castes: genetic systems explaining caste distribution among eusocial insects. *J. Theor. Biol.* **161** 111–17.

*Yamane, S. 1980. *Social biology of the Parapolybia wasps in Taiwan*. PhD dissertation, Hokkaido University (data from Suzuki 1986).

Yamauchi, K. and Kawase, N. 1992. Pheromonal manipulation of workers by a fighting male to kill his rival males in the ant *Cardiocondyla wroughtonii*. *Naturwissenschaften* **79**: 274–6.

Yamauchi, K., Itô, Y., Kinomura, K. and Takamine, H. 1987. Polycalic colonies of the weaver ant *Polyrhachis dives*. *Kontyû* **55**: 410–20.

Yamauchi, K., Furukawa, T., Kinomura, K., Takamine, H. and Tsuji, K. 1991. Secondary polygyny by inbred wingless sexuals in the dolichoderine ant *Technomyrmex albipes*. *Behav. Ecol. Sociobiol.* **29**: 313–19.

Yanega, D. 1988. Social plasticity and early-diapausing females in a primitively social bee. *Proc. Natl. Acad. Sci. USA* **85**: 4374–7.

Yanega, D. 1989. Caste determination and differential diapause within the first brood of Halictus rubicundus in New York (Hymenoptera: Halictidae). *Behav. Ecol. Sociobiol.* **24**: 97–107.

Yanega, D. 1992. Does mating determine caste in sweat bees? (Hymenoptera: Halictidae). *J. Kansas Entomol. Soc.* **65**: 231–7.

Yokoyama, S. and Nei, M. 1979. Population dynamics of sex-determining alleles in honey bees and self-incompatibility alleles in plants. *Genetics* **91**: 609–26.

Zchori-Fein, E., Roush, R.T. and Hunter, M.S. 1992. Male production induced by antibiotic treatment in *Encarsia formosa* (Hymenoptera: Aphelinidae), and asexual species. *Experientia* **48**: 102–5.

Zimmerman, R.B. 1983. Sibling manipulation and indirect fitness in termites. *Behav. Ecol. Sociobiol.* **12**: 143–5.

Species index

Note: page numbers in *italics* refer to figures and tables.

Subject index

Note: page numbers in *italics* refer to figures and tables.